Chronopoetics

Media Philosophy

Series Editors: Eleni Ikoniadou, Lecturer in Media and Cultural Studies at the London Graduate School and the School of Performance and Screen Studies, Kingston University, and Scott Wilson, Professor of Cultural Theory at the London Graduate School and the School of Performance and Screen Studies, Kingston University.

The Media Philosophy series seeks to transform thinking about media by inciting a turn toward accounting for their autonomy and "eventness," for machine agency, and for the new modalities of thought and experience that they enable. The series showcases the "transcontinental" work of established and emerging thinkers whose work engages with questions about the reshuffling of subjectivity, of temporality, of perceptions, and of relations vis-à-vis computation, automation, and digitalization as the current twenty-first-century conditions of life and thought. The books in this series understand media as a vehicle for transformation, as affective, unpredictable, and nonlinear, and move past its consistent misconception as pure matter-of-fact actuality.

For Media Philosophy, it is not simply a question of bringing philosophy to bear on an area usually considered an object of sociological or historical concern, but of looking at how developments in media and technology pose profound questions for philosophy and conceptions of knowledge, being, intelligence, information, the body, aesthetics, war, and death. At the same time, media and philosophy are not viewed as reducible to each other's internal concerns and constraints, and thus it is never merely a matter of formulating a philosophy *of* the media; rather, the series creates a space for the reciprocal contagion of ideas between the disciplines and the generation of new mutations from their transversals. With their affects cutting across creative processes, ethico-aesthetic experimentations and biotechnological assemblages, the unfolding media events of our age provide different points of intervention for thought, necessarily embedded as ever in the medium of its technical support, to continually reinvent itself and the world.

"The new automatism is worthless in itself if it is not put to the service of a powerful, obscure, condensed will to art, aspiring to deploy itself through involuntary movements which none the less do not restrict it."

Eleni Ikoniadou and Scott Wilson

Software Theory, Federica Frabetti

Media After Kittler, Eleni Ikoniadou & Scott Wilson

Chronopoetics: The Temporal Being and Operativity of Technological Media, Wolfgang Ernst, Translated by Anthony Enns.

Chronopoetics

The Temporal Being and Operativity of Technological Media

Wolfgang Ernst

Translated by Anthony Enns

ROWMAN &
LITTLEFIELD
──INTERNATIONAL──
London • New York

Published by Rowman & Littlefield International Ltd
Unit A, Whitacre Mews, 26-34 Stannary Street, London SE11 4AB
www.rowmaninternational.com

Rowman & Littlefield International Ltd. is an affiliate of Rowman & Littlefield
4501 Forbes Boulevard, Suite 200, Lanham, Maryland 20706, USA
With additional offices in Boulder, New York, Toronto (Canada), and Plymouth (UK)
www.rowman.com

This translation copyright © 2016 Rowman & Littlefield International
Originally published in German as *Chronopoetik* and *Gleichursprünglichkeit*
Copyright © Kulturverlag Kadmos Berlin, *Chronopoetik* and *Gleichursprünglichkeit*

All rights reserved. No part of this book may be reproduced in any form or by any electronic or mechanical means, including information storage and retrieval systems, without written permission from the publisher, except by a reviewer who may quote passages in a review.

British Library Cataloguing in Publication Data
A catalogue record for this book is available from the British Library

ISBN: HB 978-1-7834-8570-3
       PB 978-1-7834-8571-0

Library of Congress Cataloging-in-Publication Data
Names: Ernst, Wolfgang, 1959- author.
Title: Chronopoetics : the temporal being and operativity of technological media / Wolfgang Ernst ; translated by Anthony Enns.
Other titles: Chronopoetik. English
Description: London ; New York : Rowman & Littlefield International, 2016. | Series: Media philosophy | Includes bibliographical references and index.
Identifiers: LCCN 2015044180 (print) | LCCN 2016007247 (ebook) |
 ISBN 9781783485703 (cloth : alk. paper) | ISBN 9781783485710 (pbk. : alk. paper) |
 ISBN 9781783485727 (Electronic)
Subjects: LCSH: Mass media—Philosophy. | Time perception. |
 Technology—Philosophy.
Classification: LCC P90.E685413 2016 (print) | LCC P90 (ebook) |
 DDC 302.2301—dc23
LC record available at http://lccn.loc.gov/2015044180

∞ ™ The paper used in this publication meets the minimum requirements of American National Standard for Information Sciences—Permanence of Paper for Printed Library Materials, ANSI/NISO Z39.48-1992.

Printed in the United States of America

Contents

Preface to the Focused English Edition — vii

Foreword: Media History versus Media Archeology (Anthony Enns) — xiii

PART I: ELECTROTECHNICAL MICROTEMPORALITIES — 1

1 Time-Critical Media Processes — 3
2 Signal Transmission and Delay — 15
3 Generating Time by Technical Measuring — 37
4 The Computer as Time-Critical Medium — 63

PART II: MEDIA-INDUCED DISRUPTIONS OF THE HUMAN PERCEPTION OF TIME — 97

5 Experiencing Time as Sound: Recorded Voices, Magnetic Tapes — 99
6 A Close Reading of the Electronic "Time Image" — 123
7 The Media Timing of NonLinear Communication — 173

PART III: RE-THINKING "MEDIA HISTORIOGRAPHY" — 203

8 The Heterochronic Being-in-Time of Technical Media — 205
9 Equitemporalities in Media Knowledge — 251

Selected Bibliography — 269

Index — 277

About the Contributors — 283

Preface to the Focused English Edition

THE BIAS OF THIS BOOK

The reader is hereby presented with the translation of selected and updated chapters from twin volumes previously published in German: *Chronopoetics* (*Chronopoetik*) identifies the internal time-based and timing mechanisms of technological media, whereas *Equitemporality* (*Gleichursprünglichkeit*) deals with the disruptions in human temporal perception induced by technical media and their asynchronous being in what is known as "historical" time.[1] With this comprehensive, knowledge-oriented analysis of media tempor(e)alities in mind, the present translation of selected chapters ranges from minute technical descriptions to the analysis of media-induced phenomenological time perception, up to the final temporal re-positioning of such technologies against the historicist imagination. This structure unfolds both in a linear and in a recursive way: as successive arrangement (Parts I–III), and as recurrent topics within the subsections of the individual chapters themselves.

This book identifies technical instantiations of time(s). The guiding assumption is that the essence of technical media is revealed only in the processuality of their temporal operations. Such media do not simply exist *in time* but result in *timing* agencies. Opening the "black box" of technological media reveals a microcosm of temporal configurations. Micro-temporal events within technical devices are critical for the success of media processualities like the electronic cathode ray image. Such time-criticality deserves a precise description in its value for cultural knowledge beyond engineering departments. An analysis of the temporal autonomy of media leads to the discovery of a machine agency, which enriches the traditional philosophy of time. Electronic and digitized media not only mirror traditional time concepts but generate original figures of temporal processuality; they are thereby allowed to be called "chronopoetical." The knowledge of such time-giving technologies enriches the age-old tradition of philosophical inquiry into the nature of *chronopoiesis*.

This book brings together what often remains contained in separate fields of study. As close as possible for an author originally educated in the humanities, the technological analysis of media time processes is biased by the theoretical reflection of their epistemological implications. The distinct chapters "open access" to the delicate micro-temporal dramas performed within (and induced by) technological media by radically broadening the horizon beyond mere electro-technical expertise and computing science.

In alliance with that subject, the book aims to provide insight into the current status of media studies and the philosophy of technology. The intersection between the fields of media analysis and the philosophy of time not only produces new questions but offers unexpected outcomes. Media archeology as an innovative method is applied here to the subject of time-based and time-providing technologies in a blend of media-philosophical inquiry with the analytic fascination of technological devices themselves.

Media culture deserves a close analysis of its technological and techno-mathematical conditions. Different from discourse analysis in cultural and socio-technological studies, media archeology—in alliance with critical media philology such as platform and code studies—always starts from (and returns to) the "close reading" of the technologies involved. However dilettante this knowledge might appear when compared to expert knowledge (which is hereby invited to feed back necessary corrections), it is the epistemological bias that allows this artifactual knowledge to be deciphered in different ways. The media-archaeological approach is driven by the ambition to create sparks of insight that deserve to be expressed as general knowledge.

By tracing the proper tempor(e)alities *(Eigenzeit)* of technical configurations, the overall epistemological ambition of this publication is a representative discussion of the ways in which technical media operate in cultural and non-human time. After introducing the notion of time-critical media, the argument starts with a close media-archaeological and media-philological reading of the minutest artifactual signal events and temporal impulses, then progresses to the human-machine-coupling of temporal sensation, and finally discusses chrono-epistemological alternatives to writing traditional media historiography.

The choice of core chapters from the German twin volumes keeps intact their argumentative bridge. The succession of topics in the German edition was meant to be comprehensive (even if surely not all-encompassing); therefore, the voluminous *Chronopoetics* and *Equitemporality* discuss time-related cultural techniques; the human psycho-physiology of time; previous time machines, such as the phonograph, the cinematograph, and the mechanical clock; and further technical as well as mathematical details as well. The present selection rather decisively concentrates on the temporal epistemology of electronic and techno-mathematical media systems, which concern the extended "time window" of the present culture.

While *Chronopoetics* and *Equitemporality* diagnostically cover the major variances of the relation between technical media and time, in its decisive choice of core chapters for translation, this present book is necessarily a sampled presentation of the media chronopoetics under discussion. *Theory sampling* by means of exemplary micro-studies is a qualitative method of media archeology.

This is certainly not the first attempt to trace the relation between technology and time, but it is determined in its application of the media-archaeological method. Whereas the inquiry into the nature of time belongs to the most venerable issues of philosophy, the relationship between time and technology in a stricter sense has only been discussed for about a century—such as in Henri Bergson's critique of cinematographic temporality around 1900, which has been refreshed by Gilles Deleuze's volumes on cinema (*The Movement-Image* and *The Time-Image*). Martin Heidegger's definition of the nontechnical essence of technology was inherently connected to the question of being-in-time, which has been continued by French philosophers of technology like Gilbert Simondon (*On the Mode of Existence of Technical Objects*) and more recently Bernard Stiegler (*Technics and Time*). In the English-speaking academic world, some recent studies have addressed the techno-temporal question, e. g. Timothy Scott Barker's *Time and the Digital* and Mark B. N. Hansen's *New Philosophy of New Media*, which deals with media temporality from the phenomenological point of view. Jussi Parikka's recent *A Geology of Media* is important for the analysis of the deep time of media materialities. The methodological approach of this book is also close to that of Erkki Huhtamo and Jussi Parikka's anthology *Media Archaeology: Approaches, Applications, and Implications* and Siegfried Zielinski and David Link's anthology *Variantology 2: On Deep Time Relations of Arts, Sciences and Technologies*.

To readers educated in the elegant Anglo-American style of academic argumentation, the author's somewhat *modular* writing style will seem more*staccato*-like than a melodic narrative. Media archeology stays close to the fragment while at the same time trying not to lose sight of its epistemological ground. After all, the media temporality identified in this book tends to be more time-discrete than time-continuous, challenging the familiar linear concept of historical narrative and inviting rather discontinuous media archeography.[2]

SUMMARY OF CONTENT BY CHAPTER

PART I deals with *Electrotechnical Microtemporalities*. With the aim of sharpening the awareness of media tempor(e)alities, chapter 1 ("Time-Critical Media Processes") defines a fundamental feature of contemporary media culture: micro-temporal configurations within electronic media that deserve knowledge-oriented analysis and description. The differentiation between *chronos* and *kairos* as temporal categories escalates in the notion of "time-critical media" beyond the traditional category of "time-based media."

Then follows the analysis of "Signal Transmission and Delay," i.e. the description of minute electronic time configurations, such as the resonant circuit and delay lines. The following chapter deals with the mechanisms of "Generating Time by Technical Measuring" with such devices as chronophotography and its critique in terms of Bergsonian time and the kymograph as epistemological object and its resulting time diagrams. "The Computer as Time-Critical Medium" focuses on the computer as a time machine and digital computing as an "algorhythmic" mechanism. "Emulation" will be discussed as an ahistorical practice of re-enacting past computers.

PART II deals with *Media-Induced Disruptions of the Human Perception of Time*. The chapter "Experiencing Time as Sound: Recorded Voices and Magnetic Tapes" discusses disembodied voices and the temporal shock induced by time-axis manipulation. Then follows "A Close Reading of the Electronic 'Time-Image,'" which focuses on the temporal essence of the electronic image (television, video) and its chronopoetic aesthetics. The chapter "The Media-Timing of Non-Linear Communication" attempts a time-critical reading of telecommunication from telegraphy up to the Internet and its temporal economy.

The final PART III *Re-Thinking "Media Historiography"* first unfolds "The Heterochronic Being-in-Time of Technical Media"—that is, the proper time (*Eigenzeit*) of technical apparatuses. A Heideggerian reading of technological time is applied to a case study: the endurance of radio. The final chapter, "Equitempor(e)alities in Media Knowledge," proposes a critique of conventional media historiography and proposes non-narrative modalities of media time—alternative figures of media in time that are emerging in terms such as "equitemporality," "resonance," and "recursions."

ACKNOWLEDGMENTS

My thanks for making the book possible go to the editors of the *Media Philosophy* series, esp. to Eleni Ikoniadou's driving enthusiasm for the translation project. This brings me to the translator Anthony Enns, who simply turned out to be the perfect choice for transposing such "German school of media studies" argumentation into a text for readers in English. Further gratitude goes to the Kulturverlag Kadmos, publishing house in Berlin (with its *Berliner Programm einer Medienwissenschaft* series) for liberally granting permission to publish an abridged translation with Rowman & Littlefield.

NOTES

1. Wolfgang Ernst, *Chronopoetik. Zeitweisen und Zeitgaben technischer Medien* (Berlin: Kulturverlag Kadmos, 2012); Wolfgang Ernst, *Gleichursprünglichkeit. Zeitwesen und Zeitgegebenheit technischer Medien* (Berlin: Kulturverlag Kadmos, 2012).

2. See Wolfgang Ernst, "Media Archaeography: Method and Machine versus History and Narrative of Media," in *Media Archaeology: Approaches, Applications, and Implications*, ed. Erkki Huhtamo and Jussi Parikka (Berkeley/Los Angeles/London: University of California Press, 2011), 239–55.

Foreword

Media History versus Media Archeology: German Media Theory and Wolfgang Ernst's Chronopoetics

Anthony Enns

In his introduction to the anthology *Digital Memory and the Archive* (2013)—the first collection of Wolfgang Ernst's writings in English—Jussi Parikka explains that "the purpose of this book is to give voice to one of the names that have appeared in recent waves of media-theoretical debates in Germany ... [who] is not yet so well known in English-speaking academia."[1] Happily, this is no longer the case, as *Digital Memory and the Archive* has been followed by two additional books—*Stirrings in the Archives: Order from Disorder* (2015) and *Sonic Time Machines: Explicit Sound, Sirenic Voices, and Implicit Sonicity* (2016)—as well as numerous articles and book chapters in English. *Chronopoetics* is an abridged translation of two books that were published in quick succession in Germany: *Gleichursprünglichkeit. Zeitwesen und Zeitgegebenheit technischer Medien* (*Equiprimordiality: The Temporal Essence and Condition of Technical Media*, 2012) and *Chronopoetik. Zeitweisen und Zeitgaben technischer Medien* (*Chronopoetics: The Temporal Modes and Specifications of Technical Media*, 2012). If there is a sense of urgency to these publications, it is because that they are attempting to defend a particular approach to media theory that has come increasingly under fire for its technological determinism. Anglo-American critics frequently apply this label to the work of Friedrich Kittler (1943–2011), who is widely known as the "father of the so-called German media theory,"[2] and the continued persistence of such accusations has encouraged many German theorists to distance themselves from the anti-humanist aspects of his work.[3] Ernst's most recent books challenge this trend by arguing for the continued relevance of Kittler's technical *a priori*, and they thus attempt to lay claim to a certain territory in the landscape of contemporary German media theory that has been left vacant in Kittler's absence. To introduce the present volume, I will provide a brief overview of this context, an outline of Ernst's key concepts and methods, as well as an argument for the importance of his admittedly controversial approach to media theory.

GERMAN MEDIA THEORY

As Geoffrey Winthrop-Young points out, there is no such thing as a unified German media theory, as media theorists in Germany employ a wide range of approaches and perspectives (as they do elsewhere). The term gained currency in the 1990s, however, because of the tremendous international success of Kittler's work, and it has gradually become one of Germany's most successful intellectual exports.[4] It is not possible to provide a detailed overview of Kittler's work here, but in the context of the present volume, it is helpful to remember that Kittler's approach was largely inspired by the work of Michel Foucault—in particular his *Archeology of Knowledge* (1969), which argues for an approach to historiography that emphasizes epistemic ruptures or shifts in systems of knowledge. Kittler employs a similar approach in his own work, as David E. Wellbery explains "Kittler's discourse analysis follows the Foucauldian lead in that it seeks to delineate the apparatuses of power, storage, transmission, training, reproduction, and so forth that make up the conditions of factual discursive occurrences."[5] The problem with Foucault's approach, however, is that it fails to address the media technologies through which these discourses are inscribed, stored, and transmitted. In Kittler's words, Foucault "merely describes the *production* of discourses. There are, for example, no descriptions in Foucault's book of the *source* of these discourses, of the *channels* or the *receivers* of discourse."[6] Kittler thus criticizes Foucault's archaeologies of knowledge for failing to take into account how media technologies influence the formation of systems of knowledge.[7] Instead of focusing exclusively on discourses, Kittler is more interested in their material substrates, which he famously describes as "inscription systems" (*Aufschreibesysteme*).[8] As Wellbery explains, "these technologies are not mere instruments with which 'man' produces his meanings. ... Rather, they set the framework within which something like 'meaning,' indeed, something like 'man,' become possible at all."[9]

Unlike Foucault, Kittler also focuses on the nineteenth and twentieth centuries, and he is particularly interested in the contemporary shift toward digital media. What is essentially new about the contemporary moment, according to Kittler, is that man's position as the being who does the inscribing is gradually being displaced by the computer:

> What I keep dreaming of and what people don't like to hear because they believe that technology and science are mere tools made for people in the street . . . is that machines, especially the contemporary intelligent machines as conceived

by Turing in 1936, are not there for us humans . . . but that nature, this glowing, cognitive part of nature, is feeding itself back into itself.[10]

Kittler also explains the underlying reason for Foucault's prediction "that man would be erased, like a face drawn in sand at the edge of the sea"[11] through reference to the rise of new media technologies: "The end of 'man' postulated by Foucault is brought on by a mechanism that writes writing."[12] Winthrop-Young similarly notes that Kittler's "move into media studies served to corroborate his programmatic as well as polemical emphasis on ruptures. Foucault spoke of epistemes that changed without rhyme or reason; Kittler was able to ascertain that these changes were related to media shifts."[13] According to Kittler, these media shifts ultimately lead to the erasure of man, as "the ultimate subject of history is technology, understood . . . as the processing of nature that for an extended period of time was dependent on human intermediaries, but that now, with the arrival of digital technology, is closer to a self-processing of nature that leaves humans behind."[14] Kittler thus concludes that the concept of man was itself made possible by a medial shift—an argument that reverses the traditional approach of Anglo-American cultural studies, which assumes that humans are the driving force behind technological development.

One of Kittler's early students, Bernhard Siegert, identifies this shift in emphasis from discourse to media as a pivotal moment in the formation of German media theory: "The focus on the materiality and technicality of meaning constitution prompted German media theorists to turn Foucault's concept of the 'historical a priori' into a 'technical a priori' by referring the Foucauldian 'archive' to media technologies."[15] Kittler's work thus represented a challenge to the humanist tradition as well as the practice of hermeneutic analysis: "The insistence on these media reference systems, designed as an attack on the reason- or mind-based humanist reference systems, was guided by a deeply anti-humanist rejection of the tradition of the Enlightenment and the established discursive rules of hermeneutic interpretation."[16] However, Kittler's focus on this technical *a priori* soon led to accusations that his own work was a symptom (if not the cause) of anti-humanist tendencies within the humanities. In one of the earliest articles on German media theory in English, for example, Michael Geisler criticizes Kittler's approach for ignoring both the content of media and the practices of media users:

> [T]he great advantage of dealing with the media as paradigm-forming technologies is that one need not concern oneself with representation. . . . Similarly, hypermedia are dealt with in apocalyptic or celebratory metaphors, or as a giant scam perpetrated on the naive curiosity of unsuspecting intellectual. . . . No one in Germany ever seems to ask what people actually *do* with the internet.[17]

In other words, Kittler's work was criticized for its exclusive focus on the material properties of media technologies, which seemed to exclude any analysis of content or reception, which is the main focus of Anglo-American cultural studies.

Kittler's emphasis on the agency of media technologies also made him vulnerable to accusations of technological determinism. Lisa Gitelman argues, for example, that Kittler ignores the complex ways in which media technologies come into existence:

> It is as if Kittler doesn't need to persuade his readers of details about why or how phonographs were invented because he already knows what phonographs are, and therefore he knows what (and particularly how) they mean. ... That is to make a medium both evidence and cause of its own history. ... Media are more properly the results of social and economic forces, so that any technological logic they possess is only apparently intrinsic.[18]

According to Gitelman, therefore, Kittler's work ignores the social, political, and economic context surrounding the emergence of new media technologies, and his approach thus endorses the notion that media obey an "intrinsic technological logic"[19] that remains outside the sphere of human thought and action.[20]

CULTURAL TECHNIQUES

According to Winthrop-Young, the anti-humanist tendencies underlying German media theory had "run its course ... by the mid-1990s, when Kittler's own apocalyptic anti-humanism had passed its peak."[21] The generation of media theorists who followed Kittler, including Bernhard Siegert, Cornelia Vismann, and Markus Krajewski, gradually moved away from this position by developing new approaches that were largely centered on the concept of "cultural techniques." According to Siegert, this term represents a "mediating third" between nature and culture, as it remains outside this binary opposition yet it is also the structural element that makes this opposition possible in the first place. The term can thus be applied to virtually any technique that establishes such a distinction: "Humans *as such* do not exist independently of cultural techniques of hominization, time *as such* does not exist independently of cultural techniques of time measurement, and space *as such* does not exist independently of cultural techniques of spatial control."[22] The concept of "cultural techniques" thus provided a way for German media theorists to move away from the anti-humanist tendencies in Kittler's work and to focus instead on cultural practices, as Winthrop-Young explains:

"The culture-technical approach offers a viable alternative or escape route. To speak of operations and connections allows those inspired by the Kittler effect to speak of practices without saying society; to readmit human actors allows them to speak of agency without saying subjects."[23] Kittler's anti-hermeneutic stance is thus transformed "into a less intransigent post-hermeneutic approach involving certain notions of praxis and limited human agency that Kittler was prone to eschew."[24] Bernard Dionysius Geoghegan similarly notes that the concept of "cultural techniques" has its roots in agricultural practices, and these "agricultural connotations . . . allow for an introduction of those questions of life and bios that . . . Kittler scrupulously avoided . . . but which have recently reasserted themselves as problematics for critical reflection in 21st century philosophy and media theory."[25] The concept of "cultural techniques" thus reemerged at this particular moment in order to overturn "the anti-biologism that prevailed in nearly all Kittlerian analysis."[26]

While this focus on cultural practices has become increasingly popular in recent years, Winthrop-Young adds that it has not been universally embraced—particularly by Kittler's most staunch followers: "Some of Kittler's more dedicated and hence less original disciples . . . continue to write like it's 1999 and indulge in ever more detailed readings of ever more arcane technologies. Media theory can forfeit its relevance in many ways; one of the safest is to engage in increasingly stale artifactualism."[27] He also describes this "artifactual" approach as a "bid to out-Kittler Kittler," which "is most clearly on display in the media-archaeological work of Wolfgang Ernst."[28]

MEDIA ARCHEOLOGY

Like German media theory, the term "media archeology" has been employed in different ways by different theorists. The term first emerged due to widespread discontent concerning the exclusionary nature of the accounts written by media historians. For instance, Eric Kluitenberg argues that media history often marginalizes technological failures due to its "implicit construction of a unitary narrative of progress."[29] Huhtamo and Parikka similarly describe the "negligence" and "ideological bias" inherent in the accounts of many media historians.[30] Michael Goddard also argues that media historians are "not empirical enough" and that they should "get closer to the materiality" of media technologies.[31] Media archeology was designed to address these concerns by focusing on the materiality of media technologies and rejecting traditional accounts of media history as a narrative of technological progress.[32]

These concerns are also at the heart of Ernst's own particular brand of media archeology. Trained as a classicist and a historian, Ernst completed his

Ph.D. on museology in 1989. His work started to take a medial turn in the 1990s (largely due to the influence of Kittler's work), as he began to develop his own Foucauldian reading of media archives, which culminated in such works as *M.edium F.oucault* (2000), *Das Rumoren der Archive* (*Stirrings in the Archives*, 2002), and *Im Namen der Geschichte. Sammeln—Speichern—Er/zählen* (*In the Name of History: Collecting—Storing—Re/counting*, 2003). In these books, Ernst extends Kittler's focus on archival technologies by arguing that media are the true archivists of the past: "It is worth remembering that the archive as the condition for our knowledge of history becomes dependent on the media of its transmission."[33] However, his approach differs somewhat in its explicit critique of historiography. Like Kittler, for example, Ernst embraces Foucault's model of epistemic ruptures, but unlike Kittler he emphasizes that this model is primarily opposed to the discourse of history: "The archaeology of knowledge, as we have learned from Foucault, deals with discontinuities, gaps and absences, silence and ruptures, in opposition to historical discourse, which privileges the notion of continuity."[34]

Ernst eventually turned to media archaeology as a Foucauldian method of critiquing the discourse of history, as it primarily focuses on "the mechanisms that regulate entry into the discourse of history or exclusion from cultural memory."[35] Instead of extending the historical discourse,[36] therefore, Ernst employs media archaeology as an alternative way of writing—or rather *not* writing—media history:

> Media archaeology . . . is interested in procedures and events that are not "historical" (i.e., narratable) but rather consist of "autochthonic transformations" (Foucault) within the realm of machines and their symbols. Words and things happen within the machine (computers) as logic and hardware. The media-archaeological gaze, accordingly, is immanent to the machine. Human beings, having created logical machines, have created a discontinuity with their own cultural regime.[37]

In other words, Ernst's media-archeological approach is based on a fundamental awareness that media do not process cultural signs but rather technical signals, which are imperceptible to the human sensorium. As a result, "media themselves, not exclusively humans anymore, become active 'archaeologists' of knowledge."[38] Although Foucault's understanding of archeology as a method of discourse analysis clearly diverges from the more materialist approach of media archeology, Ernst emphasizes that the idea of media as archeologists is distinctly Foucauldian, as it represents a vivid realization of his desire to "define a method of historical analysis from the anthropological theme . . . a method of analysis purged of all anthropomorphism."[39] Just as Kittler sought to explain how media technologies influence the formation of systems of knowledge, Ernst thus sought to explain how media technologies influence the construction of history—including the history of media themselves.[40]

Ernst also argues that media archeology "introduces a third element to the Promethean dichotomy of culture and nature,"[41] much like Siegert describes "cultural techniques" as a "mediating third" between nature and culture. Unlike "cultural techniques," however, media archeology does not make the nature/culture distinction possible; rather, it represents a way out of the nature/culture binary, as a technical medium is "based on cultural knowledge—but it is still of a physical nature because there are electro- or even quantum-physical laws at work that are not solely dependent on the respective cultural discourse."[42] In other words, media are not simply cultural products, but rather they represent the implementation of physical and mathematical laws that transcend human culture: "Media technology thus emerges from culture as an autonomous entity."[43]

In 2003, Ernst became the first professor of media theories in the seminar for media studies, which was later incorporated into the Institute for Musicology and Media Studies at Humboldt University in Berlin. In the same year, he also created the Media-Archaeological Fundus, a collection of historical (and often obsolete) media technologies, and the Signal Laboratory, which focuses on the study of less obsolete artifacts, such as computational hardware and software. These projects are a natural extension of his media-archeological approach and his call to open up media technologies to better understand their technical operations. Unlike museum objects, the devices collected in the Fundus are fully functional. In contrast to initiatives like Bruce Sterling's "dead media" project (a collection of obsolete, forgotten, and dysfunctional technologies),[44] the Fundus is thus informed not by a nostalgia for historical media, but rather by a desire to operationalize them, which reveals the complex entanglement of the past and the present:

> Media archaeology understood as an analysis of epistemological configurations (both machinic and logic) does not simply seek a redemption of forgotten or misread media of the past, nor is it confined to a reconstruction of the crude beginnings and prehistories of technical media. Rather than being a nostalgic collection of "dead media" of the past, assembled in a curiosity cabinet, media archaeology is an analytical tool, a method of analyzing and presenting aspects of media that would otherwise escape the discourse of cultural history.[45]

The operationalization of historical media thus reveals that media technologies possess their own intrinsic temporalities or "proper times" (*Eigenzeit*), which typically elude media-theoretical analysis—a concern that also informs his most recent work.

TIME-CRITICAL MEDIA

Ernst's most recent work has turned even further in a technomathematical direction by focusing on the temporal processes of "time-critical media" (*zeitkritische Medien*)—a term that has several meanings:

- Time-critical media are technologies in which time plays a crucial role (beginning with chronophotography and phonography, but escalating with the development of electronic transmission media and digital computing).
- Time-critical media do not simply refer to the axis of time (like time-based media), but are instead capable of time axis manipulation.
- Time-critical media introduce new modes of measuring and recording time, and they thus create their own unique tempor(e)alities.
- The tempor(e)alities of time-critical media disrupt abstract notions of human time, and they thus separate the time of technical media from the time of human culture.
- The tempor(e)alities of time-critical media can only be understood according to their own operative dimensions through an analysis of technical (time) signals as opposed to cultural (semiotic) signs. In other words, the study of time-critical media necessarily disregards content and focuses instead on temporal processes.

The chapters presented in this volume provide detailed descriptions of the tempor(e)alities of a wide range of time-critical media, including storage media like photography, phonography, and magnetic tape, as well as transmission media like radio, television, and computers, which each represent autonomous temporal forms. The book also outlines three basic stages in the development of time-critical media:

- Mechanical storage media, like photography, phonography, and cinematography, disrupt the human sense of time by introducing discontinuities and ruptures.
- Electronic transmission media, like radio and television, intensify these temporal disruptions, as they operate at the level of microtemporal processes that are no longer perceptible to the human senses.
- Digital computers operate at the smallest temporal intervals, and the shift from analog to digital represents a shift from signal delay times to calculated time windows. Computers thus introduce the concept of digitally calculated "real time," which is fundamentally dynamic.

It is important to note that these chapters are not organized chronologically (as this brief overview might suggest). While the non-chronological organization of the book might seem confusing at first, it is fundamental to the book's underlying argument that an understanding of the tempor(e)alities of time-critical media undermines the linear narrative of media history. By moving backward and forward in time, Ernst's analysis thus performs its own time axis manipulation, which allows him to draw connections between the operations of media from different time periods—connections that fundamentally challenge the concept of technological progress: "[Media archeology], understood in this way, hints at a non-linear relation between past and present media technologies, a short-circuiting of media tempor(e) alities which escapes traditional, narrative history of technology. Instead of one media system resulting from another, there are sudden recursions."[46] For example, "The Gutenberg Era, which starts with the Greek discrete symbol-based alphabet and escalated with the printed book, was radically challenged by signal recording and signal transmission media. . . . Then all of a sudden the Gutenberg Era returns as an alphanumeric and discrete alphabet, which consists of two letters: zero and one. But it's an alphabet."[47] These "recursions" are not anchored in "human culture as a historical figure, but rather exclusively in the autonomy of electrophysics, logic and mathematics."[48] The concept of "recursions" thus offers an alternative form of time-writing, which represents a way of *not* writing media history: "What is at issue, therefore, is not *another* way of writing media history, but rather *the other* of writing media history itself."[49]

One of the major implications of Ernst's concept of time-critical media is that the focus of media analysis shifts from the discursive register of signs to the non-discursive register of signals, which are algorithmic and thus techno-mathematical processes. The signals recorded on magnetic tape, for example, "do not exist as character strings in the sense of printed texts or notes, but rather as a magnetic latency that only takes place in the inductive moment of the passing machine, the unwinding, when it is converted into a signal."[50] In other words, the latency of magnetic signals pushes media history from semiotics to mathematics—from telling to counting:

> Events that occur in the domain of micro-seconds can no longer be consciously perceived by human senses, and they thus fall into the realm of pure measuring electronics. The anthropological narrative of time thus comes to an end, and it is replaced by the concept of the human as an ensemble of computable numbers (the difference between narrativizing and counting data).[51]

The concept of time-critical media not only illustrates how media technologies disrupt the human sense of time and thereby challenge the "anthropological narrative of time" (i.e., history), but it also represents a rejection of humanism: "Let us try for a moment to suspend the voluntary self-restriction of the human temporal horizon by means of the category of history. Thus, the face of the historical human being does not disappear like a figure drawn in sand at the edge of the sea, but rather like the sand in an hourglass."[52] Ernst thus presents this project as a logical extension of Kittler's approach, as it focuses on the intrinsic logic of media technologies and their role in the gradual disappearance of the human. It also potentially represents an update of Kittler's approach, as the "historical human being" is not simply displaced by media technologies; rather, human time is displaced by the tempor(e)alities of technical media. Ernst thus revises Kittler's famous claim that "media determine our situation"[53] by revealing that it is actually the *time-critical processes* of media that determine our situation.[54]

The other key concept informing Ernst's recent work is "equiprimordiality" (*Gleichursprünglichkeit*),[55] which also has several meanings:

- The operations of time-critical media always take place in the present (when these media still function), and they are thus always equiprimordial (i.e., they make no distinction between the past and the present). This is one of the major differences between media-archeological and classical-archeological artifacts: "[N]ot only is the past still present as a condition in the remnant of a technical antiquity, but the technical antiquity also creates the present in its processuality."[56] One of the examples that Ernst frequently employs is the *Volksempfänger*—a radio built in Germany during the National Socialist regime that was originally used to broadcast propaganda speeches: "[It still] receives radio programs when operated today, because the stable technological infrastructure of broadcasting media is still in operation. There is no 'historical' difference in the functioning of the apparatus now compared to then. . .; rather, there is a media-archaeological short circuit between otherwise historically clearly separated times."[57] While Ernst is clearly familiar with the history of this device and its political significance, he concludes that such radios also perform time-critical functions that transcend historical time, as technical signals always take place in the present and they thus blur the distinction between history and actuality.
- This technical differential also results in the blurring of storage and transmission. Ernst argues, for example, that when listening to analog radio "you cannot tell whether what you hear is from a record being played at the radio station or whether it's actually being performed live. The medium of radio, and electromagnetic wave-based media, turns everything into signal transmission . . . and the signal itself [does] not change."[58] In other words, it makes no difference whether technical signals originate in the past (tape) or in the present (radio), as they remain temporally indistinguishable (i.e., equiprimordial). The concept of a "historical" past is thus foreign to

technical apparatuses, as sounds and images from the past are pure presence in the moment of playback.
- This sense of continued presence is a result of the fact that media technologies always obey the same physical and mathematical laws—regardless of the time of their invention or implementation: "Every medium that is operative today also *realizes* at the same time transhistorical iterations of technological infrastructures, symbolic circuits, and electrophysical laws."[59] This means that media from different eras are still equiprimordial with regard to their basic time-critical operations.

One of the major implications of this concept is that media technologies only reveal their temporal essence *in implementation*—that is, a technical apparatus can only be considered a medium when it is operative. This is yet another difference between media-archeological and classical-archeological artifacts:

> [W]hat drastically separates an archaeological object from a technical artifact is that the latter discloses its essence only when operating. While a Greek vase can be interpreted by simply being looked at, a radio or computer does not reveal its essence by monumentally being there but only when being processed by electromagnetic waves or calculating processes. If a radio from a museum collection is reactivated to play broadcast channels of the present, it changes its status: it is not a historical object anymore but actively generates sensual and informational presence.[60]

Media technologies thus not only exist *in* time but also consist *of* time, and their temporal essence can only be understood through an analysis of their underlying tempor(e)alities.

This concept also represents an implicit critique of the curatorial practices employed in museums of technology. According to Ernst, media technologies need to be displayed *in implementation* in order for them to be understood ("otherwise a medium like a TV set is nothing but a piece of furniture"),[61] yet this is impossible for most museums of technology. Ernst's Fundus can thus be understood as an essential tool for modeling his media-archeological approach, and the difference between museum-displayed technologies and the operationalization of technical media in the Fundus becomes particularly evident when Ernst explains that the work of a media archeologist more closely resembles that of an engineer than a historian.

TECHNOLOGICAL DETERMINISM

As Parikka explains, "Ernst's way of differentiating *Medienwissenschaft* [media studies] from . . . *Kulturwissenschaften* [cultural studies] lies in the resolute demand that if we study media, we really need to study their modes of technical epistemology and how they process signals in a channel."[62] By focusing on the technical aspects of media, however, Ernst often leaves out people: "Just as with Kittler, there is not much room for humans in the media ontology of Ernst."[63] Like Gitleman, Parikka also points out that Ernst's media-archeological approach avoids any discussion of the sociohistorical contexts surrounding the emergence of new media technologies, which makes it seem as if the technologies themselves have no history: "At the same time that this idea tries to complexify the idea of history as nonreducible to human cultural realms . . . it simplifies exactly the other bit, that despite their nonhistorical nature, such technologies cannot avoid being embedded and entangled in such human temporalities as well."[64] Ernst's emphasis on the technical *a priori* thus makes his work vulnerable to the same criticisms that were once leveled at Kittler.

While German media theorists often defend Kittler's work from accusations of technological determinism, this appears to be a label that Ernst openly accepts and even promotes: "Media archeaology is driven by a certain 'Berlin school of media studies' obsession with approaching media in terms of their logical structure (informatics) on the one hand and their hardware (physics) on the other—and thus is different from British and U.S. cultural studies, which analyze the subjective and social effects of media."[65] Ernst thus freely admits that his media-archeological approach differs from Anglo-American cultural studies, but he describes this difference as an advantage by emphasizing the potential dangers of focusing exclusively on content and context rather than technical operations:

> [A] reconciliatory analytical merger of the technologies with the people who created, consumed, and imagined them blurs the decisive differences that are at work here. From a decisively non-humanist perspective, inventors do not figure as the primary agents of a so-called media history, but their creations turn out to be governed by non-relativistic agencies such as external technological laws.[66]

Ernst thus justifies his approach by reiterating the idea that the development of media technologies is driven by their own intrinsic logic rather than human agency. He even compares this principle to the concept of "memes": "In the incubation phase of new technologies, humans themselves sometimes become the 'media' of a knowledge that anamnetically insists on manifesting. Technologically implicit knowledge is particularly compliant with cultural practices and human curiosity—much like the concept of the 'meme,' which makes use of humans and machines as hosts for the purpose of its transmission."[67]

Ernst is clearly aware of the theoretical assumptions underlying his approach, just as he is aware that media technologies are historically and culturally situated. Indeed, he openly concedes that "media archeology is certainly aware of the irreducible historicity of technical knowledge"[68] and "it seems obvious that all media innovations are culturally determined."[69] He quickly adds, however, that there is more to media technologies than this cultural and historical layer of meaning—namely, there are also the physical and mathematical laws that govern their operations, which are far more crucial to an understanding of their temporal essence. He thus argues that "*there are* media in which something always occurs that does not depend on human-cultural knowledge,"[70] and therefore "this chiastic historical model calls for a supplement: the assumption of an inner logic of media development."[71] Instead of rejecting the charge of technological determinism, Ernst more often assumes the opposite approach by insisting on the importance of precise media-archeological analysis, which reveals the degree to which media technologies operate as autonomous entities. He also notes that Anglo-American media theorists are gradually coming to the same conclusion, as his approach closely resembles that of software studies and media forensics:

> What looked like an antithetical configuration in German hardware-oriented and Anglo-American socially and culturally oriented media studies for a long time nowadays seems "sublated" by a Hegelian trick ("List") of media-theoretical reason. So-called software studies (which acknowledges the increasing virtualization of programming and its phenomenologization as "apps") and a refreshed materialist (forensic) approach link both cross-Atlantic schools.[72]

Ernst certainly occupies a unique position in the landscape of contemporary German media theory (post-Kittler), and he has clearly established his own particular brand of media theory, which he proudly describes as the "Berlin school of media studies." It is safe to say that no other theorist working today focuses more intently on the technical *a priori* and demands such extensive media competence. While it remains unclear whether Anglo-American theorists will develop the requisite skills to perform such a rigorous analysis of media-technical operations and temporal processes, Ernst emphatically argues that they are neglecting the most crucial element of media technologies if they fail to take up this challenge.

NOTES

1. Jussi Parikka, "Archival Media Theory: An Introduction to Wolfgang Ernst's Media Archaeology," in *Digital Memory and the Archive*, by Wolfgang Ernst (Minneapolis: University of Minnesota Press, 2013), 2.

2. Geoffrey Winthrop-Young, "Material World: An Interview with Bernhard Siegert," *Artforum International* 53.10 (Summer 2015): 324.

3. Kittler describes his approach as "information materialism," as it focuses on the merging of information and material media: "Information has been transformed into matter and matter into information." Friedrich A. Kittler, "Media Wars: Trenches, Lightning, Stars," in *Literature, Media, Information Systems: Essays*, ed. John Johnston (Amsterdam: OPA, 1997), 126.

4. Winthrop-Young refers to this as the "Kittler effect." See Geoffrey Winthrop-Young, *Kittler and the Media* (Cambridge: Polity, 2011), 143–46.

5. David E. Wellbery, foreword to *Discourse Networks 1800/1900*, by Friedrich A. Kittler, trans. Michael Metteer and Chris Cullens (Stanford: Stanford University Press, 1990), xii.

6. Armitage, John. "From Discourse Networks to Cultural Mathematics: An Interview with Friedrich A. Kittler," *Theory, Culture & Society* 23.7-8 (2006): 19.

7. Following this method, critics are now beginning to examine how Kittler's own work was influenced by the rise of new media technologies in the late twentieth century, such as the personal computer. See, for example, Claus Pias, "Kittler und der 'Mißbrauch von Heeresgerät': Zur Situation eines Denkbildes 1964–1984–2014," *Merkur: Deutsche Zeitschrift für europäisches Denken* 69.791 (April 2015): 31–44.

8. This term, which was originally borrowed from Daniel Paul Schreber's *Memoirs of My Nervous Illness* (1903), is usually translated as "discourse networks."

9. Wellbery, foreword, xii.

10. Friedrich A. Kittler, *Short Cuts*, ed. Peter Gente and Martin Weinmann (Frankfurt am Main: Zweitausendeins, 2002), 270.

11. Michel Foucault, *The Order of Things: An Archaeology of the Human Sciences* (London: Routledge, 2002), 422.

12. Wellbery, foreword, xxx.

13. Winthrop-Young, *Kittler and the Media*, 64.

14. Ibid., 80.

15. Bernhard Siegert, "Cultural Techniques: Or the End of the Intellectual Postwar Era in German Media Theory," *Theory, Culture & Society* 30.6 (2013): 50.

16. Ibid., 52.

17. Michael Geisler, "From Building Blocks To Radical Construction: West German Media Theory Since 1984," *New German Critique* 78 (Fall 1999): 104–5.

18. Lisa Gitelman, *Always Already New: Media, History, and the Data of Culture* (Cambridge: MIT Press, 2006), 10.

19. See Geoffrey Winthrop-Young and Michael Wutz, "Translators' Introduction: Friedrich Kittler and Media Discourse Analysis," in *Gramophone, Film, Typewriter*, by Friedrich A. Kittler, trans. Geoffrey Winthrop-Young and Michael Wutz (Stanford: Stanford University Press, 1999), xiv.

20. Despite these criticisms, critics like Nicolas Gane actively promote the techno-determinist aspects of Kittler's approach. See Nicolas Gane, "Radical Posthumanism: Friedrich Kittler and the Primacy of Technology," *Theory, Culture & Society* 22.3 (2005): especially p. 40.

21. Geoffrey Winthrop-Young, "Cultural Techniques: Preliminary Remarks," *Theory, Culture & Society* 30.6 (2013): 14.

22. Siegert, "Cultural Techniques," 57.
23. Winthrop-Young, "Cultural Techniques," 14.
24. Ibid., 15.
25. Bernard Dionysius Geoghegan, "After Kittler: On the Cultural Techniques of Recent German Media Theory," *Theory, Culture & Society* 30.6 (2013): 75.
26. Ibid., 77.
27. Winthrop-Young, "Cultural Techniques," 14.
28. Ibid., 15.
29. Eric Kluitenberg, "On the Archaeology of Imaginary Media," in *Media Archaeology: Approaches, Applications, and Implications*, ed. Erkki Huhtamo and Jussi Parikka (Berkeley: University of California Press, 2011), 51.
30. Erkki Huhtamo and Jussi Parikka, "Introduction: An Archaeology of Media Archaeology," in *Media Archaeology: Approaches, Applications, and Implications*, ed. Erkki Huhtamo and Jussi Parikka (Berkeley: University of California Press, 2011), 3.
31. Michael Goddard, "Opening up the Black Boxes: Media Archaeology, 'Anarchaeology' and Media Materiality," *New Media & Society* (April 28, 2014), 2.
32. Huhtamo and Parikka describe media archeology as a discipline that constructs "alternate histories of suppressed, neglected, and forgotten media that do not point teleologically to the present media-cultural condition as their 'perfection.'" Huhtamo and Parikka, "Introduction," 3. Geert Lovink similarly describes media archeology as the "reading of the 'new' against the grain of the past, rather than a telling of the history of technologies from past to present." Geert Lovink, *My First Recession: Critical Internet Cultures in Transition* (Rotterdam: V2/NAi, 2004), 11.
33. Wolfgang Ernst, "Let There Be Irony: Cultural History and Media Archaeology in Parallel Lines," in *Digital Memory and the Archive*, ed. Jussi Parikka (Minneapolis: University of Minnesota Press, 2013), 42.
34. Geert Lovink, "Archive Rumblings: An Interview with Wolfgang Ernst," in *Digital Memory and the Archive*, ed. Jussi Parikka (Minneapolis: University of Minnesota Press, 2013), 196.
35. Ibid.
36. Projects that extend the historical discourse in relation to media culture include, for example, the *Graduate Research Program "Media of History—History of Media," which was funded by the German Research Foundation from 2005 to 2013.* See http://www.mediale-historiographien.de.
37. Wolfgang Ernst, "Media Archaeography: Method and Machine versus the History and Narrative of Media," in *Digital Memory and the Archive*, ed. Jussi Parikka (Minneapolis: University of Minnesota Press, 2013), 69–70.
38. Ibid., 55. Ernst even argues that "media archaeology is more akin to the gaze of the optical scanner than to that of the anthropological observer." Ibid., 67.
39. Michel Foucault, *Archaeology of Knowledge* (New York: Pantheon, 1999), 16.
40. While Kittler admits that "Wolfgang's concerns are not my concerns," he also notes that this approach influenced his later work: "Wolfgang . . . is trying to get out of narrative approaches to history, a project that I accept and like very much. . . . [W]hat I have learned, in part from Wolfgang, is that we have to stop narrating the history of writing, computing, mathematics or music as linear history. So this is what I

have been attempting for some time now." Armitage, "From Discourse Networks to Cultural Mathematics," 32.

41. Wolfgang Ernst, "From Media History to *Zeitkritik*," *Theory, Culture & Society* 30.6 (2013): 133.

42. Wolfgang Ernst, "Experimenting with Media Temporality: Pythagoras, Hertz, Turing," in *Digital Memory and the Archive*, ed. Jussi Parikka (Minneapolis: University of Minnesota Press, 2013), 185.

43. Ernst, "From Media History to *Zeitkritik*," 137.

44. Bruce Sterling, *The Dead Media Project*, http://www.deadmedia.org.

45. Ernst, "Media Archaeography," 56.

46. Lori Emerson, "Archives, Materiality and the 'Agency of the Machine': An Interview with Wolfgang Ernst," *Library of Congress* (8 February 2013), http://blogs.loc.gov/digitalpreservation/2013/02/archives-materiality-and-agency-of-the-machine-an-interview-with-wolfgang-ernst.

47. Ghislain Thibault, "What We Used to Call 'Media History': A Feature Interview with Wolfgang Ernst," *Amodern* (27 January 2013), http://amodern.net/article/ernst-media-history.

48. Wolfgang Ernst, "Kittler-Time: Getting to Know Other Temporal Relationships with the Assistance of Technological Media," trans. Yuk Hui and James Burton, in *Media After Kittler*, ed. Eleni Ikoniadou and Scott Wilson (London: Roman & Littlefield, 2015), 61.

49. Chapter 9 in this volume.

50. Chapter 5 in this volume.

51. Chapter 3 in this volume.

52. Ernst, "From Media History to *Zeitkritik*," 144.

53. Friedrich A. Kittler, *Gramophone, Film, Typewriter*, trans. Geoffrey Winthrop-Young and Michael Wutz (Stanford: Stanford University Press, 1999), xxxix.

54. Ernst also describes this project as an extension of Kittler's approach, as the concept of time-critical media was already implicit in Kittler's own insights into the nature of time-measuring media. See Ernst, "Kittler-Time," 54.

55. Martin Heidegger employs this term to refer to elements that exist together without being placed in a hierarchy of primacy. In other words, equiprimordial elements cannot be derived from each other and cannot exist without each other, as they are equally fundamental and mutually dependent. This term can also be translated as "co-original." See Martin Heidegger, *Being and Time*, trans. Joan Stambaugh (Albany: State University of New York Press, 1996), 124.

56. Chapter 8 in this volume.

57. Ernst, "Media Archaeography," 56–57.

58. Thibault, "What We Used to Call 'Media History.'"

59. Chapter 8 in this volume.

60. Ernst, "Media Archaeography," 57–58.

61. Emerson, "Archives, Materiality and the 'Agency of the Machine.'"

62. Jussi Parikka, "Afterword: Cultural Techniques and Media Studies," *Theory, Culture & Society* 30.6 (2013): 154.

63. Jussi Parikka, "Operative Media Archeology: Wolfgang Ernst's Materialist Media Diagrammatics," *Theory, Culture & Society* 28.5 (2011): 60.

64. Parikka, "Archival Media Theory," 13–14.
65. Lovink, "Archive Rumblings," 196.
66. Wolfgang Ernst, "Media Archaeology as a Transatlantic Bridge," in *Digital Memory and the Archive*, ed. Jussi Parikka (Minneapolis: University of Minnesota Press, 2013), 24.
67. Chapter 9 in this volume.
68. Chapter 9 in this volume.
69. Ernst, "From Media History to *Zeitkritik*," 133.
70. Chapter 9 in this volume.
71. Ernst, "From Media History to *Zeitkritik*," 133.
72. Ernst, "Media Archaeology as a Transatlantic Bridge," 26. Despite his criticisms, Parikka similarly notes that "because of his controversial suggestions his theoretical insights are exactly the food for thought we need in our current debates in digital humanities and new media theory." See Parikka, "Archival Media Theory," 3.

BIBLIOGRAPHY

Armitage, John. "From Discourse Networks to Cultural Mathematics: An Interview with Friedrich A. Kittler." *Theory, Culture & Society* 23.7-8 (2006): 17–38.
Emerson, Lori. "Archives, Materiality and the 'Agency of the Machine': An Interview with Wolfgang Ernst." *Library of Congress* (8 February 2013). http://blogs.loc.gov/digitalpreservation/2013/02/archives-materiality-and-agency-of-the-machine-an-interview-with-wolfgang-ernst.
Ernst, Wolfgang. *Digital Memory and the Archive*. Edited by Jussi Parikka. Minneapolis: University of Minnesota Press, 2013.
———. "From Media History to *Zeitkritik*." *Theory, Culture & Society* 30.6 (2013): 132–46.
Foucault, Michel. *Archaeology of Knowledge*. New York: Pantheon, 1999.
———. *The Order of Things: An Archaeology of the Human Sciences*. London: Routledge, 2002.
Gane, Nicolas. "Radical Post-humanism: Friedrich Kittler and the Primacy of Technology." *Theory, Culture & Society* 22.3 (2005): 25–41.
Geisler, Michael. "From Building Blocks To Radical Construction: West German Media Theory Since 1984." *New German Critique* 78 (Fall 1999): 75–108.
Geoghegan, Bernard Dionysius. "After Kittler: On the Cultural Techniques of Recent German Media Theory." *Theory, Culture & Society* 30.6 (2013): 66–82.
Gitelman, Lisa. *Always Already New: Media, History, and the Data of Culture*. Cambridge: MIT Press, 2006.
Goddard, Michael. "Opening up the Black Boxes: Media Archaeology, 'Anarchaeology' and Media Materiality." *New Media & Society* (April 28, 2014).
Heidegger, Martin. *Being and Time*. Translated by Joan Stambaugh. Albany: State University of New York Press, 1996.
Huhtamo, Erkki, and Jussi Parikka, eds. *Media Archaeology: Approaches, Applications, and Implications*. Berkeley: University of California Press, 2011.

Ikoniadou, Eleni, and Scott Wilson, eds. *Media After Kittler*. London: Roman & Littlefield, 2015.

Kittler, Friedrich A. *Discourse Networks 1800/1900*. Translated by Michael Metteer and Chris Cullens. Stanford: Stanford University Press, 1990.

———. *Gramophone, Film, Typewriter*. Translated by Geoffrey Winthrop-Young and Michael Wutz. Stanford: Stanford University Press, 1999.

———. *Literature, Media, Information Systems: Essays*. Edited by John Johnston. Amsterdam: OPA, 1997.

———. *Short Cuts*. Edited by Peter Gente and Martin Weinmann. Frankfurt am Main: Zweitausendeins, 2002.

Lovink, Geert. *My First Recession: Critical Internet Cultures in Transition*. Rotterdam: V2/NAi, 2004.

Parikka, Jussi. "Afterword: Cultural Techniques and Media Studies." *Theory, Culture & Society* 30.6 (2013): 147–59.

———. "Operative Media Archeology: Wolfgang Ernst's Materialist Media Diagrammatics." *Theory, Culture & Society* 28.5 (2011): 52–74.

Pias, Claus. "Kittler und der 'Mißbrauch von Heeresgerät': Zur Situation eines Denkbildes 1964–1984–2014." *Merkur: Deutsche Zeitschrift für europäisches Denken* 69.791 (April 2015): 31–44.

Siegert, Bernhard. "Cultural Techniques: Or the End of the Intellectual Postwar Era in German Media Theory." *Theory, Culture & Society* 30.6 (2013): 48–65.

Thibault, Ghislain. "What We Used to Call 'Media History': A Feature Interview with Wolfgang Ernst." *Amodern* (27 January 2013). http://amodern.net/article/ernst-media-history.

Winthrop-Young, Geoffrey. "Cultural Techniques: Preliminary Remarks." *Theory, Culture & Society* 30.6 (2013): 3–19.

———. *Kittler and the Media*. Cambridge: Polity, 2011.

———. "Material World: An Interview with Bernhard Siegert." *Artforum International* 53.10 (Summer 2015): 324–33.

Part I

ELECTROTECHNICAL MICROTEMPORALITIES

Chapter 1

Time-Critical Media Processes

LEVELS OF MEDIA-ARCHEOLOGICAL TIME ANALYSIS

Technological media always take place in the temporal dimension, regardless of whether they are understood through epistemological reflection. Technical constellations are only operative when actualized in time. Time-critical processes—such as delicate electronic synchronization between image senders and receivers ("television") and the exact orchestration of binary instruction cycles ("computer")—occur *in* electro-technical (commonly called "analog") and techno-mathematical (commonly called "digital") media. The signal-technical discovery of time-critical processes *through* measuring media (like chronophotography) revealed for the first time a corresponding epistemological sensitization.[1] A second level of investigation involves the temporal affects in people that are induced by the *re-play* of stored recordings. People are addressed through media in their existential (not historical) sense of time. Lastly, the question concerning the cultural ways in which media process time does not seek to prove "medial historiographies"[2] or the role of media as a history-making power acting in concert with the historical discourse; rather, it reflects on the proper time of media. Technical media are not only reconfigured repeatedly in the course of time; they also serve as models for the conception of emphatic time itself through their timing *specifications*, which is how the concept of "real time" came into our current vocabulary. While the role of media as agents in historical processes has been sufficiently examined,[3] time-critical analysis focuses on the genuine event-like nature of media on both sides of the concept of history. It thus encompasses signal-technical, electro-mathematical, and media-epistemological time series analyses. Such studies do not approach this structure historiographically; rather, they allow media-induced temporal processes themselves to be addressed.

Technical time critique reveals a microcosm of time figures that are usually concealed in media apparatuses; it is assisted by a phenomenology of the temporal affects that media induce in people. This raises the question of which representational form of the temporality of technical media is expected, and thus how *not* to write media *history*. The operative linking of these different levels of temporal knowledge, whose agents are often technical media themselves, calls for differentiations. "A micro-temporal level of physical and techno-physical processes, a meso-level of psychic-cognitive processes, and a macro-level of social systems and discursive formations as well as macro-physical processes."[4] Gilbert Simondon thereby systematically divides the temporal modes of existence of technical objects into intrinsic machine realities, human-machine relations, and the genesis of technicity.[5]

Time-critical media processes are first and foremost dedicated to analyzing the specific ways of processing time that were and are introduced into culture through techno-mathematical media. This includes the smallest temporal events, which are essential for the realization of sound, image, and computing processes as well as the cognitive disruption of the human awareness of time through media of time axis manipulation. This also includes the challenge not to write media archeology and genealogy exclusively according to the model of history, as this ignores their unique chronopoetics.[6] That "technical objects embody complex temporalities" is an insight of Simondon's philosophy of technology.[7] Instead of describing how technical media are part of cultural history, media-induced alternatives to history itself thus emerge.

A MEDIA THEORY OF THE TIME-CRITICAL

The time-critical is a field of knowledge originated by media and their analysis. It includes concepts like real time, time axis manipulation, as well as the actualization of stored time signals and the temporalizing variants of Aristotelian *metaxy*, which in this sense means not only the spatial in-between as media channel, but also media-technically the temporal in-between, the smallest memory buffers and signal delays. Time-*critical* processes, interpreted literally, *determine* the overall process and success of systems in electronics and informatics; on a functional level, the concept is familiar enough in all related disciplines. In industrial process control, the time-critical is understood simply in terms of punctuality. Heathrow Airport in London advises arriving guests to "follow the time-critical flight connection streams"; as with Internet communication, connections are not only spatial but also temporal nodal points. For a long time, however, the time-critical element, which is characteristic of operative systems, has lacked a fundamental media-epistemological meaning. Signals, defined electrophysically as genuine time events, are the chief subject of media studies in contrast to a cultural semiotics of sign relations. A premise of cybernetics, which still remains current, is thus also evoked for the technosphere. "No analysis of natural science, whether it be physics or biology, is complete unless we possess a proper analysis of its appropriate time-concept."[8] Media archeology, which is based not only on philosophy but also on the mathematical and natural sciences and thus belongs to the humanities as much as it understands itself as a science, drew the obvious conclusion from this insight. The literally critical point here is the category of time, which oscillates between micro- and macro-temporal levels, between moment and history. "We observe a temporal sequence of events, and our experiments are attempts to reproduce at various times that which we have observed at one particular time. Therefore, all the improvements and modifications which have been made in the theory of time itself are relevant in the study of all the sciences."[9] A theory of measuring media is thus always also a theory of time.

If media events are firmly understood as micro-time-critical processes, then they refer to processes on this side of the "historical" field. Media history deals with the actual implementation and thus the temporalization of logical relations in physical materiality in the double sense of techno-logy. If the implementation turns a technical-symbolic constellation into a media process, then being-in-the-world means being-in-time. This also applies to cellular automata. Every binary switch of discrete information consumes a minimal time interval, which can literally be counted on and which tends toward infinitesimally small moments. "Switching time is inversely proportional to the energy expended. . . . This theorem has consequences for the geometry of spacetime and the computing power of the universe."[10] In addition to this binary moment of time, signal analysis primarily deals with time series, which Wiener formulated for the electronic anti-aircraft predictor under wartime conditions: the lightning-fast extrapolation of measurement data for the purpose of statistical prediction, which from the outset refused to make any claim with regard to an exact predetermination of concrete events. Certainties and causalities (the event concept of history) were thus replaced by probabilities and correlations.

SIGNAL TIME

If media were for the longest time simply mechanical extensions or amputations of human organs and senses—to borrow loosely from Marshall McLuhan—electronics and their mathematization introduced a new situation: electronic media are extensions of the central nervous system itself. "The human—and also his pride: fantasy, art—is divided into physiology and data processing, which can only be reintegrated through a media theory";[11] however, this theory must be time-critical. It delineates the field in which the alliance of electronics, physiology, and data processing takes place.

Forms of time implemented in the real, thus time-critical processes, long remained undiscovered as objects of knowledge in the Western temporal economy because they were hardly measurable with human senses and mechanical instruments; Leibniz anticipated their discovery as *petites perceptions*. When light shined, it appeared as a pure emanation and not as a vibration in the electromagnetic spectrum. Reality, insofar as it consists of the smallest time-critical moments, eludes symbolic notation. Time analysis has long confined itself to historiography. "This indescribability only disappears when a time range is successfully transformed into a frequency range entirely without metaphysics or a philosophy of history," wrote Friedrich Kittler, with regard to the techno-mathematical process of fast Fourier transformation, which indeed replaces the time axis as the classical abscissa of causal chains with a frequency axis, whose units are inversely proportional to its units of time as evidenced metrologically on an oscilloscope. "On this axis everything that brought even only a trace of periodicity or regularity over time appears as ordinate values."[12] Yet, in order for all of the sampled values to be calculated at the same time within a time window, they must remain temporarily stored; real time analysis is based on latency in the present. The space of the archive and the actuality of the present are thus no longer strictly separate, but rather a mutual condition. "All of the circulating theories that seek to distinguish between historical and electronic time as between delay and simultaneity are myths."[13] In the time sense of highly technical media, space and time are dynamically relative in terms of the electromagnetic field, which is why the concept of media-induced time ratios seems more plausible than the crude concept of a media age. Pass a magnetic tape by the tape head of a magnetophone and there is precisely no immediate contact between the magnetic charge on the tape and the coil of the ring head, but rather an electrodynamic process is *induced* without contact. Waves and particles converge in technical sound. "The magnetic field lines that change with the rhythm emerge at the gap, penetrate the passing tape, and leave behind a magnetic remnant. A sine-shaped recording signal can thus be conceived as a series of infinitesimal bar magnets."[14]

During playback, conversely, this differential calculus literally becomes media-operative: on their way through the core of the tape head, the field lines cut into the coil windings and induce an electrical current in them that corresponds to the temporal changes. Hermann Minkowski's concept of spacetime entanglement thus takes place *de facto* at the most concrete level of electrodynamic processes. As a result, it is necessary to think about the time of technological media differently than in narrative figures. The temporal form of implementation is the central criterion of media-being (*Mediendasein*).

An anthology on early digital valve computers with the time-critical title *Faster Than Thought* explains: "All the operations . . . carried out by these valves could equally well be achieved by the use of ordinary switches and variable resistances, but for one thing—time. Valves can be switched on and off almost instantaneously. . . . The fastest mechanical switch is a thousand times slower than this."[15] State metaphors, which correspond to their respective current technical systems, are also formulated as temporal events. In his third letter "On the Aesthetic Education of Man," Schiller writes: "When the mechanic has the works of a clock to repair, he lets the wheels run down; but the living clockwork of the State must be repaired while it is in motion, and here it is a case of changing the wheels as they revolve."[16] This practice first became a reality as real-time programming or *live coding* (as in the programming environment SuperCollider)—a dynamic temporal mode based on the programmable memory of digital computers.

CHRONOS AND KAIROS

Time is generally conceived as chronological in Western culture: a continuous stream. On the other hand, there is a time horizon that extends between two extremes, which ancient mythology designated as antipodes to the God Chronos, who devours everything (even his own children): Aion and Kairos, the Gods of time, who are reflected in the etymological entanglement of apparently distinct categories (ancient Greek *rhein* as flow, *rhythmos* as its derivative). It is characteristic of the internal embargo on time epistemology in early Greek thought that the word *kairos* initially meant "right position" and "proportion," and in Homer, it served as the designation for a vulnerable body part. With the Pythagoreans, the basic measurement of cosmic rhythms has a geometric rather than a dynamic meaning; if anything, according to ancient Greek accounts, dynamics revealed themselves in the musical technique of glissando. The time-critical is an object of knowledge that is largely un-Greek, as it originated from extremely technical (measuring) media. However, *one* archaic concept comes close to the temporal essence (*Zeitwesen*) of technical media: the phenomenology of *kairos*. "Aion shines at the transcendental dimensions: time that stretches far, far beyond the life span of humans and planet Earth; pure time, like that of machines. . . . By contrast, Kairos's time is doing the right thing at the right moment: he is the god of the auspicious moment. . . . He challenges us to make a decision. . . . Once Kairos has passed by, it is too late."[17] In contrast to linear, arithmetical time (*chronos*), kairotic time is the time of contingent dramaturgies, narratives, decisions. However, the fleeting moment loses the non-binding character of this mythological entity as soon as he is chrono-technically mastered.

Culture first diagnosed the common temporal process: planetary orbits, seasons, "historical time." With the electromechanical measuring media of the nineteenth century, a mirror image of this process revealed itself at the micro level. Media history itself could be reformulated kairotically.

> Is there actually a suitable time and a correct place for an invention or are such . . . statements paradoxical due to their implicit belief in a directed and compulsory technical progress? . . . The adoption of the mythic or synchronic time *kairos* instead of the diachronic arrow of time *chronos* is sensible in so far as one can speak of it as *right* or better still *critical* time.[18]

Technical media emerge in this time field.

TIME AXIS MANIPULATION AND DASEIN-CRITIQUE: THE EXISTENTIAL AWARENESS OF TIME-CRITICAL PHENOMENA

Time etymologically means a cut in a continuum; the Latin *tempus*, derived from the Greek *temnein*, means parts, and thus an act of discretion, discontinuation, "critique" in its most fundamental form. However, human perception is hardly capable of consciously registering the smallest slices of time as discrete events; instead, cognitive time awareness calculates with intervals.

> Events can only be synthesized as perceived shapes after a period of roughly three seconds. The three-second segmentation is considered a neuronal correlate of attention, and indeed also with regard to the perception of aesthetic objects as aesthetic objects (hearing melodies, viewing images, reading, etc.). Aesthetic perception can serve here as a mirror of the experience of time in relation to the production of units of meaning, rhythm, time intervals.[19]

The ability to fix moments is a condition of the analyzability of temporal processes. Self-writing measuring media register fleeting moments that human senses are unable to capture. Time is transformed from a metaphysical signified to technical availability. "It is no coincidence that in the modern era the development of techniques of acceleration was accompanied by the development of techniques of controlled deceleration that serve the targeted interruption and the collective production of stasis in the temporal sense."[20] Time figures used in linguistic articulation (*lógos*), such as ellipses and pauses, have long since found their techno-logical correlate in the epoch of electromechanical and electronic recording media. Chrono-techniques like freezing, fixing, replaying, decelerating, and accelerating are practiced with the stop, play, rewind, and fast-forward buttons on tape recorders, video recorders, and their iconic emulation as software.

Media time communicates its processuality to humans. Time-critical moments are part of this pattern and thus turn out to be Dasein-critical. The time horizon extends from the knowledge of the "being-at-an-end of Dasein in death."[21] However, technical intensification shifts this question concerning ontology to the micro-temporal. The concept of time as a noun suggests the existence of a signified. Time can also be an action, such as the verb *to time, timing*. The trivial machine of the clock is reified time. In contrast, the world of electromagnetism, electronics, and computers introduced a way of processing time that developed its own temporal cosmos or chronosphere. Media archeology seeks to reveal the temporal relations of electro-technical processes and to describe how this electro-technical temporality subliminally intervenes in human experience as such. The media-archeological exploration of techno-mathematical ways of processing time involves not deducing explanations but rather exposing causes. The exposure of an event structure and its temporal conditions of possibility means not only gaining "an *ontological* understanding of *historicity*,"[22] but also at the micro-temporal level on and in electrophysics itself. Time-critical phenomena involve firstly the infra-temporal domain of the smallest moments,[23] secondly the human window of perception for "the present," and finally gradual ongoing transformations that are not noticed by people at all due to their slowness. During his trip to China, Marco Polo did not consciously experience the time difference in the smallest intervals; however, contemporary processes of technical time acceleration and scarcity become more intense, culminating in the editing frequency of music videos. Physiologist Karl Ernst von Baer's virtually pre-cinematic thought experiment concerning accelerated and decelerated time perception using the example of fighting fish and snails, respectively,[24] can be transferred to technological processes, which take place in ultra-fast computers or as long-term signal storage on magnetic tape. The neuronal time window of the present and the exposure time of photography converged for a moment; in 1866, Hermann Wilhelm Vogel answered the question of what a "photographic moment" is called with "three seconds." Yet, the gap widens when it concerns, on the one hand, the evolution of physiological organs of perception and, on the other hand, the evolution of time-critical media technologies. Since the development of phonography, cinematography, and computer games, advanced media technologies influence the human time sensor; its coupling with time-critical media challenges its ability to process reactions. The human time sense initially sustains a shock through escalating technological speeds. Time-critical processes as consciously perceived reality first became an object of knowledge through the development of measuring media that could capture and calculate them. This actually began with the clock in so far as it divided movement. This progressing discretization approaches the assumed continuum of time itself; Leibniz and Newton's infinitesimal calculus contributed to it on the symbolic level. Time is the subject and object of technologies; in an electronic time counter, the expiration time of hardware lies below the critical threshold at which it influences the measurement of time itself.

To put it pointedly, time critique applies to the world of micro-temporal processes that subliminally evade immediate human perception. The question of the critical moorings of being and time is scientific as well as technical—up until the moment when the concept of "time" as an emphatic signified shifts to the implementation level of signifiers, at which point it dissolves into a variety of time figures that render the concept of time in the singular as metaphysics.

INTENSIFICATION AS OPPOSED TO THE CONCEPT OF "TIME-BASED MEDIA"

"Time-critical media" should not be understood as mass media broadcasts that journalistically engage in political discourse. Rather, they are media events in which minimal time processes represent a critical and thus decisive criterion for medial operativity itself—from electronic video images, which strictly speaking consist solely of time, to the Von Neumann architecture of the computer, which displays its chrono-*techné* in the synchronization and timing of the smallest moments of time. Acoustic signal events in particular are only conceivable and perceptible through their extension in time, and they thus stand out in alliance with the temporal phenomenality of technical media. Their merger with networked communication worlds indicates an epistemology of streaming, which originates from the permanent transmission of the data stream itself and thereby dissembles its radical time-discrete character at the sub-technical level.

The time-critical represents an escalation with respect to the established concept of *time-based media*, such as the status of electronic images, which—unlike the cinematic frame as the basis of the illusion of motion in the cinema—consist solely of time or the motion of signals, or acoustic processes, which circumvent the reaction time of human perception. "Time-critical" does not mean simply that media operations are time-based—that is also true of other cultural techniques. Rather, this concept means that medial operations under the conditions of digital signal processing must be processed in strictly predefined time windows in order for them to succeed and for a message to materialize at all. Time is thus no longer simply a physical parameter, but rather an intelligent operator—"a paradigmatic shift from semiotic sign-processes to simulation-controlled signal-processing."[25] There is a (time-) critical difference between *time-based* and *timing media* (analogous to the active concept of the *imaging sciences*). The phono-technical concept of the *track* cuts straight to the point: on the one hand, it is a trace *of* something (as a recording), and on the other hand, it is a pathway (channeling the sound scanning).

There are forms of *aisthesis* that are not consciously perceived as signs (and thus aesthetically coded), but rather almost instantly interconnected in neuronal processing as signal chains. This momentariness is more pronounced in technical-operative media, where minimal time operations remain below the human perceptual threshold and thus subliminally generate a world according to its own law. One-second intervals originated from the quantization of time in clockwork; the measurement of movement was visualized as chronophotography. In the electronic domain, however, this measuring takes place at high frequencies, which circumvent human perception so completely that they—productively turned from analysis to synthesis—imperceptibly dominate the high-tech communication of the present. The power of media is that their temporal process becomes imperceptible.

In the classical humanities, the objects of research are usually not analyzed in terms of their chrono-technical production; in textual worlds, temporal moments play an important role at best in the eye movements that occur while reading. In fact, the operations of the alphabet and writing have not been time-critical for the longest time. In his treatise *Laocoon* (1766), Gotthold Ephraim Lessing called for the representation of dramatic moments in the visual arts, so that the observer would be made to interact; the critical event was thus supposed to be represented not directly, but only in its approach. Lessing coined the concept of the "transitory"[26] for this—a temporalizing variant of what McLuhan defined for "cold media," namely the inducement to active participation. "The phenomenon of the moment can *in principle not* be clarified in terms of the *now*," wrote Martin Heidegger in connection with Edmund Husserl's concept of the internal protention and retention of present awareness.[27] In the move from narrative to time-critical practices, this theme is literally played out in current computer worlds.[28]

The horizon of media-temporal processes ranges from the time sensitivity of hearing to the calculations of the computer; it includes Lessing's *Laocoon* theory from 1766 as well as the temporal nature of video scan lines, Hermann von Helmholtz's measurements of nerve impulses, and the mathematical modeling of temporal ergodicity in cybernetics. The specific focus on signal processes goes beyond the crude concept of *time-based media*; time-critical processes constitute the theoretical-epistemological and technical-practical signature of contemporary media culture. Processes that develop in the smallest windows of calculated time ("real time") are now analytically and phenomenologically central. Media archeology can demonstrate the operative mathematics of symbol-processing media as well as the shift from discursive and cultural-technical to non-discursive and algorithmic processes through an analysis of media processes in which the smallest moments of time play a decisive role. As Konrad Zuse said concerning the memory programming of calculating machines, which he had already envisaged in 1939. "Memory programming only works . . . when the memory is very fast, thus electronic, and requires a huge storage capacity. . . . This is no problem for an electronic machine, but it was impossible with a relay machine."[29] In contrast to pure Boolean logic as a form of thinking,[30] its realization in actual circuits is highly time-sensitive, as it takes places in the real electrophysical world. Since the development of the sequential Von Neumann architecture, computer history has been above all a history of faster switching.[31] The switch from zero to one is actually micro-temporal. A switching organ "spends only very little time transiently in the intermediate states that form the connecting continuum."[32] Norbert Wiener called this the *time of non-reality*.[33] Operative mathematics, brought into the world as computers, is thus overtaken by time. Time critique and twentieth-century media machines are technologically linked. Through their media-cultural usage in the form of human-machine interfaces, time-critical reactions are deeply ingrained—a legacy of the physiological stimulus-response research of the nineteenth century. Electronically accelerated calculating technologies and the total mobilization of communication media culminate in this one message: the intensification of the temporal moment.

NOTES

1. See Axel Volmar, ed., *Zeitkritische Medien* (Berlin: Kulturverlag Kadmos, 2009).

2. A graduate program at the Bauhaus University in Weimar researched this concept for years; see *Mediale Historiographien*, http://www.mediale-historiographien.de.

3. See Fabio Crivellari, Kay Kirchmann, Marcus Sandl, and Rudolf Schlögl, eds., *Die Medien der Geschichte. Historizität und Medialität in interdisziplinärer Perspektive* (Konstanz: UVK, 2004).

4. Martin Donner, "Rekursion und Wissen. Zur Emergenz technosozialer Netze," in *Rekursionen. Von Faltungen des Wissens*, ed. Ana Ofak and Philipp von Hilgers (Munich: Fink, 2010), 112f.

5. See John Hart, foreword to *On the Mode of Existence of Technical Objects*, by Gilbert Simondon (London: University of Western Ontario, 1980).

6. See Ana Ofak and Philipp von Hilgers, eds., *Rekursionen der Wissensgeschichte* (Munich: Fink, 2010).

7. N. Katherine Hayles, "Komplexe Zeitstrukturen lebendiger und technischer Wesen," in *Die technologische Bedingung. Beiträge zur Beschreibung der technischen Welt*, ed. Erich Hörl (Frankfurt am Main: Suhrkamp, 2011), 217.

8. Norbert Wiener, "Time, Communication, and the Nervous System," *Annals of the New York Academy of Sciences* 50 (1948), 197.

9. Ibid.

10. Oswald Berthold, "Computational Universe" (lecture presented at the Technical University of Berlin, 10 October 2009). See also Seth Lloyd, "Computational Capacity of the Universe," *Physical Review Letters* 88 (2002).

11. Norbert Bolz, "Computer als Medium—Einleitungm," in *Computer als Medium*, ed. Norbert Bolz, Friedrich Kittler, and Christoph Tholen (Munich: Fink, 1994), 9.

12. Friedrich Kittler, "Real Time Analysis, Time Axis Manipulation," in *Draculas Vermächtnis. Technische Schriften* (Leipzig: Reclam,1993), 200.

13. Ibid., 201.

14. Karl-Heinz Finke, *Bauteile der Unterhaltungselektronik* (Berlin: VEB Verlag Technik, 1980), 102f.

15. B. V. Bowden, ed., *Faster Than Thought: A Symposium on Digital Computing Machines* (London: Pitman, 1971), 42.

16. Qtd. in Martin Fontius, "Produktivkraftentfaltung und Autonomie der Kunst," in *Literatur im Epochenumbruch*, ed. Günther Klotz et al. (Berlin/Weimar: Aufbau, 1977), 486.

17. Siegfried Zielinski, *Deep Time of the Media: Toward an Archaeology of Hearing and Seeing by Technical Means*, trans. Gloria Custance (Cambridge, MA: MIT Press, 2006), 30.

18. Robert Dennhardt, *Die Flipflop-Legende und das Digitale. Eine Vorgeschichte des Digitalcomputers vom Unterbrecherkontakt zur Röhrenelektronik 1837–1945* (Berlin: Kulturverlag Kadmos, 2009), 52.

19. Andreas Gelhard, Ulf Schmidt, and Tanja Schultz, eds., *Stillstellen. Medien—Aufzeichnung—Zeit* (Schliegen: Argus, 2004), http://www.editionargus.de.

20. Ibid.

21. Martin Heidegger, *Being and Time*, trans. Joan Stambaugh (Albany: State University of New York Press, 1996), 216.

22. Ibid., 344.

23. Information aesthetics defined the possible measure of temporarily changeable and thus time-variant structures in the present window of perception as 16 bit/s. See Herbert W. Franke, *Phänomen Kunst. Die kybernetische Grundlage der Ästhetik* (Cologne: DuMont, 1974), 121f.

24. On the epistemological assumption and media-technical measurement of the length of the "moment," see Karl Ernst von Baer, "Welche Auffassung der lebenden Natur ist die richtige? und wie ist diese Auffassung auf die Entomologie anzuwenden?," in *Zeitkritische Medien*, ed. Axel Volmar (Berlin: Kulturverlag Kadmos, 2009), 45–59.

25. Martin Carlé, "Psychoacoustics and Simulation: Breakdown and Reconstruction of the Bicameral Mind" (paper presented at the 10th Transatlantic Dialogue conference on *Sound Politics*, New York, April 2005).

26. Gotthold Ephraim Lessing, *Laocoon: An Essay on the Limits of Painting and Poetry*, trans. Ellen Frothingham (Boston: Roberts Brothers, 1887), 17f.

27. Heidegger, *Being and Time*, 311.

28. See Espen Aarseth, "Aporia of Epiphany in *Doom* and *The Speaking Clock*: The Temporality of Ergodic Art," in *Cyberspace Textuality: Computer Technology*

and Literary Theory, ed. Marie-Laure Ryan (Bloomington: Indiana University Press, 1999), 31–42.

29. Bernhard J. Dotzler, "'. . . anderes Zahlensystem, das bedingt ja andere Rechenverfahren'—Interview mit Konrad Zuse," in *Schreiben und Rechnen. Eine Technikgeschichte der Informationskultur* (Regensburg: Universitätsverlag, 2009), 67.

30. George Boole, *An Investigation of the Laws of Thought on Which Are Founded the Mathematical Theories of Logic and Probabilities* (New York: Dover, 1958).

31. Claus Pias, "Elektronenhirn und verbotene Zone. Zur kybernetischen Ökonomie des Digitalen," in *Analog/Digital—Opposition oder Kontinuum? Zur Theorie und Geschichte einer Unterscheidung*, ed. Jens Schröter and Alexander Böhnke (Bielefeld: Transcript, 2004), 306.

32. Ralph W. Gerard, "Some of the Problems Concerning Digital Notions in the Central Nervous System," in *Cybernetics: The Macy Conferences 1946–1953*, ed. Claus Pias (Zürich: Diaphanes, 2003), 1: 177.

33. Norbert Wiener, "Sensory Prosthesises," in *Cybernetics: The Macy Conferences 1946–1953*, ed. Claus Pias (Zürich: Diaphanes, 2003), 1: 158f.

Chapter 2

Signal Transmission and Delay

TEMPORAL MODES FOR LIGHT AND SOUND

Acoustic spatial perception is a function of the delay time of sound. The dimension of space thus transforms almost completely into a function of overlapping time events (waves, frequencies, and impulse responses). The concept of the "audio-visual" is divided by different delay times; with regard to the time-critical, it breaks into both of its separate channels. Because the delay time of sound can be consciously perceived by human senses, while the delay time of light cannot, every tentative analogy between sound and light waves comes to an end in terms of phenomenology. At a temperature of 20°C, sound waves propagate through the air relatively slowly at a rate of 343m/s; in contrast, the speed of light has always been the ultimate measure of the present time. William Stern introduced the concept of present time as the interval between two stimuli that are experienced as a unified impulse rather than two disjointed sensations.[1] Acoustic stimulus processing in humans is hardly able to distinguish between impulse sequences under a tenth of a second; the auditory threshold from a crackle to a sound lies in this range. On the other hand, optical stimuli are already perceptible at a duration of approximately 0.04 seconds.[2]

As soon as light is made audible by media-technical means—that is, transposed into the delay time of sound—the human experience of time is disrupted. The performances of Fournier d'Albe's optophone unsettled the certainty of a naturally ordered division of labor between the senses with respect to spatial and temporal perception. This was described with ironic commentary in the London *Pall Mall Gazette* in 1912.

An ingenious Birmingham scientist has turned the element of selenium to account by making light audible, and we are to be dazzled and deafened both at once. Sunlight makes a roaring sound, and lightning, presumably, anticipates its concomitant thunder. All we require now is to increase the anticipative process, and then daylight will awaken us every morning a couple of minutes before it arrives.[3]

Long before the speed of light revealed itself in the early modern period through the interaural time difference between the lightning and thunder of cannon fire, Aristotle (with his proto-media-theoretical instinct) had already approached sonic events with a keen auditory sense, which was more measuring-technical (and thus media-archeological) than music-aesthetical. In his discussions on objects of perception (*Parva Naturalia*), he anticipated Leibniz's later concept of *petites perceptions* as time-critical processes. The secret of audio vision is actually a (mostly technically constrained) synchronization of image and sound. The existence of a contributing transmission medium as a spatial-acoustic function of the delay time of sound could thus be discerned, which led Aristotle to develop his concept of media *avant la lettre*. Time is only directly accessible to people when it circumvents or surpasses their sensory perception; all other concepts of time are imaginary constructs (like "history"). The power of high-tech media over human perception lies in the subliminal domain of time. Based on empirical observation, Aristotle concluded that sound and vision require an intervening medium, so media-theoretically there must be an intermediary.[4] The difference between light and sound, however, is time-critical: while sight appears to be immediate for human senses, the acoustic echo proves the delay of (signal) time. The idea of light emanations as immediate was also revised following the discovery of a time barrier; however, the speed of light remained hidden from ancient measuring capacities and aesthetics and was only realized and then conceptualized in the early modern period. Scientists have since discovered that sound and light are not the only phenomena that occur in time; James Clerk Maxwell (through mathematical calculations) and Heinrich Hertz (through the performance of media experiments) also discovered this temporality in electromagnetic induction, which contradicted Newton's assumption of an immediate and virtually timeless *actio in distans*.[5] Once light was first identified as a temporal transmission channel, it became possible to conceive of time travel.

While Aristotle identified the irreducible role of an intermediary (*to metaxy*) through the delay time of spatial-acoustic impulses, this method of processing time was instrumentalized in a twentieth-century positioning technology called the "echolot," which was patented in Germany by Alexander Behm in 1916. The echo of the voice has always already deconstructed all phonocentric metaphysics of presence as a temporal extension or even retention of the present through pure propagation delay. In Shannon's communication model, the channel (which he explicitly declared the actual *medium*) functions as an interference-critical factor in the transmission of messages; in the echolot, however, the delaying medium for sound waves is water, which becomes the condition and source of information itself—an extreme shift in emphasis from the channel model. The media-anthropological function of Valdemar Poulsen's telegraphone, which was patented in 1898—namely, the recording and time-delayed reproduction of telephone conversations and announcements using magnetized wire coils—was also used as prosthetic memory in Samuel Spitz's echolot system, which was patented in the United States in 1918. By connecting a magnetic loop between the sender and the receiver of a sound, the time dilation of an extremely brief signal event became measurable and analyzable, just as the later magnetophone enabled the deceleration of playback for human senses. The media-archeological condition of possibility for this invention was the amplification of electric signals, which were otherwise too weak—Spitz first connected a telephone receiver and a microphone, but later used an electron tube. Although it was largely impractical, this method was nevertheless excellent from a theoretical standpoint—particularly from the perspective of media epistemology.[6]

In 1909, Max Dieckmann envisioned the possibility of transmitting images over distance using Braun's cathode ray tube to circumvent human nature. "If we only had apparatuses that worked rapidly enough . . . it would help us come to terms with our relatively low consciousness threshold for spatial and temporal sensitivity."[7] Electronic time is time-critical not only in the technical sense; it also epistemologically separates the time of human culture from the dynamic, phase-sensitive ways of processing time of high-tech media, just as the mechanical time of clocks separated the naturally perceptible time cycles of earlier cultures from the literally modern period.[8] The difference between machine time and electronic media time also breaks with the traditions of writing: while measurement information is inscribed and thus stored by registering apparatuses like the clockwork-driven kymograph (Ètienne-Jules Marey's graphic method), a signal change in the oscilloscope causes an immediate change in representation and overwrites the previously signaled condition. In the micro-temporal domain, the medium articulates itself. "In this sense, technology has a speed problem. . . . Every long-distance line limits the impulse sequence due to its capacity, and every receiving apparatus . . . proves to be far too slow."[9] Dieckmann then refers to the measuring medium that marked the beginning of the epoch of electronics, which is defined as the controllability of the flow of electrons: Ferdinand Braun's electron picture tube used as an oscilloscope. "It is highly significant that the cathode rays follow the deflecting forces instantaneously, so there is no detectable slowness."[10] Nevertheless, a physically conditioned slowness—namely, the luminescence in the phosphorous layer of the picture tube—must come into play in order for the comparatively slow human sense of sight to grasp the process in the first place. Through electr(on)ic measuring media, the micro-world of the time-critical becomes analyzable as a form of time. For their part, measuring media like the electronic oscilloscope—a device that Braun developed explicitly to measure the *temporal* flow of electric currents[11]—are electro-technical and material formations in the physical world (i.e., in time). In the oscilloscope, the time deviation with the signal to be measured, which is slowed down while passing through the line, is synchronized through an antagonistic delay line.[12]

In the archaic phase of digital computers, special buffers served "to accommodate the fundamental difference in speed between electronic and electromechanical technologies"[13]—interfaces of heterogeneous tempor(e)alities. The function of delay lines is also employed successfully in color television in order to produce visual impressions for sensitive human eyes. The basis of Walter Bruch's phase alternating line (PAL) color television system is a micro-temporal figuration—the phase change of each line, a quasi-musical principle that is familiar from sonic worlds. Its media-archeological background is Henri de France's idea not to modulate the two color signals necessary for the colorization of the image using a single carrier wave, as in the American NTSC system, but rather to transmit alternatingly only one of these signals for the duration of a television line and to retain the first signal in short-term memory until the second signal arrives and enables the superimposition necessary for color. The technical as well as chronopoetic name of the SÉquentiel Couleur A Mémoire (SECAM) system results from the alternating transmission and the signal buffer, "which must produce from 'memory' what it previously registered."[14] In 1961, Bruch made an important modification to the SECAM system. What goes on here is time-critical communication not between human and machine, but rather within the electronic apparatus itself. "Mathematical considerations led to the idea of forcing the system to display in consecutive lines the unavoidable transmission errors in reverse polarity, to make them coincident by storing them both in memory as with SECAM, and to throw out the mistakes through electronic communication."[15] In order to delay the chrominance signal in the PAL system to a line duration of 64 μs, electromagnetic conduction or even sound conduction is employed. The electric signal to be delayed is first transformed into an ultrasonic signal using an appropriate electromechanical converter; after following a predetermined route through the delaying medium of glass that corresponds to the desired delay time, the acoustic wave is then reciprocally retransformed into an electrical signal.[16]

Inductivity or capacity, the familiar parameters (and problems) in the transmission of electromagnetic waves, can be quickly manipulated in the form of continuous or discrete pulse-forming networks; in principle, every transmission line can be converted into a delay line. The lengths of electrical lines can be measured through propagation speed and impulse responses. Spatial segments (as in ancient Greek geometry) are thus replaced with temporal intervals. Since 1983, fractions of seconds have defined the length of a meter as temporal duration, namely as a light path oriented toward the oscillations of the cesium atomic clock. Frequencies must be measured in order to calculate the wavelengths of laser light—time is here inversely proportional to the number.

Every electro-physical transmission is characterized by the finite nature of its speed—like light itself. This also constitutes the time barrier in the design of complex circuits on computer chips. In long connection lines, a signal occasionally does not arrive punctually at a distant switching element, so a false value is processed—these so-called hazards are found in asynchronous switches. "This is solved . . . through the introduction of synchronized timing, whose cycle time must be measured adequately enough that the different delay times have no impact."[17] First, the object of measuring perception becomes literally mechanical time delay and thus the slowness of decreasing or oscillating material. Galileo Galilei constructed an inclined plane (a ramp) in order to make gravity experiments understandable on the laboratory bench—and that means in a decelerated time window. In the case of light, however, human senses are incapable of distinguishing between the impression of promptness and a minimal dilation (because it remains below the perceptual-critical threshold); the formula $t = \Delta t$ applies here in the sense of the aesthetics of infinitesimal calculus. "If a shorter path requires less time than a longer one, then the propagation is temporal."[18] As Aristotle discovered the existence of a resistant intermediary as quasi-medium based on the propagation delay of sound, this was still conceived from physical material; in contrast, Michael Faraday's observations of electromagnetic induction and James Clerk Maxwell's calculations of the same using differential calculus established a much more radical media temporality. The media channel dissolves into dynamic transit time.

Time-critical moments emerge at the interface between humans and machines. The exposure time of the photographic apparatus, for example, can be below the light stimulation threshold of the human senses. While Nicephore Niépce, in the primal scene of modern photography, still used ten-hour exposures, daguerreotypy already decreased this to ten minutes. Richard Leach Maddox's gelatin-coated dry plates, which were further developed by Charles Harper Bennett in 1878, reduced exposure times to 1/25 of a second, which subsequently made chronophotography and then cinematography possible.[19] The media-operative time field thus evades immediate human perception; with highly developed mechanical cameras, analog photography brought the interval with 1/1000 of a second virtually to the present moment. This moment is never without extension, however, and it thus reveals the implicit ideality of the concept. Time critique is also a critique of metaphysical media theory.

In geophysical earthquake research, the term "critical point" refers to the intersection point between the velocity-time curves of the direct wave and the seismic frontal wave. Submarine sonar and radar calculate spatial images as a human interface using the delay times of acoustic or optical echoes. The ear is actually capable of a highly temporal resolution of different frequencies, yet only the eye is capable of precisely differentiating the simultaneity of diverse movement information on a two-dimensional plane. The so-called acoustic camera transforms the delay times of sound into spatial-visual locations by mathematically extrapolating spatial distances from the delay times of acoustic signals. Through corresponding delays as interpolations, a signal received through distributed microphones is once again brought into phase (beamforming); conversely, a spatial image can be literally calculated from different sound sources because this analysis passes as time-discrete AD conversion in the sampling of arriving signals. The sampling rate is time-critical because the precision of the spatial location, which can be determined to the nearest centimeter, is a direct function of this time information. Time here becomes geometric space, and the analysis of consecutive time events leads to the localizations of objects.[20] Scanned in differently weighted pixels, this sound field is then also visually perceptible for humans.

The delay times of sound and light belong to entirely different time regimes. However, the discovery of the finite nature of light made it (despite the difference between physical compression waves and electromagnetic waves) similar in character to acoustic processes as time events. This resulted in Christian Huyghens' analogy concerning the spread of light and sound waves. Even though the speed of light is much faster than the speed of sound, a propagation speed—regardless of how unimaginably fast it is—is nevertheless anything but momentary; "the difference between that and this is the same as the difference between the finite and the infinite."[21] This implicitly articulates the Christian-theological sensitization for finite processes, which deviated from the cosmic thought of ancient Greece. At the same time, the message is a temporal *to metaxy*, or in physical symbolic language: Δt.

In order to verify the existence of an ether wind experimentally, Albert Michelson constructed an instrument called an interferometer that was designed to prove the assumed differences in the delay times of light through reflection. It thus artificially extended rays of light through an arrangement of mirrors, telescopes (employed here, in contrast to Galileo, for physical microstructures), and light sources. However, this experiment actually led to the discovery (*ex negativo*) that the ether did not exist and that a medium was not necessary for the transmission of light, unlike the transmission of sound. The apparently immediate effect of light refraction through different media thus proved to be *temporalized* (for analog as well as for electromagnetic dynamism): the propagation speed of light changes when it crosses over into another medium. Every camera obscura, every photographic apparatus thus performs a time-critical function that precedes the temporality of the exposed substrate. The different nature of acoustic and optical waves is also emphasized here: the former is on the side of mechanics (elasticity), while the latter is on the side of the electromagnetic spectrum, of which it constitutes a small segment. However, electromagnetism could only be described through the media concept of classical physics, and it required the field concept developed in the mathematical analysis of Maxwell's equations. The connection between electrical and magnetic fields expressed in these equations brought speed time-critically into play. The value c, which is familiar as the speed of light, is also identical to the speed of electromagnetic waves in open space. The interferometer constructed in Potsdam in 1881 resulted in delay times caused not by the refraction of light in the "ether" but rather by vibrations from the traffic in neighboring Berlin.[22]

MAGNETIC TAPE AND TIME DELAY

The critical element of delay times became aesthetically productive in the time-linear storage medium of magnetic tape. In the Sun Records studio of music producer Sam Phillips, they became influential for a specific strain of early rock 'n' roll—namely, the recordings of Elvis Presley beginning in 1953. For the song "That's All Right," which was recorded before multitrack technology, two tape recorders were played with the same recording minimally delayed. This resulted in the so-called tape echo, which was to become the first temporal signature of that form of pop music and which developed different effects depending on the acoustic articulation (on the one hand voices and on the other hand instruments). Presley himself internalized the genuinely media-induced temporal mode of *slapback sound*, which shaped the hiccup-phrasing of his syllabic singing style; the model for human articulation thus became the technical medium itself, and in this moment the singer became a tape recorder. While western music as an event was thus far strictly linked to the presence of the musician or rather the instruments (live), the presence-generating power of their electronic performances was already a function of technical storage.[23]

For the purpose of recording and replaying linearly in time, magnetic tape evolved into a one-dimensional media channel. Acoustics and sound thus became virtually identical, yet this was shattered by the combined concept of audio vision once it involved the recording of electronic images. While the sound track on video tape still remained linear—as did the track for the synchronization pulses, which made the question of temporal modes an inner object—the technology of video recording advanced from the originally obvious parallel recording of individual television lines (also known as "longitudinal recording" or "direct recording"—a term that is striking in a media-epistemological way)[24] to transverse recording. Another process that fell between these two methods was presented to the public in October 1952 by the laboratory of Bing Crosby Enterprises in the United States: on the basis of the high frequency range of the electronic image signal, the recording was divided into several channels (partial frequencies) and written on the magnetic tape in parallel tracks using the same number of magnetic heads. However, the synthesis was to a large extent time-critical. "In order to reproduce the original image signal all of the tracks had to be scanned and the partial frequencies recombined. . . . However, the defects with the interference-free reassembly of partial components into the original video signal were considerable."[25] The smallest differences in delay time disrupt the phase-correct reassembly, which is indispensable in order to achieve the successful image effect and thus the intention of the technical nature of the image; the physiological sensory channel is highly sensitive to such deviations in time. "The visible results of the replay of images were that they demonstrated the so-called shutter effect. That is, continuous horizontal bars appeared on the screen, and the images cast shadows and flickered due to a lack of stability."[26] This time-critical challenge is already familiar from the related process in the sonic domain—namely, the technology of the vocoder, which divided the frequencies of the human voice and thus made them transmittable and resynthesizable in parallel channels.

In March 1955, the Ampex Corporation, which went from tape recorder technology to the development of video, successfully presented the transverse process in the quadruplex system. A technical metamorphosis was also at work here in the temporal essence or mode of the video recorder. "The significant technical change that the process had undergone was the replacement of amplitude modulation, which was normal in magnetic tape technology, with frequency modulation."[27] This was a radical temporalization in the formation of the transmission medium, although it was familiar from VHF tank radio in World War II. The public premiere of this new image recording system was the broadcast of a CBS news program entirely from magnetic tape on November 30, 1956. The program was previously produced in New York, where it was broadcast live, and sent via transcontinental cable to CBS Television City in California to be recorded on video and then broadcast around the country, keeping in mind the three-hour time difference from east to west. What began as a delay-critical question was therefore updated here as micro-temporal phase shifting. The mastery of time-critical processes at the media-archeological level (i.e., in the apparatus) is rewarded with time sovereignty, but it also leads to the disruption of temporal certitude. "The replay of the program was free of interference, and its quality could hardly be distinguished from that of a live transmission."[28] The result was a Turing test of human perception from a time-critical perspective; event time, broadcast time, and receiver time converged in the electronic semblance of temporal immediacy. In contrast to the time-consuming process of developing film, the interval between these points in time (Δt) tends toward zero. Marshall McLuhan interpreted the electronic time regime as similar in nature to "acoustic space." The contrast between the technical minimization of the time delay in video and the rhythm of cinematic newsreels made television the medium of the present based on the model of radio. It was thus close to the temporal essence of the transmission speed of electromagnetic waves as the actual medium as the time-critical event. The reverse of the superficial appearance of the television and video image is the media-archeological insight into its temporal modes.

ELECTRONIC MEMORY

The rectification or rather smoothing of alternating current (a condition of the audibility of radio reception) is accomplished through the use of condensers as buffers: accumulators that become charged "fast as lightning"[29]—exactly as fast as electricity—whenever the tube lets current flow through and then emits the stored current if necessary to the receiver. As part of the wiring of an operational amplifier, a condenser is able to offset memory against time itself in that the charging current for the condensers is divided and only a fraction of the current reaches the condensers—whereby "the circuit simulates substantially larger capacities than are actually available."[30] An ideal simulator of such processes is the RC element. "A condenser is charged 'slowly' via a resistor until (after a theoretically infinite period of time) it is charged to its final condition."[31] The output voltage follows the input voltage with a delay. The tempo-spatialization that philosophy identifies with the act of writing as opposed to the apparent immediacy of the spoken word[32] is implemented by electronics as a time-critically intensified *différance*.

This intermediacy is familiar not only to the world of analog media electronics. The electromechanical or electronic features of a calculating machine made of bistable circuits, "which can temporarily store a number,"[33] are registers in the techno-mathematical (not archival) sense. The essence of this memory lies in the transition. The high cost (of electron tubes) leads to their restricted use only for short-term, temporary storage. The same arrangement thus functions as register or counter. Time and number stand side by side in Aristotelian equitemporality. Computer memory is close to the archival *dispositif*, but extends it with the parameters of time and timing. Magnetic storage with a moving recording medium is literally an "archive in motion."[34] "In most processes a clock track is recorded parallel to the storage track so that the recorded impulses can be assigned to their temporally correct place values when replayed."[35] The place value system as a condition of discrete calculation already implies temporal seriality in symbolic mathematics. Temporal regularity is time critique; memory is made dynamic to the smallest temporal moment, like word recognition on magnetic tape. If delayed clock pulses move the content of the shift register in the direction of lower place values, then number, time, and memory become entangled. The value of the word counter indicates which number is contained in the shift register, and it thus generates the so-called address of the stored number.

Ferrite core memory—the working memory of early computers—uses the physical property of hysteresis of a ferromagnetic material for the purpose of remagnetization, whereby the symbolic binary values zero and one are realized in the real world as magnetic field strengths. In ferrite core memory, however, zero and one materialize as signals in different time processes.[36]

In the flip-flop circuit of a pair of electron tubes (triodes), a minimal time delay also comes into play at the level of the smallest informational (memory) unit, the *bit*: like a neuron in the nervous system, the stimulation must exceed a certain threshold value in order for a pulse to be released (commonly known as the so-called Schmitt trigger). Switching hysteresis is a genuine time figure of electronics at its elementary operative and thus media-archeological level: the switchover of tubes or rather transistors occurs later with increasing voltage (hysteron) than with decreasing voltage. Binary switches always necessarily operate with the smallest time moments of the circuit itself; here, the delay times are no longer infinitesimally minimized, such as through interpolation. In the "image" of magnetic core memory (as it involves a two-dimensional matrix), reading simultaneously means erasing—as an invasive, media-archeological gaze. Reading and rewriting are thus time-critically correlated, which in the temporally distributive field of process execution (scheduling) supports the chronotechnology of batch processing.

> Core storage was an improvement on the Williams tube and not only in terms of reliability. It is unusual in that the memory is non-volatile—you can remove the power from a core store and the data will remain intact. Until, that is, you read the data. Reading data from a core memory is destructive; a read resets the store to zero. So with a core memory the critical time is not just the time taken to read the memory, but the time to read the data, and then restore it, so that it can be read again.[37]

The smallest dynamic time information is media-operationally implemented in conjunction with short-term memory, as in a German radio location device from World War II—the phase direction finder[38]—which determined locations by calculating time differences. The privileged human organ of such temporal relations is the ear; the direction finder acoustically succeeds when it is able to discern weak signals from background noise. However, such time signals were only truly understandable when they were played back at slower speeds; in other words, they were only understandable as technical recordings, and a magnetic record was therefore attached to the device. In this way, media time was translated into human time.

TEMPO-SPATIALIZATION: SHORT-TERM MEMORY

The electronic relay circuit, which keeps an actual binary condition stabile, constitutes a momentary storage and thus the smallest electronic memorial monument. Norbert Wiener associated it with the desire "to have special apparatus to retain an impulse which is to act at some future time"[39]—a suspended and deferred time, a time in latency, and a virtual time that awaits actualization.

> A very important function of the nervous system, and . . . a function equally in demand for computing machines, is that of *memory*, the ability to preserve the results of past operations for use in the future. . . . There is first the memory which is necessary for the carrying out of a current process, such as a multiplication, in which the intermediate results are of no value when once the process is completed.[40]

This is realized through static or dynamic buffers, like the construction of a short-term memory that keeps a sequence of impulses traveling around a closed circuit until this form of time-based buffer is cleared by intervention from outside and a new sequence is fed in. Wiener assumes that "this happens in our brains during the retention of impulses, which occurs over what is known as the specious present"[41]—very much in the sense of the neurological concept of a *present time window*. In addition, he also discusses delay memory in early computers, such as the Williams–Kilburn tube, which was able to store not each individual bit statically—like a flip-flop circuit—but rather thousands of such choices dynamically.[42] What is commonly known in cinematographic physiology as the "afterimage" is implemented here in the temporary storage of digital information as the afterglow of electron radiation on the phosphorous layer (0.2 s). While the afterimage produces the impression of continuous movement in humans, the purpose of the Williams–Kilburn tube consists precisely in retaining the binary discreteness of the pixel through amplification-based refresh cycles, which is the entire difference between analog and digital media time. Any information that is not stored or renewed on the tube in 0.2 s is lost to the computer; the finite nature of electronic-immaterial signifiers makes emphatic memory the operational storage of the present itself—a delicate, processual time window that depends on synchronization.

What is commonly known in the logistical movement of goods as so-called chaotic storage also occurs micro-archivally on computer hard drives. The data stored on a compact disc (CD) is nested; in other words, it is stored not serially in the temporal playback sequence, but rather dissipatively. This reduces the risk of losing entire bytes of a sound signal through scratches. A marketable CD player reads and sorts the datasets on the disc, so that every second 44100 words (16 bits each) emerge in the correct temporal sequence—in the sense of a reconversion of the optical and thus visual storage of the CD into the temporal sequence of a quasi-sonic (ac)count. The emphatic time of archival memory is undermined by the micro-event of electronic storage media.

"DELAY TIME"—A TIME-CRITICAL CONCEPT

Delay time is a familiar concept from physical acoustics to electronics that describes the effective time consumption of every act of signal transmission; mechanical and electronic signal transmissions are both fundamentally affected, albeit to different degrees. In contrast, informatics introduces logical delay time. The time span within which a software application is executed in working memory can be measured through the asymptotic delay time of an algorithm, which estimates the amount of time a program needs to solve a problem depending on the input length.

In the microwave range, the performance of space charge tubes decreases as a result of the delay time effects of electrons—at the limits of the speed of light. Yet this temporality was made technologically productive: for ultra-high frequency ranges, different velocity-modulated tubes are employed "with which the limited delay time of electrons is consciously utilized to amplify and generate high-frequency electromagnetic vibrations."[43] The magnetron, the secret of allied radar reconnaissance in World War II, generates diverse resonance frequencies from the flow of electrons and cavity resonators, which are induced to oscillate at their natural frequency; the interaction space thus functions as a delay line. Delay time devices make it possible to calculate (electro)physical delay times *qua* buffering in that they sample the input voltage with an arbitrary clock frequency, convert the sampled value into binary-coded text, and store the values in a shift register. The analog signal is then reproduced at the outputs via the digital-to-analog converter. The device can be used to transform the time of unique events, and it thus represents a veritable time machine.

A variant in the realm of digital audio and video transmission is run length encoding, which was designed for the purpose of loss-free data compression. This process replaces a series of identical symbols repeated multiple times in the string, such as homogeneous color areas that are sampled and transmitted line-by-line, with the description of the singular value and the number of its repetitions. In the sonic realm, this means that calculated sequences of silence in audio files are temporal modes that economize storage and transmission capacity.

Circuit technology models not only logical systems, but also their temporalness, which manifests in time processes and their denials (superpositions, delays, and time invariance). The expression $t \rightarrow t1$ indicates a temporal delay to the right; the culture-technical effect of reading direction thus converges here with the thermodynamically justified arrow of time. Time invariance is a system whose behavior remains the same regardless of such time delays; in acoustic space, this reaction uniformity can be perceived by human senses. The form of the output signal is thereby independent of the transformation of the signal. There is an internal time dependence, but it is independent of the activation time (temporal invariance). In technical systems, time delays are caused by "internal processes"[44] in the active and passive components; line lengths in the circuits (even though they are extremely compact in microprocessor architectures) and temperature fluctuations come into play here. Processes are identified as stationary when their average size does not change, like the technical alternating currents. Quasi-stationary processes with weak time dependence, like the electronic drift experienced sonically in short-wave radiophone reception and in the calculating space of analog computers, constitute a special case. A virtually poietic bouquet of dynamic transfers unfolds between stationary conditions: skip functions, starting actions, and pulse values. Transient analysis or differential equations serve to describe such events, which are intensified in the time domain.

BUFFERING MODES

The concept of time delay as a form of "analog storage" is at work not only in classical electronics but also in a media-epistemologically distinctive and decisive way in the digitization of analog signals (as "sample and hold"). A handbook explains this by means of the difference between analog and digital oscilloscopes.

Unlike conventional oscilloscopes, the waveform display shown on the cathode-ray-tube (CRT) screen is not directly produced by the input signal voltage. The beam is controlled digitally from information previously recorded in memory. The information stored in memory was received from an analog-to-digital converter which measured the input signal at typically 4096 times following the start of the sweep.[45]

Digitally buffered signals thus allow their manipulation as time event; suspended in memory, time becomes mathematically available.

> One difference in the recording and display processes results from the desirability of showing each signal waveform as it occurs, at times; and at other times to hold a particular waveform for prolonged examination by the operator. . . . In this instrument every signal is recorded. . . . Each new signal is measured and the voltage information replaces in memory that of the previous signal, in this mode. The stored information is read from memory and displayed continuously, so if signals do not occur frequently the display is nevertheless ready.[46]

With digital computers, a temporal mode of delay comes into play whose mathematical-discrete nature is radically different from the constant delay time required by analog technology. "A disadvantage of a digital method of solution lies in the relatively longer amount of computing time required for each solution," it says in the analysis of a computer-based method of missile air defense[47]—therefore in a context where small delays in prediction can be deadly if the calculated result lags behind the target event (a critical oscillation between protention and retention). The numerical integration of the flight data to be determined and the sampling of measured data require—albeit the smallest—computing time intervals;[48] in a concrete situation, "this would result in a running time of about eight times real time."[49] It first became possible to achieve sufficient precision in computing applications through the hybrid combination of the analog computer, which was powerful in the temporal domain due to parallel processing, with the virtues of the digital computer; "the solution could be run at about one-tenth real time."[50] The actual battle here is with time.

In contrast to the live transmission of classical electronics, in which the signals represent pure physical time functions, the irreducible delay times in digital communications media are of another nature. The time-critical deceit of digital electronic media is that they are able to dissimulate the actual delay in computing time with algorithmic intelligence. It is remarkable that this computing time (as well as the data to be calculated from the present or from storage) is not bound to a one-to-one locatable carrier; rather, it can in principle take place at diverse locations, culminating in "distributed computing" with its cascades of buffers. "Only a medium that always runs over a discrete period of time as intermediate storage instead of simply sending signals makes it possible to replace the arduous process of adjusting each individual device with digital automatic self-calibration."[51] Since the introduction of the interrupt signal, the communication between input, calculation, and output units distinguishes itself through a form of time-critical (self-)observation, whose condition of possibility is intermediate data storage or the holding of information. Signal distortions are particularly noticeable in the temporal domain; special buffers (jitter buffer and dejitterizer) serve here to restore a continuous signal or data stream. In the course of the international Phobos mission of 1988, the R3m video image storage system developed in the former GDR was employed in space. This system stored digital image data on magnetic tape, as the data produced by satellite cameras normally cannot be transmitted continuously to Earth and intermediate storage is needed until the appropriate moment for broadcasting.

In analog media, the delay as a time-critical component of technical communication is not like the "run-time" of executed programs or the delay time of algorithms; rather, it is merely the delay of signal transmission, which is decidedly a function of the materiality of cables. The analog/digital difference as the difference between live transmission (delay time) and real-time transmission (digitally calculated time window) can thus be fixed to the concrete parameter "t" in the domain of Δt. Material transmission channels, like wires between electronic components, always already implies a delay time—even with minimal distances, as Hermann von Helmholtz already identified for communication in humans in his time-critical measurements of nerve impulses as intermediate times (and thus literally as *media* time *avant la lettre*).[52] The temporal interval in which human consciousness "lags behind the present," which von Helmholtz called *temps perdu*, was illustrated by means of graphic curves.[53] From a cultural perspective, this concept builds a bridge to the literary work of Marcel Proust, but in the world of actual media, it more closely corresponds to the *terminus technicus* "dead time." What was employed positively in the delay time memory of early computers was made noticeable as interference in closed circuits, such as control loops:

> The time it takes for material to travel from one point to another can add dead time to a loop. If a property (e.g. a concentration or temperature) is changed at one end of a pipe and the sensor is located at the other end, the change will not be detected until the material has moved down the length of the pipe. The travel time is dead time. ... The distance may only be an arm's length, but a low enough flow velocity can translate into a meaningful delay.[54]

With sampling as a process of AD conversion, this assumes a techno-mathematical form, as the digital necessarily depends on moments of intermediate value storage. "Control loops typically have 'sample and hold' measurement instrumentation that introduces a minimum dead time of one sample time, T, into every loop. This . . . indicates that every loop has at least some dead time."[55]

The power and the limits of techno-mathematics are evident in this context. In agonal situations, the deployment of an "anti-aircraft predictor" was able to compensate for foreseeable dead time—a process that was employed in artillery as well as air defense and that inspired mathematical cybernetics as well as computers.[56] Claude Shannon was for this reason fascinated with the challenge of juggling automata and the real-time processing involved.[57] The transmission and reaction time are thereby omitted from calculation, but this can only be achieved in the macro-physical temporal domain. A complex analog signal path—insofar as it was not generated synthetically from deterministic signals and thus mathematical functions $x(t)$—can hardly be predicted accurately at the level of its concrete micro-temporal eventfulness and thus can only be corrected through an artificial delay in limits. However, the transmission and processing of coded signals can be calculated as symbol strings; mathematical probability prediction and other forms of calculated time intelligence are thereby accessible, especially when the time channel itself is time-discrete.

Technical "delay" also applies in other time domains. As systems theory emphasizes, (technical) memory *is* not; rather, it must be implemented as a difference, as a given fact. Generative memory machines record vectorial time indices in stored datasets—logical-archival, not historical-hermeneutic time. Georg Wilhelm Friedrich Hegel also emphasized that stored memory and memory-time represent different modes of existence. The storage-archival network stands more on the side of logistics than emphatic temporality; it is subject to time at best as computing time for access, as the digital computer is *clocked*. This timeless, archival (because addressable) space is radically present as latency as well as actualization. As opposed to storage, emphatic time and human memory (commonly known as the "historical") are external, improper functions.

THE TIME CHANNEL AS MEMORY: ACCESS TIMES, DELAY MEMORY

Every transmission of a signal, no matter how fast, is a temporal process—even when it coincides in digitized telecommunications with the moment of switching itself. In delay memory, time is understood not as a point, but rather as technical implementation; this also includes macro-technical delay time memory. Physical (unlike time-logical or so-called virtual) storage can be realized in the computer mechanically, electromechanically, electrostatically (Williams–Kilburn tube), magnetically (magnetic drum storage), acoustically (mercury delay line), optically, or chemically. The decisive criterion for static or dynamic memory is chronotechnical. "The application of different qualification levels depends on the speed at which the telecommunication system manages the accruing amount of information. The choice is dictated by the 'level of intelligence' of such systems, . . . 2–200 msec for magnetic drum storage, for example, and 1–10 sec for magnetic tape storage."[58] Memory-programmable computer architectures in their classic formulation require temporary storage areas for the short-term memory of values, starting with the registers and continuing in working memory. The technical-operative diagram of an archaic delay memory, the mercury delay line, features a bit-series transformed into acoustic pulses and intervals, which moves through the mercury in order to be converted back into current pulses that are fed into the circuit as information or sent to further intermediate storage renewed in the delay line.[59]

The condition for memory hierarchies in computers is the access time or the interval that elapses from the activation of a memory cell through the control unit to the output of the stored code word. "If the memory is only realized with triggers, then the access time is equal to zero."[60] The alternative delay memories follow another mathematically economical calculus in that the disadvantage of the longer access time is offset by the advantage that they record more information using less material and energy. In such dynamic memories, the data words are in constant motion and need to be addressed with temporally infinitesimal precision. Memory here becomes an object of techno-mathematical analysis in every sense. In the early buffers of electronic computers, magnetostrictive delay lines functioned analogously to mercury delay lines. These were metal rods with coils attached to the ends. The electromagnetic vibrations that resulted from the passage of pulses through the input coil caused mechanical vibrations in the rod, which were converted back into electromagnetic vibrations and pulses from the output coil—a special kind of oscillating circuit.

In addition to the actual arithmetic logic units at the center of the central process units of computers, however, are the so-called registers—a concept adopted from administration and archival practices to refer to the filing of interim data in computations. They merit a closer look: registers are an archival part of the operative administration and its immediate working memory as so-called "old registers"; they do not become part of the actual archive until after they are sorted. It also holds true for computers that working memory means the removal of data from the operative register, not the temporary storage of data at the register level. Registers serve here in general as momentary buffering and thus constitute part of the present window rather than part of memory in the emphatic sense—even if the content of CPU registers can now be preserved for longer periods without being refreshed.

These small operative memories as extensions of the present were at work in a very direct form in Colossus, England's first fully electronic computer, which was developed to decode German military communications in World War II. "The machine added bits together in specialised valves called pentodes. But the real innovation was the memory. In order to calculate deltas"— and thus the folding of bit strings onto themselves, like the ancient double writing tablet *deltion*, for the purpose of cryptographically detecting repeated character combinations—"Colossus had to 'remember' a bit for a split second until its neighbor arrived. For this task, it used a bank of capacitors which it charged up and discharged as needed."[61] At the same time, the power of this new intermediate archive was time-critical in the sense that the smallest temporal moments were crucial for data processing. The Mailüfterl—one of the first fully transistorized computers, which was built at the Vienna University of Technology in 1957—operated in terms of a dynamic circuit technology, which used the slowness of the transistors for time delay and storage. A weakness here becomes an opportunity.

"Process dead time" lies under the sign of Thanatos. "Dead time is the delay from when a controller output (CO) signal is issued until when the measured process variable (PV) first begins to respond. The presence of dead time, Θp, is never a good thing in a control loop."[62] In the core process of digitization, namely in the "sample and hold" with the digital measuring of analog signal currents, a minimal amount of dead time per "sample time" is unavoidable. It is this dead time that separates the real world from the information world. With signal and data processing, however, a line-conditioned dead time is sometimes not a defect, but rather a media-temporal quality.

> If a property (e.g. a concentration or temperature) is changed at one end of a pipe and the sensor is located at the other end, the change will not be detected until the material has moved down the length of the pipe. The travel time is dead time. . . . The distance may only be an arm's length, but a low enough flow velocity can translate into a meaningful delay.[63]

NOTES

1. William Stern, *Die differentielle Psychologie in ihren methodischen Grundlagen* (Leipzig: Barth, 1911).

2. Ernst Pöppel, *Grenzen des Bewußtseins. Wie kommen wir zur Zeit, und wie entsteht Wirklichkeit?* (Frankfurt: Insel Verlag, 2000), 35ff.

3. Qtd. in Cornelius Borck, "Blindness, Seeing, and Envisioning Prosthesis: The Optophone between Science, Technology, and Art," in *Artists as Inventors—Inventors as Artists*, ed. Dieter Daniels and Barbara U. Schmidt (Ostfildern: Hatje Cantz, 2008). See also E. E. Fournier d'Albe, *The Moon-Element: An Introduction to the Wonders of Selenium* (New York: D. Appleton, 1924), 98–99.

4. Aristotle, *Über die Seele*, § 419a (19–25).

5. An example is Felix Eberty's *Die Gestirne und die Weltgeschichte* (Berlin: Rogoff, 1923), which was first published anonymously in 1846.

6. See Harvey C. Hayes, "Measuring Ocean Depths by Acoustical Methods," *Journal of the Franklin Institute* 197.3 (March 1924): 323–54.

7. Max Dieckmann, "Fernübertragungseinrichtungen hoher Mannigfaltigkeit," *Prometheus* 20.1010 (March 1909): 337.

8. See Peter Berz, "Uhrwerk und Zeitgetriebe," in *Zeitreise. Bilder, Maschinen, Strategien, Rätsel*, ed. Georg Christoph Tholen, Michael Scholl, and Martin Heller (Basel/Frankfurt am Main: Stroemfeld/Roter Stern, 1993), 172f.

9. Dieckmann, " Fernübertragungseinrichtungen hoher Mannigfaltigkeit," 337.

10. Ibid., 339.

11. Ferdinand Braun, "Ueber ein Verfahren zur Demonstration und zum Studium des zeitlichen Verlaufes variabler Ströme," *Annalen der Physik und Chemie* 60.1 (1897): 552–59.

12. Dietmar Benda, *Wie misst man mit dem Oszilloskop?* (Poing: Franzis, 2004), 29.

13. William Aspray, introduction to *Proceedings of a Symposium on Large-Scale Digital Calculating Machinery*, ed. Howard Aiken (Cambridge: MIT, 1948), xvii. The section on "Storage Devices" discusses an entire bouquet of alternative delay memories.

14. Walter Bruch, *Eines Menschen Leben* (Mittweida: Hochschule Mittweida, 2008), 390.

15. Ibid., 391.

16. Rudolf Schiffel and Artur Köhler, eds., *Elektronik-Arbeitsblätter: Fernsehen I* (Munich: Franzis, 1974), 35.

17. Jörg Pflüger, "Wo die Quantität in Qualität umschlägt. Notizen zum Verhältnis von Analogem und Digitalem," in *Hyperkult II. Zur Ortsbestimmung analoger und digitaler Medien*, ed. Martin Warnke, Wolfgang Coy, and Georg Christoph Tholen (Bielefeld: transcript, 2005), 45.

18. Heinrich Hertz, *Über sehr schnelle elektrische Schwingungen. Vier Arbeiten* (Leipzig: Akademische Verlagsgesellschaft, 1971), 108.

19. See Jens Schröter, "Die Macht der Stillstellung. Zur technologischen Abtastung und Verfolgung am Beispiel der Fotografie und des Computers," in *Stillstellen. Medien—Aufzeichnung—Zeit*, ed. Andreas Gelhard, Ulf Schmidt, and Tanja Schultz (Schliengen: Edition Argus, 2004), 60–74.

20. Felix Pfeifer, "Echtzeit Schallvisualisierung, Ortung und Quellentrennung mit Mikrofronarrays und Beamforming" (paper presented in the seminar *Semantic Media—Semantic Audio*, University of Potsdam, October 2009).

21. Christian Huyghens, *Abhandlung über das Licht*, ed. E. Lommel (Leipzig: Engelmann, 1890), 11.

22. Silvio Bergia, *Einstein. Das neue Weltbild der Physik* (Heidelberg: Spektrum der Wissenschaft Verlagsgesellschaft, 2002), 14.

23 Thanks to Jens Gerrit Papenburg for drawing my attention to this fact.

24. Siegfried Zielinski, *Zur Geschichte des Videorekorders* (Berlin: Wissenschafts-verlag Spiess, 1986), 72.

25. Ibid., 72f.

26. Ibid., 73.

27. Ibid., 83.

28. Ibid., 104.

29. Eduard Rhein, *Wunder der Wellen. Rundfunk und Fernsehen dargestellt für jedermann* (Berlin: Deutscher Verlag, 1935), 80f.

30. *Handbuch zum Lernpaket: Elektronik mit ICs* (Poing: Franzis, 2008), 41.

31. Ibid., 29.

32. Jacques Derrida, "Die différance," in *Randgänge der Philosophie* (Vienna: Passagen, 1988), 29–52.

33. A. Huber, "Programmgesteuerte elektronische Rechenmaschinen," *Funk-Technik* 24 (1957): 828.

34. See Eivind Røssaak, ed., *The Archive in Motion: New Conceptions of the Archive in Contemporary Thought and New Media Practices* (Oslo: Novus, 2010).

35. Huber, "Programmgesteuerte elektronische Rechenmaschinen," 829.

36. See Heinz Billing, "Datenverarbeitungsanlagen," in *Technik der Magnetspeicher*, ed. Fritz Winckel (Berlin/Heidelberg/New York: Springer, 1977), 307.

37. David Morton, "Memory Lanes," *Personal Computer World* 15.2 (February 1992): 312f.

38. See Hans Schellhoss, "Die Funkortung beim Heer, Tagung Landsberg," http://www.cdvandt.org/navigation-tagung.htm.

39. Norbert Wiener, *Cybernetics or Control and Communication in the Animal and the Machine* (Cambridge, MA: MIT Press, 1985), 120.

40. Ibid., 177f.

41. Ibid., 178.

42. Frederic C. Williams and Tom Kilburn, "A Storage System for Use with Binary-Digital Computing Machines," *Proceedings of the Institution of Electrical Engineers* 96.40 (1949): 81–96. See, in particular, figure 2: snapshots of screen displays as dynamic binary data buffers for 1024 and 2048 bits.

43. Jörg Bretting, *Technische Röhren. Grundlagen, Funktionen, Anwendung* (Heidelberg: Hüthig, 1991), 162.

44. Horst Völz, *Information I. Studie zur Vielfalt und Einheit der Information. Theorie und Anwendung vor allem in der Technik* (Berlin: Akademie, 1982), 153f.

45. Operating Manual for the Explorer Digital Oscilloscope Models 1090A and 1090AR (Madison: Nicolet Instrument Corporation, 1975), 1.

46. Ibid., 1f.

47. For more on the "missile intercept problem," see Harold K. Skramstad, "Combined Analog-Digital Techniques in Simulation," *Advances in Computers* 3 (1962): 283.

48. The so-called "memristor" is distinctive in this regard, as its internal resistance varies depending on the electric current, and it is able to hold this value constant when the current is switched off.

49. Skramstad, "Combined Analog-Digital Techniques in Simulation," 283.
50. Ibid.
51. Friedrich Kittler, "Gleichschaltungen. Über Normen und Standards der elektronischen Kommunikation," in *Interface 1. Elektronische Medien und Künstlerische Kreativität*, ed. Klaus Peter Dencker (Hamburg: Hans-Bredow-Institut, 1992), 180.
52. Hermann von Helmholtz, "Über die Methoden kleinste Zeittheile zu messen und ihre Anwendung für physiologische Zwecke," *Königsberger naturwissenschaftliche Unterhaltungen* 2 (1851): 169–89.
53. Henning Schmidgen, "Die Helmholtz-Kurven. Auf der Spur der verlorenen Zeit," (Berlin: Merve, 2010). See also Henning Schmidgen, "Mind the Gap: The Discovery of Physiological Time," in *Film 1900: Technology, Perception, Culture*, ed. Klaus Kreimeier and Annemone Ligensa (New Barnet: John Libbey, 2009), 53–66.
54. Doug Cooper, "Dead Time Is the 'How Much Delay' Variable," http://www.controlguru.com/wp/p51.html.
55. Ibid.
56. David A. Mindell, *Between Human and Machine: Feedback, Control, and Computing before Cybernetics* (Baltimore: Johns Hopkins University Press, 2004).
57. On Shannon's "mathematical theory of little juggling clowns," see Axel Roch, Claude E. Shannon. Spielzeug, Leben und die geheime Geschichte seiner Theorie der Information (Berlin: gegenstalt Verlag, 2009).
58. Fritz Winckel, foreword to *Technik der Magnetspeicher*, ed. Fritz Winckel (Berlin: Springer, 1960), ix.
59. Wolfgang Hagen, "Der Stil der Sourcen. Anmerkungen zur Theorie und Geschichte der Programmiersprachen," in *Hyperkult*, ed. Wolfgang Coy, Georg Christoph Tholen, and Martin Warnke (Basel: Stroemfeld, 1997), 50.
60. Lew Pawlowitsch Teplow, *Grundriß der Kybernetik* (Berlin: Volk und Wissen, 1966), 294.
61. Barry Fox and Jeremy Webb, "Colossal Adventures," *New Scientist* 1081 (10 May 1997): 41.
62. Cooper, "Dead Time."
63. Ibid.

Chapter 3

Generating Time by Technical Measuring

COUNTED TIME UNDERMINES NARRATIVE TIME

No matter which dramatic form it assumes, the writing of media history is trapped in a "vicious circle between the act of narrating and temporal existence."[1] Focusing on time-critical processes and couplings between human and machine signal processing at the micro-temporal level offers a way out of this circle. This culminates in digital signal processors and corresponding algorithms, where the calculated and thus *counted time* undermines *narrative time*. A narrative—whether historiographic or literary—*simulates time* at the referential level; it is only *in time* at the performative (human) and operative (media) level. The bored patients in the sanatorium in Thomas Mann's novel *The Magic Mountain* gradually begin to fill their days with measurements; they go hiking in the mountains with thermometers. Hans shows Joachim how slowly time passes when it occurs as temperature readings.[2] The perception of time is coupled with the act of measuring and measuring media, whether clocks or laboratory devices. For their part, clocks (since Huyghens' pendulum) are considered the *time-critical* reason for scientific precision in the first place.

With the acceleration of emphatic time perception—expressed philosophically as "history"—the epoch of industrial and political revolutions also subtly increased awareness of the smallest temporal moments. Hermann von Helmholtz even discovered intermediate moments of signal transmission in nerves as negative time, which is known among engineers as the so-called "dead time."[3] The temporal interval thus became an event, much like the interruption of the image in cinematography and electronic television. When electrical impulses cause a frog's leg to twitch and an organic element thus becomes part of an electromagnetic circuit, the smallest impulse delays and their differences become galvanometrically measurable. Unlike mechanical recording processes, which are still immediately familiar to people, electricity itself here becomes media-technical time writing. Technical components (relays and indicators) were also taken from the experimental setup of contemporary telegraphy, which communicated in Morse code not simply with two pulse lengths, but also with the pause—the blank space as empty time.

Controlled experiments in the natural sciences primarily reproduce the identical time figure. The radical time-writing experiments of the nineteenth century, based on the indicator of steam engines and the "wave-writing" kymograph, focused on microevents that were emancipated from the narrative regime of history as the dominant time discourse. On the basis of such measurements, Karl Ernst von Baer defined the human awareness of the present as time-critical, *quasi* cinematic. "In one second we have on average roughly six living moments, or at most ten."[4] Von Baer correlated living time and the timing of perception, so a compression of human life into 29 days would give rise to a thousandfold increase in nerve signal time. Humans would thus be in a position to observe a passing bullet, whose movement would otherwise micro-temporally evade the slow human retina. Von Baer's thoughts can be extended from the *dispositif* of the measuring technologies possible at that time to the domain of electromagnetic waves. "If our lives were shortened to a millionth of their actual duration, our hearing ability would begin far above our current perceptual threshold. We would hear light, as our ears would be able to hear everything in the chaos of high-frequency vibrations in which they would be immersed. And we could finally hear radio."[5] Radio here refers to high-frequency electromagnetic carrier waves in contrast to that technically obsolete program content that actually emanates from loudspeakers after demodulation as low-frequency language or music.

THE BIRTH OF TIME CRITIQUE FROM TECHNICAL MEASURING MEDIA

The emphatic signifier "time" has always been an object of thought in philosophy. Nevertheless, the discovery of time-critical moments as specific objects of knowledge was a direct function of highly sensitive measuring media. The "critical aesthetic" defined by Abraham Moles—that is, making spatial and temporal relations that are not accessible to human perception visible or audible through technical observation—is a genuine *time-critical* aisthesis.[6] Time-sensitive measuring media, which are capable of time-critical intensification through their lightning-fast (or ultra-slow) processuality, were necessary to make time-critical processes tangible as events that long surpassed the human sense of registering time (both under and in the time scale).

Since the late nineteenth century, physiology has chosen the concept of the moment as the measure of the perceptible present and thus the critical experience of time (von Baer, von Uexküll). This becomes a media-technical event in cinematography: changes to objects and situations that occur faster than the time specified by that interval (Δt) can no longer be directly perceived in their temporal flow. By employing electrophysical measuring devices, it became possible "to improve this 'biological limit' of time measuring accuracy by an order of magnitude of 10 or at most 15."[7] In contrast to a totalizing concept, time reveals itself here as a collective singular, for its operative reality occurs on diverse levels: the macro-physical (for which historiography is responsible), the meso-physical (the domain of human narratives), and the micro-physical (signals and electronic processes). In his *Confessions*, St. Augustine already discovered the world of the smallest temporal moments in the act of speaking—namely, long and short syllables.[8] Conceived in linguistic phonemes, the speech act is always already speaking *in time*, and its cinematographic image is the reading of vowel sounds. A glider floating through the air is slightly different from a propeller, which divides the air into countable frequencies. Yet only high-precision measuring media are able to perceive infinitesimally small temporal moments. Events that occur in the domain of microseconds can no longer be consciously perceived by human senses, and they thus fall into the realm of pure measuring electronics. The anthropological narrative of time thus comes to an end, and it is replaced by the concept of the human as an ensemble of computable numbers (the difference between narrativizing and counting data).[9]

In the time-critical domain of neuronal events, latency (commonly known in the optical domain as the photochemical intermediate state of the exposed image) means *per definitionem* not concealment, but rather the interval between stimulus and response—a physiological phase shift. According to von Helmholtz's experiments with human nerve reactions, the threshold value is one-tenth of a second, which separates the humanities from the natural sciences as well as the experiential from the experimental sciences:

> The realm of the humanities is founded on data with a temporal differential threshold larger than one tenth of a second. ... Media begin where the humanities stop. ... According to Dilthey, what cannot be experienced and consequently also not understood in history is the real, which can only be recorded by technical media. It exists in writing, but not in narrative: the "noise of battles, the positioning of the opposing armies, the effects of their artilleries, the influence of the terrain on the decision."[10]

This non-discursive tumult represents articulations of the real, which can only be registered by technical measuring media as they evade the writing of history (and thus the concept of historical time). Étienne-Jules Marey initially derived his graphical method from the musical-symbolic practice of Guido von Arezzos, who arranged mensural notation in a line.[11] Léon Scott's phonautograph, a self-writing machine developed in 1857 to measure voice frequencies, supplied "microscopies of time." Von Helmholtz called such apparatus "chronographs,"[12] and he developed special measuring media that had to be electromechanically tuned to frequencies such that the smallest stimuli and delay times could be made perceptible and thus calculable in physiological signal processing. As measuring rather than representational media, these devices effectively generated data. This introduces a time-critical difference between machine and psyche: that which is not on time in a machine normally expires; with humans, it remains suspended.[13]

CHRONOI: THE MEASURING OF LIFE IN THE SMALLEST TEMPORAL UNITS

It makes a difference if a philological text is devoted to *Goethe's Life from Day to Day* and thereby symbolically discretizes over a 24-hour cycle or if this biographical sampling escalates from the macro-temporal to the subliminal signal sequence of 24 images per second, which are not consciously perceived by human senses as discrete. Goethe himself wrote *Day- and Yearbooks*. Like the Ken Burns effect in editing, however, camera movements, zooms, and other cinematic tricks breathe the semblance of life into chronological entries through the narrativization of photographs.[14] "Life consists of day after day," Goethe wrote to Johann Heinrich Voß on 22 July 1821. In other words, life does not occur as organic continuity but rather as discrete jumps. "The optics of snapshots makes it possible to capture the most subtle and secret developmental moments."[15] As a complete chronicle, 24 hours per day is hardly possible in the medium of writing, such as the *écriture de soi* technique in Michel Foucault's sense or the total observance of a biographer. The *individual* first becomes identifiable when divided into his smallest units of movement. Yet, the concept of "optics of snapshots" already articulates that technical medium that circumvents all human (and thus individual) perceptual thresholds: no longer the 24-hour daybook, but rather the 24 images per second film. Even more time-critical, however, is the human perception of lightning-fast vibrations (such as radio flashovers), where the eye sees through the integration of apparently parallel patterns that actually alternate one after the other. The optics of the camera in radio-cinematography offers a better insight into ultra-time-critical processes, as human perception is then made possible through slow motion playback. There are time-critical phenomena that only arise from the time of their media-technical recording and playback. The secret of the electronic image (television and video) is scan line signals, which are time- and value-continuous but are forced by sawtooth voltages to skip lines and finally also change screens—a technological hybrid of continuous and digital time processes, which appears to humans as pure motion. As editing (continuity and montage), this temporal hybrid returns dramaturgically to the second order.

The slowness of the human retina allows every optical impression to linger for roughly one sixteenth of a second; retinal images are thus similar to photographs, but distinguish themselves through their transience. For this reason, the retina is exactly between the memoryless *camera obscura* and photography. What was already described by Ptolemy from Alexandria in 150 AD was technologically realized in 1832 with the phenakistoscope (the "wheel of life"), a stroboscopic device for animating drawn images that was later combined with the *laterna magica* for cinematographic projection. The chronotechnical condition for Eadweard Muybridge's series photographs was no less time-critical; a galloping horse was recorded by 12 and then 24 sequentially arranged cameras. Out of "The Horse in Motion" then came cinema. Vision became even more high frequency with Nipkow's patent in 1884, which enabled the dissection, scanning, and reconstruction of an image by means of a spiral-shaped perforated disc. This process repeats itself at least 16, today 25 times per second, "so that the observer does not notice that it involves not real-time images but rather image components that are sequentially assembled."[16] The effectiveness of technical media thus begins where the time resolution of human nerves ends[17]—the media-archeological moment of transition. Von Helmholtz anticipated this when he described the technical methods of measuring the smallest physiological time intervals. In the nineteenth century, measuring rather than representational media revealed a world in which the smallest temporal processes played a decisive role in perception without themselves being perceived—a time-related variant of the *petites perceptions* identified by Leibniz. Media technology here lies in acts of measuring, with which the human constitutes an empirical-transcendental object not of self-observation, but rather of the machinic observation of others. The humanities and natural sciences converge not in anthropology, but rather in the practices of engineers and mathematicians.

In his monograph *Unconscious Memory*, Samuel Butler, who is famous as the writer of the techno-utopian novel *Erewhon*, responded to Hering's physical justification for cerebral memory:

> Every point of a medium through which a ray of light passes is affected with a succession of periodical movements, recurring regularly at equal intervals, no less than five hundred millions of millions of times in a second; that is by such movements communicated to the nerves of our eyes that we see. . . . Yet

the mind that is capable of such stupendous computations as these so long as it knows nothing about them, makes no little fuss about the conscious adding together of such almost inconceivably minute numbers.[18]

The term "computations," which also applies to the processing of acoustic signals in the human ear, as they are calculated using Fourier analysis, is here applied to optical signal processing.[19] From then on, a treatment of physiological phenomena was considered "in the spirit of mathematics," meaning an examination of the correlation of variable quantities.[20] The computability of time-critical processes lies at the origin of an epistemology of the computer in its current form. What is crucial here is the relation between number and time, whose privileged aisthetic phenomenon (the media channel that enables such analyses to be perceived) is acoustics. Logical-machinic processes thus become perceptible to the senses, which explains why video artist Bill Viola describes the gramophonic model of optical pixels as "the sound of one line scanning."[21]

The graphical method in the nineteenth century epistemologically brought about the chrono-cybernetic equivalence of humans and machines. In parallel, a genuinely mathematized graphical method was developed for Vannevar Bush's analog computer, which was called the differential analyzer.[22] From the graphical method to the analog computer to the mechanical integration of differential equations, the problem remains precision—an entire world thus lies between the *analog* and the *digital*. The ideal of mechanical objectivity overlooks measuring inaccuracies;[23] in the time-critical domain of intelligent weapons, this tolerance can be deadly.

Narratives are always used when the countability of data eludes perception and is delegated to measuring instruments—a media-anthropological division. Signal processing in the microsecond domain is only still perceptible via measuring cinematography, such as the myographical curves from attempts to determine the propagation speed of nerve stimuli. Von Helmholtz described time-writing machines as "microscopies of time" or chronographs.[24] Explicitly, the *so-called* living phenomena literally dissolve into measurements, transitively. Du Bois-Reymond summarized this idea in his speech at the opening of the Physiological Institute in Berlin on November 6, 1877. "The perception of the size of so-called living phenomena as a function of variables and the so to speak bodily recording of their behavior in curves combined to create an entirely new way of handling old tasks."[25] Media archeology means here not simply the scientific-historical research of such relations, but also the operations of which the apparatus alone are capable. Marey personally compared the work of the physiologist who employs the new graphical recording automata with the work of the archeologist who deciphers the traces of past cultures: through their hermeneutic look, "which as such deciphers, collects, compares, catalogues, and thus gives meaning,"[26] they produce sensible inscriptions of the recording of the noisy real in the symbolic order. But since sensory data is no longer only computed from human nerves, as this computation has consequently been delegated to calculating machines, the graphical method has exhausted itself in favor of discrete signal processing.

THE MECHANIZATION OF PHOTOGRAPHIC TIME: CHRONOPHOTOGRAPHY

In a time-aesthetic field located between annalism and chronophotography, Claude Monet painted from 1888 to 1893 a series of haystacks at different times of the day and the year (*Meules*). "The representation of an object from different perspectives in space was abandoned in favor of a representation of the same object at different times"[27]—a transformation of the painterly perspective of space into the time axis according to momentary illumination. Technical measuring media seek to take into account the depth of time in both of its extremes. "It should be noted that the difficulty a movement affords to perception is not always caused by its high speed; some movements also evade our perception through their slowness. The hand of a pocket watch, for example, seems immobile . . .; however, chronophotography also possesses a means of analyzing these movements."[28] If such a media-technical discretization of movement actually grasps its temporal being, then is the so-called flow of time perceived directly or only indirectly? "Imagine a series of instantaneous photographs to be taken. . . . No matter how closely they follow one another, there is no more motion visible in any one of them than if they were taken at intervals of centuries."[29] The more precisely time-critical processes are analyzed, the stranger the so-called historical time appears. From a media-archeological perspective, extremely contracted and extended moments are technically controllable time functions. Charles S. Peirce had a processual concept of perception and comprehension, which bears an affinity to the essential function of technical media:

> Consequently, we cannot reason without having already the idea of time. A greater difficulty is that an instantaneous photograph, though it may contain a symbol of time . . . can certainly not contain a true likeness of time. . . . The other proposition, that the instants of time are so crowded as to merge into one another and lose their distinct existence, seems to be involved in the conception of the "flow" of time. For this phrase likens time to a homogeneous fluid in which the "particles" are mere creations of the mind, made for convenience of calculators.[30]

An excursus in Thomas Mann's *The Magic Mountain* (1924) asks: "Would there be no time if there were no movement? No movement, if there were no time?"[31] Chronophotography with stationary plates lets a movement pass by using a slotted wheel and discretizes *kinesis* analytically as the simple derivative of time, in which it is already imminent (this also applies to Nipkow's disc in electromagnetic television). This process media-technically determines a concept of time with epistemological implications: the conventional interval becomes controllable "intermediate time,"[32] in which the medial moment itself (the Aristotelian *to metaxy*) is time. In chronophotography "with moving film" (eventually celluloid), on the other hand, the photosensitive element itself changes its position before the focal point of the lens and thus freezes there for a short exposure time.[33] The concept of "partial images"[34] suggests that a second temporality is induced by the *chrono*photographic apparatus (in contrast to earlier photography), as the image sequences become a variable of time and thus open the field to scanning and derivation. When movement is analyzed not simply statically but rather dynamically, then time is no longer the object but rather the subject of the media process. It suddenly makes sense that the process is not called "photochronography," as Marey initially intended, but rather "chronophotography."[35]

The chronomechanical "determination"[36] corresponds to the logic of clockwork—culminating in Claude Shannon's concept of "hindrance" in relay-based circuits, which first granted the binary symbolism its temporicity.[37] The apparatus of chronophotography constitutes the reciprocal value of the wheel clock with an escapement: time-critical measuring media are the answer to time-critical moments. In the mechanical intermediate time of images, the critical "time of exposure" (one of Marey's subtitles) is an interval that becomes evident typographically with a hyphen (instead of just *spatium*). "In that event the mechanism must be stopped in the between-time."[38] In addition to the discrete recording apparatus, the media-archeological condition of the photographic manageability of fleeting moments is a correspondingly sensitive chemical emulsion, which yields new objects of knowledge. The temporal mode of chronophotographic processes is intensified through tempo-spatial extension. "It is the nullification of the transmission time between a phenomenon and the photographic plate."[39] Discretized by the apparatus in regular and thus periodic exposures, the photochemical mark of this event time becomes media-technically countable—a sampling of analog kinetics *avant la lettre*.

Marey identified the intensification of observation methods as a shift from the point axis to the time axis: in addition to instruments that measure static values there are also recorders capable of micrographically representing shape and location changes as well as continuities "in the form of multiple undulating curves."[40] Alongside the symbolic ordering of time—language-based historiography, which claims to be capable of emphatic temporal processes but is not able to take place on its own volition—emerges a genuine form of time writing for the analysis of microevents. It is a graphism interposed between culture and technology, which is authored by nature itself and which thus circumvents the traditional Aristotelian separation of *physis* (inherent movement) and *techné* (sterile artifacts). "Generally speaking, with chronophotography all of the movements of bodies that are under the influence of different forces register themselves on the fixed plate."[41] The movements of people as well as machines can thus be analyzed—the *mathematical* and therefore nonvisual hypothesis of kinematics.[42] Chronophotography is even able to kinematize geometry itself—the return of a question that has been raised repeatedly since the times of ancient Greece. Marey concludes that this emphatic time figure stands more on the side of "natural history" than classical history.[43] Indeed, "it is another nature which speaks to the camera rather than to the eye,"[44] and this other nature is that of time. Whoever gives an account of the movements of people with natural senses is not able "to say anything about that fraction of a second when a person *starts to walk*." The past (*passé*) and passing (*passer*) here blur together. "Photography, with its devices of slow motion and enlargement, reveals the secret. It is through photography that we first discover the existence of this optical unconscious"[45]—namely, in its dynamics. As a function of technical apparatus, this other perception is not only "a thoroughly historical variable,"[46] but it also brings about new time objects in the critical domain. A form of writing that is able to *write time transitively* distinguishes itself from the *intransitive description* of (historical) time figures, just as mathematical time distinguishes itself from Bergsonian time (duration).

Does the media-specific temporal mode of motion photography stand between human and world? The exact periodic time sequences of chronophotography, which each fix complete recorded images, is different from the processing of moving images in humans, which resembles current digital image sequence compression processes insofar as it does not store each complete image but rather only the differences to each previous perception (the remainder is interpolation). Mechanic cinematography is thus farther from human image processing than the electronic-digital. The brain processes not optical and acoustic signals as such, but rather their conversion into anaisthetic nerve impulses,[47] which is entirely consistent with coded data processing in electronic computers. Inner time consciousness here becomes comprehensible in information-technical neutrality.

Chronophotography in the strictest sense is the moment when a point in time as object and as apparatus of photography are overlaid and roughly converge, so that the time of photography does not simply represent the unchanging variable as opposed to a fleeting object, but rather it is carried along in the time-critical domain itself (relative time vs. Newtonian time). This is evident when a sundial is the object of a snapshot and the hastening shadows of the gnomon inscribe their temporal moment both indexically and symbolically onto the photograph. The shortening of photographic exposure times (which were initially quite long) and thus the compression of the temporal interval or rather the present window (Δt) resulted more covertly than consciously in a new mode in the concept of the present itself. This led to a way of scanning points in time, a virtual sampling of the present, a discretization that enabled the storability and computability of its previously fleeting, constantly receding nature. This still involved the scanning of two-dimensional objects, unlike early electromechanical and electronic television, where an image in the present was scanned line-by-line by a continuous point—initially a hole in a rotating disc (Nipkow's patent), then the "flying spot" of an electron beam (Von Ardenne's patent)—which thus generated the image signal. However, Marey approximated this process through the enlistment of the dynamograph, which enabled the discrete-amplitude analysis of the foot pressure of a walking motion.[48] In the snapshot's time-critical concentration on the (virtual) point—namely, the photography of microscopic movements—chronophotography already reaches the aporia of Heisenberg's uncertainty principle. Photographing the ultra-fast movements of organisms under the microscope required magnification, but "the magnification of the images entailed a corresponding reduction in the intensity of light, which affected every point on the photographic plate. In order to obtain sharp pictures of very fast movements, it was also necessary to shorten the exposure time considerably"; on the other hand, the heat of the strongly focused light killed the organisms to be observed.[49] The solution was noninvasive illumination through synchronously triggered electric sparks—a process that was later used by Ernst Mach and Peter Salcher. The electrification of mechanical chronophotography embodies the media-epistemological autonomization of the time-critical. The atomic force microscope, for example, detects the miniscule mechanical deflection of the scanning tip over the angular refraction of a diode laser beam, and it thus only exists as a calculation. The media-epistemic thing—the electron microscope—brings forth objects that are no longer concrete but rather temporal beings. Ultra-microscopic objects, like electrons and photons, can be described not as isolated objects, but only as time-critical events; as the results of techno-empirical knowledge, they are more like analytical models. "In the world of microphysics, the individual loses its substantive properties"[50] in favor of its microtemporal existence. It is no longer a tangible elementary particle, which is able to give empirical information to the observer,[51] but rather imprecise eventfulness.

The implicit mathematics of chronophotography is differential calculus in the sense of analysis, which makes it possible "to derive the general relationship between the form of the movement-tools and their characteristic function."[52] Henri Bergson criticized this techno-mathematical discretization of time as the misapprehension of the phenomenological being of time: duration. However, Marey occupied a middle position: the physiological perspective. Just like Hermann von Helmholtz's *Sensations of Tone*, movement first takes place in the interlocking of the perceptual apparatus and cognition. The chronophotographic process does not actually provide a coherent representation of the reproduced change; "however, the recorded images can be strung together so closely that the missing stages are mentally added in a thoroughly admissible way."[53] The sequences of pictures resulting from this media-technical process provided art with "models for an iconography of the subjective visualization of time." They thus circumvented Lessing's *Laocoon* theory, which defined the spatially coexistent quality of literally still images (which includes sculpture and painting as well as photography) as essential. "Marey contrived to decompose movement into a multiplicity of equal and discrete units"[54]—a discretization and thus reduction of movement to space-time (*temps-espace*), which was already thematized in Zeno's paradox of the arrow that always stands still at the moment of analysis and which Bergson vehemently contradicted by arguing that temporal reality constitutes a fluid continuum (the sculptor Rodin also insisted that motion sequences are continuous). "Chronophotography are images not of movement through time, according to Bergson, but of position and succession."[55] The mathematical answer (since Leibniz and Newton) is infinitesimal or rather fluxion calculus, but the sensory and neurophysiological answer is the slowness of optical perception in the human eye and the cognitive fusion effect (afterimage and phi effect) as a condition for the impression of constant movement. In the pre-photographic epoch, the image analysis mechanism called the "zoetrope" employed drawn images; 16 images per second were sufficient to produce the impression of constant movement in the eye. Marey's photographic gun was only based on 12 images per second. After 150 years of media training and in the age of rapidly edited music video clips, viewers have developed a sharpened sense of speed perception; after an epoch of media-technical habituation and acceleration, is a higher image frequency required to achieve the same effect? Such perceptual frequencies vary with their media cultures.

If chronophotography makes temporal processes accessible to discrete scientific analysis, then it involves the taming of previously fleeting real time. Chronophotographs were perceived as real images of moving life, as moving images. The registering medium is thus passive in relation to time and only later assumes an autonomous, shaping form in the cinematographic "time image."[56] Actually, Muybridge already laid out the media-technically arbitrary operation of the time image by *retroactively* ordering the individual photographs into series—a construct that followed the epistemological dictate of progression (the time-linear entanglement of *passer* and *passé*, of progress and the past). A chronophotographic "plot" (a discrete series) could thus be connected to a possible history (narrative); Muybridge arranged fragments of the world in temporal sequences, which could then be dramatically transformed into histories. In contrast, Marey's gaze is more analytical in a media-archeological sense; he went against the tradition of perspective and thus introduced a genuine media aesthetic. His quasi-stroboscopic overlays of photographic moments through multiple exposures on the same negative (virtually optical "noise") did not offer any material for narration, but rather condensed a process into the real-time image.

COUNTING TIME: CINEMATOGRAPHY AND BERGSON

Bergson rejected the Aristotelian concept of time, which triggered the scientific parameter of countable time. Physics actually operates with basic quantities defined by measuring apparatus like the oscillation of a pendulum; values are then derived, which are calculated. The time response of a physical experiment must be repeatable—in this respect, it is close to cinematography. The necessary apparatus for periodic time-discrete measuring is the timing clock. By reference to the measuring of athletic times, Walter Benjamin analyzed (almost media-anagrammatically) how *agonal* relations are replaced by *analog* measuring through physical standards; as a generalization, this also means that the evaluation of human time through historical and thus human-made time is replaced by the technically discrete parameter t. "It is not without reason that Nurmi said he ran against the clock."[57] When human movement, acceleration, and deceleration are measured by a chronotechnical apparatus like the stopwatch, the relationship between humans and time turns from signified to signifier. Clock time is human-made, but it becomes a time order of a second nature.

In the name of philosophy, Bergson opposed the mathematical-physical concept of time and identified the being of time as "the experienced present as pure duration."[58] This duration is like an irreversible flow or a rolling snowball, and it is thus subject to entropy—unlike clock time (in the sense of Newton's mechanics), which in principle represents a reversible process. Technical storage media literally emerge between this dichotomy in that a section of time is removed from the continuous flow of duration and becomes time-identically reproducible. In 1895, shortly after the first public demonstration of film, Bergson devoted himself to moving images in *Matter and Memory*; after cinema became a mass medium, he turned to this topic again in *Creative Evolution* (1907). According to Bergson, the relationship between the understanding of movement in ancient Greece and today is the same as the relationship between the noting of the phases of a movement by the eye and the much more complete recording of these phases by instantaneous photography. It is the same cinematographical mechanism in both cases, but it reaches a precision in the second that it cannot have in the first. ... Instantaneous photography isolates any moment; it puts them all in the same rank, and thus the gallop of a horse spreads out for it into as many successive attitudes as it wishes, instead of massing itself into a single attitude, which is supposed to flash out in a privileged moment and to illuminate a whole period.[59]

Infinitesimal signal processing, or the perception of time in "phases," is on the side of duration. In the case of electric lights connected to alternating current, however, what appears to perceptually slow eyes as continous light is actually a function of lightning-fast on/off switching; the moment of the photographic exposure is nothing else than the time-critical extreme of such flashing. It is no coincidence that the same Faraday who discovered the phenomenon of electromagnetic induction also experimented with cinematic effects involving turning wheels. Every viewer of western films is familiar with "aliasing," or the apparently backward movement of spoked wheels on carriages despite the fact that time is moving forward. The difference between the visual and acoustic addition of the individual oscillations of a complex sound is similarly perceptible. Unlike the spectral analysis of individual tones seen in the visual interface of the oscilloscope, in which the time window appears as a segment of space, the auditory sound impression is more than the sum of its parts because the ear is able to recognize a temporal extension.

In an article on the representation of horses in art, which was published in 1878, Marey referred admiringly to the representation of a galloping steed on the ancient Parthenon Frieze in Athens; in opposition to the spatializing, dissecting, and measuring concept of time, the sculptor Phidias created an image of living, continuous, flowing duration.[60] The concept of duration actually more closely resembles the stochastic streams of electrons in a vacuum tube than cinematography. Yet, such time courses first became calculable through the diagrammatic geometrization of technical measuring media. The film apparatus *analyzes* movement in the same way that the phonetic alphabet analyzes spoken language—namely, discretely. And what is technically analyzed in this way can then be technically synthesized. The trick of the cinematograph is not merely technical, but also media-epistemological, as it resonates with the cognitive apparatus in humans themselves. "We hardly do anything else than set going a kind of cinematograph inside us."[61] Human perception virtually constitutes a differential memory.[62] In contrast to Bergson's philosophical critique of techno-mathematical time, media archeology happily engages this technical formalization in order to discover its own perception of time.

Like Bergson's time critique, Sigmund Freud's analysis of Jensen's novel *Gradiva* also foregrounds "not the kinetic nymph, the motion study of Gradiva. . ., but rather the resurrection of antiquity—not film, but rather the still image."[63] In his description of Michelangelo's *Moses*, however, Freud for the first time breaks the sculpturally compressed moment of tension into film-like sequences; only at the end does the statue become a still image, a filmic *still*. As Siegfried Kracauer insisted, "there is a fundamental analogy between the writing of history and photographic media."[64] The historian as cameraman seeks to preserve the passing moment and break it down into its light colours like "a spectral analyst."[65] The transition from *passé* to *passer* is the media-archeological site where the past is negotiated in the present. In *The Creative Mind* (1934), Bergson criticized all attempts to construct a living reality out of rigid concepts (or technologies). For Bergson, therefore, the Aristotelian interlinking of time and number is a misleading approach. The diverse can only be represented in space as juxtaposition. Does the cinematographically counted sequence neglect the being of time, or does it define it for the first time?

Futuristic painting was a direct cultural-aesthetic effect of chronophotography as well as an update of Lessing's *Laocoon* theory under media-technical conditions. The *Futurist Manifesto* became the mouthpiece of technical acceleration and thus the message of new media (at that time). Time and space thereby point to an ever-present speed in the sense of what Paul Virilio later called "dromology." The static world image is replaced by a dynamic approach. In 1911, Bocchioni painted a picture with the title *The Noise of the Street Enters the House*; however, the fourth dimension (time) also enters the picture. The aesthetic of futurism was influenced by contemporary scientific developments (relativity and quantum theory); Bragaglia introduced photodynamism to counter the discrete character of chronophotography. Muybridge's sequential photography failed due to the length-contraction of relativity. In contrast, Albert Einstein chose another method. In his theoretical model, chronophotography was replaced by clocks themselves—differential time.[66]

LIGHT SPEED: TIME-CRITICAL MEASURING INSTRUMENTS

Günter Schabowski's articulation of the answer "immediately" to a follow-up question at a press conference in East Berlin on new travel regulations between East and West Germany on November 9, 1889, triggered an almost *immediate* reaction from the population due to the fact that it was transmitted live by electronic media. Electricity first opened up that world of approximate immediacy, in which the measurement of time-critical moments is crucial. "Measuring devices with small time coefficients are necessary for regulation";[67] the delay of the measuring device to the values being measured should aim infinitesimally toward zero. The problem is common in fluid mechanics: almost all currents are non-staionary, meaning time-dependent; mechanical measuring probes are unable to track relatively fast or erratic changes (like shock waves), or they are only able to track them slowly; the measuring of real events instead of models (which had previously been necessary) was first made possible by laser velocimetry.[68]

Once time is metrologically analyzed, it can consequently also be synthesized. The observation of time turns into the technical enforcement of simultaneity, like electrotechnical synchronization in television transmission. Synchronization is an act of (electro)technical force—an act of violence that combines two media processes; simultaneity, on the other hand, appears to be a natural phenomenon. However, from the perspective of physics, which is always also the perspective of media (studies), there is actually no such thing as simultaneity; in fact, the theory of relativity defines simultaneity as a connection through a time-limited beam of light.

A toothed, rotating disc scans the zone between the discrete and the continuous, which the human senses are optically as well as acoustically led to believe. What is allegorically and iconologically still "the tooth of time" here becomes chrono-technically concrete.[69] The notches of a gearwheel cannot be perceived when it turns rapidly; what appears instead is "an apparently continuous transparentness, which consists of the successive appearance of teeth and the gaps between them."[70] The impression of continuity thus turns out to be the dissimulation of the discrete in the implementation of time. This effect, which is familiar to all clockmakers, here acquires an epistemological dimension.

In the middle of the nineteenth century, Hippolyte Fizeau attempted to measure the speed of light. Measuring processes no longer take place simply on the basis of a given time axis, but rather time itself is the object of measurement (in the sense of relativity theory). The measuring principle, and thus the *arché*, here becomes the technical framework: a beam of light falls through the notches of a rotating wheel, is reflected in a mirror at a sufficient distance, and then returns to the wheel. "The rotation speed is then adjusted so that the beam of light arrives back at the wheel precisely when it has turned one tooth width further, and therefore the returning light can no longer be seen."[71] This is an intensification of something that is in itself already time-critical; for a certain type of experiment, such as research on hydro- and thermodynamic processes, the moment of reading is "a critical parameter."[72] Certain homeostatic conditions and the moment of dynamic equilibrium require "a pronounced and trained time feeling" in order to be observed.[73] In such experimental settings, the human is placed in specific temporal relationships and is thus subject to a technologically induced proper time field. Scientific experimentation thus becomes a function of a chronotechnical semantics with its own law. Fizeau was able to calculate the speed of light through the number of teeth on the wheel, its rotation speed, and the distance between the measuring point and the mirror. The speed of light as an object of time-critical measurement is not subject to historical time.

Turned into the time-function, the stroboscopic disc later served to calibrate the rotation speed of records. However, the sonospheric perception in humans is not aware that the ear actually analyzes the sound of the sea on the beach in its individual wave breaks—the calculating instinct of *petites perceptions*.[74] By converting complex signals into the frequencies of their individual oscillations, dynamic temporal beings (or modes) become accessible to mathematical treatment and complete numerical determination according to their recorded measurements (and thus in genuine media analysis). This mathematization initially happens at the expense of temporal indexicality; the point in time when the individual frequencies of a non-stationary signal emerge is no longer evident from their reciprocal value to the spectrum. Temporally "windowed" variants (like short-time Fourier transforms, Gábor transforms, or wavelet transforms) first expressed the frequency response as well as the temporal delicacy of each signal event, and thus analytically moved *with the times*.

MICRO-CHRONOMETRIES

In the tradition of the (electro)technologies developed by von Helmholtz to measure the "smallest fractions of time" below the perceptual threshold, the world of micro-times requires time-critical measuring media—"an operation in the dispositional space that Frank Haase called the 'telecommunicative apriori.'"[75] Even further below this time window lies the analysis of ultrashort temporal moments that are hardly still measurable, like the time it takes light to pass through the nucleus of a hydrogen atom. In order to expand such moments for the purpose of analysis—in other words, to temporally dissect them—physics shifts from direct measurements to models. Since mathematics was temporalized by the computer, it is now possible to mathematically slow down physical time processes that are not measurable.[76] In the nineteenth century, laboratories were a privilege of physics; as a result, only the classical humanities were able to reflect on this possibility, but not to develop the operative knowledge. The position of media studies looks different today: signal-processing machines with programming environments like SuperCollider as well as powerful modeling tools like MATLAB® and Simulink® now put the philosophical faculty in a position to run through such experiments. Media theory thus moves from the philosopher's room to the laboratory (and back).

Micro-temporal measuring devices "make temporal differences measurable by transforming them into spatial differences"[77]—a genuine scientific operation in the Aristotelian tradition, which defines time as the index of movement between before and after. In this sense, Feddersen's radio-micrometer also functions as cinematography. If the oscilloscope visualizes the change of an electrical signal over time, it is not narrative but rather counting cinema. "A rotating concave mirror serves to spatially represent and project an objective image of the changes that occur in the electric spark before it. A photographic plate placed at the image makes it possible to calmly observe and spatially measure the duration of the momentary appearance, free from every subjective illusion."[78] Marey's chronophotographic analysis of the flight of birds also employs time axis manipulation. "Instantaneous pictures of flying projectiles and air movements, . . . illustrate the principle of *time magnification*, which is applied in these cases."[79] Data collection is time-critical as soon as it takes place in temporal succession. With this emphasis, media archeology goes beyond the anthropology of the senses as defined by Helmut Plessner, who distanced himself from traditional aesthetics and proclaimed an *aesthesiology*.[80] Media archeology instead examines the aesthetics of data itself. Data is acquired through measuring apparatus. A recording device functions as an automated secretary, "in that it compares and counts—and shows what it has counted. It is therefore possible to say that facts are verified, but data is produced."[81] By resolving the old aesthetic question as to whether galloping horses had been correctly painted, Eadweard Muybridge and Jules-Étienne Marey's chronophotography already pushed between the human perception of movement and "life." Ernst Jünger identifies in sports the "tendency to determine the record in exact numbers down to the smallest spatial and temporal fractions"[82]—time and number in a critical combination. However, there is a significant difference between technical clocking (which manifests in the indispensable use of high-precision timing impulses in industrial measuring practices and computing processes) and "aesthetic synchronization" as the almost unnatural production of a close parallel correspondence between image and sound.[83] The time parameter is crucial to most technical media, which resulted in the concept of "time-based media." Aesthetics as art was for a long time primarily based on spaces, surfaces, and volumes, but in media art, these are joined by an aesthetic of time-critical processes.

THE TIME AXIS BECOMES WRITABLE AS AN EVENT: KYMOGRAPH, DYNAMOMETER, OSCILLOSCOPE, TIME DIAGRAMS

Cinematography is associated *a posteriori* with the optical reproduction of motion. Its basic approach is more fundamental, however, so media archeology does not simply regard it as the prehistory of cinema.[84] There is a consciously heuristic reason for this "negation of the historical"[85]: the object of research is momentarily suspended from history, which makes it possible to contemplate its specific temporal being and its other temporal modes. The concept of temporal flow belongs to the archaic cultural figures of time writing, so it was already operative as a linguistic figure, but it suddenly escalated in 1846 when Carl Ludwig patented the kymograph—an apparatus that literally wrote waves through the continuous registration of physiological changes. The kymograph, and then later the phonograph, encapsulated the flow of time through signal storage, and it was thereby implicitly made reproducible. In the beginning, though, the kymograph represented the condition of possibility for the analyzability of physiological processes by freezing the flow of time. Once they are recorded, temporal processes become analytically manipulable through acceleration or deceleration without being destroyed. Time is thus not only the object of media-technical recording, but also at the same time the subject. Recorded time is a function of the apparatus; the kymograph is the basic model of self-writing (time) machines. Time is no longer solely an object of arbitrary speculation. The dynamometer (and later the ergograph), which was used to study physiology in the nineteenth century, was a force-measuring device and a medium for numerically encoding people that was also strictly linked to physical data streams. Self-writing machines perform the direct analog imaging of temporal processes through mechanical transmission (*dynamis*), which is poised on the verge of a truly mediographic time critique.[86] However, a media-epistemological difference still remains: the dynamometer converted movements into numerical values on a discrete, disjointed scale, while the ergograph put the graphical method into effect. Based on such technical measurements of the smallest temporal events, Charles Féré formulated the "psycho-motor rapport"—a physiological term from the time-critical episteme of electromagnetic fields.[87]

The drum of an ergograph is driven by clockwork and is thus timed; the discrete here becomes rhythmic, metronomic. The obtained measurements first emerge in a temporal interval—a signal conversion like the conversion of current into voltage values, which is common today in digitization. The dynamometer is a temporalized notation function. The media-archeological gaze (like that of every camera) is aisthetic, not aesthetic, and its reification in the form of measuring apparatus treats bodies and machines from the same perspective of signal processing. While the first apparatus-mediated graphic curves still charted the perceptible domain of the senses, this process actually became time-critical with the measuring of nerve stimuli in the subliminal domain.

The time axis is an artificial time introduced by the clockwork-driven cylinder of a kymograph or the running paper strip. However, transitory time mechanics actually measures not instantaneous values, but rather the smallest time intervals. The electromechanical method of representing changes in physical measurements over time involves converting moving coil instruments into chart recorders by replacing the pointer with a pencil that draws the values on a moving strip of paper. However, the mechanical slowness of writing in the tradition of graphic inscription fails to yield results in the micro-temporal domain. Electronics therefore replaced the inertially loaded pointer, which is always subject to mechanical hysteresis, with the fully electronic indicator, which opened up another world of immediacy, like radio photography. When the conventional oscilloscope displays analog time signals or the logic analyzer shows the instantaneous values "1" or "0" of computer signals, time first appears here in its technical representations.

In the use of measuring media, the time axis—a genuinely diagrammatic and only superficially visual phenomenon[88]—is assumed to be an observable parameter; however, a technically realized time derivative is dilatory and thus never an absolute reference. Its prelude lies in mechanical devices; this is where the *terminus technicus* time axis was developed. In contrast to the rotating mirror, which registered and analyzed temporal deviation imperfectly, Ferdinand Braun's assistant Zenneck represented time events on the luminescent screen of the electron tube itself. "This is done by installing beside the horizontal coil, through which the variable current passes, a second vertical coil with a current whose intensity is directly proportional to time."[89]

The temporal deviation on the oscilloscope is arbitrarily created with respect to the time signals from the real electrophysical world that are registered on the vertical axis. The latter are chronotechnically modulated—that is, they are channeled along a time axis that is fractured by technological time. By representing a graph of a temporal process as a cathode ray image, the oscilloscope effectively generates an operative time diagram, which is only possible in a technical medium. This form of time writing first becomes visual through the emission of light caused by the appearance of electrons on the luminous screen; graphical curves thus no longer serve only to record a temporal process; rather, they themselves only come about as a bombardment of time-critical moments. A short impulse is prolonged through the application of an artificial time axis (time base); through the media-technical creation of a stretch of time the actual physical point in time is thus artificially dilated, out of which its proper temporality emerges. In highly precise electrotechnical acts of measuring, however, time-critical interferences inevitably occur, and the time set and the measured event are thus blurred.

Oscillography serves to demonstrate periodic and non-periodic processes with cathode rays. Yet, nothing is recorded here, as recording is storage; the oscilloscope does not store, but rather realizes a fleeting, transient representation of vibrations and impulses. *Theoría* became time-critical as it discovered the time world inherent to implicit media implementation: with oscillography, "the temporal flow of state changes can be immediately visualized in a uniquely vivid and comprehensive way."[90] The temporalization of the momentary glance is thus time-critically intensified.[91] "As a continuous line, the connection of individual momentary values, which are always shifted a degree corresponding to the elapsed time, results in a curve"—it is thus the function of a time-dependent process.

With a writing speed of over 50,000 km/sec, a high-frequency high-voltage oscillograph is sufficient for practically every short-term process that occurs in electrical technology. The time here becomes immanent to electronics. Television is nothing else than the illusionary visual application of such electronic time deviation: "the time area thus becomes the line grid of television technology."[92] As an epistemic object, the Braun tube constitutes both the media-archeological measuring instrument and the mass-media television picture tube. Media culture emerges at such transition points, as microelectronics becomes the cultural semantics of images.

As John Herschel shaped the concept of "photography" in 1839,[93] he emphasized that this process should be understood as a medium of not only quasi-painterly light images but also light measurements, as the physical event reveals itself indexically. The message is here in the medium itself entirely in the sense of the caricature of a glowing light bulb on the original cover of McLuhan's *Understanding Media* in 1964. "Photography should reveal the truth about light not through what it shows, but rather through what it is."[94] In an ontological sense, the being of the technical medium lies in the physical implementation of signals. The cathode ray oscillograph is also pure transmission—an analytical television—as it involves measuring, not representation. A visual epiphany nevertheless stands out on radar screens during fog, such as the figure of a coastline. Radar constituted the immediate precursor to the so-called Williams tube—the oscilloscope as buffer memory for digital signals. Radar enables the differentiation of what is displayed according to time, so constant or near-constant echoes disappear and only moving echoes are shown. "It is irrelevant whether the movement shown indicates advancing targets or losses. In this way, all of the signs that have not changed within the scanning period are eliminated and only moving processes are displayed."[95]

Time critique is written not exclusively in an alphabetic form, but also in a genuinely technographic form. In his lectures *On the Phenomenology of the Consciousness of Internal Time*, Husserl sketched a "diagram of time." As an alternative to the metaphysical discourse of "appearances," he struggled for concepts to describe "running-off phenomena" or "modes of temporal orientation."[96] Time diagrams made it possible to visualize the mechanism of the time-integral, but they could not perform it themselves. They thus had to be replaced by a technical medium of time, like the analog computer. On

September 1, 1805, Goethe had an insight. "If the eye were not sun-like, it would not see the sun"; it is also true of measuring media that they themselves have a time-critical sense, and in order to grasp the temporal beings (or modes) of nature they must be time media: oscillating, resonant. In contrast to hand-made diagrams and the graphic curves of the kymograph, oscillograph images as well as video images consist of pure time writing. The oscilloscope display, which is analog in an electrophysical sense, can also be digitally realized through signal-processing software. The digital variant of signal processing is physically authentic and not simply animation because on the basis of techno-mathematical scanning theory its inherently discrete moments engage the temporal modes of analog signals themselves. The earthly chronosphere and time-critical media thus converge.

NOTES

1. Paul Ricoeur, *Zeit und Erzählung* (Munich: Fink, 1989), 1: 90.
2. Thomas Mann, *Der Zauberberg* (Berlin: Aufbau, 1953), 767.
3. Hermann von Helmholtz, "Über die Methoden kleinste Zeittheile zu messen und ihre Anwendung für physiologische Zwecke," *Königsberger naturwissenschaftliche Unterhaltungen* 2 (1851): 169–89. See also Henning Schmidgen, *Die Helmholtz-Kurven. Auf der Spur der verlorenen Zeit* (Berlin: Merve, 2009).
4. Karl Ernst von Baer, *Schriften* (Stuttgart: Greiner und Pfeiffer, 1907), 141.
5. Bernhard Siegert, "Das Leben zählt nicht. Natur- und Geisteswissenschaften bei Dilthey aus mediengeschichtlicher Sicht," in *Medien. Dreizehn Vorträge zur Medienkultur*, ed. Claus Pias (Weimar: Verlag und Datenbank für Geisteswissenschaft, 1999), 177.
6. See Abraham Moles, *Kunst und Computer*, ed. Hans Ronge (Cologne: DuMont, 1973).
7. Otto-Joachim Grüsser, "Zeit und Gehirn. Zeitliche Aspekte der Signalverarbeitung in den Sinnesorganen und im Zentralnervensystem," in *Die Zeit. Dauer und Augenblick*, ed. Jürgen Aschoff et al. (Munich: Piper, 1992), 98.
8. St. Augustine, *Bekenntnisse*, trans. Herman Hefele (Jena: Eugen Diederichs, 1921), 245.
9. Translator's note: there is a play on words here in the original text, as the German word for "narrative" (*Erzählung*) closely resembles the word for "counting" (*Zählung*).
10. Siegert, "Das Leben zählt nicht," 175. Siegert here refers to Wilhelm Dilthey, "Die Abgrenzung der Geisteswissenschaften. Zweite Fassung," in *Gesammelte Schriften, Band VII* (Stuttgart: Vandenhoeck & Ruprecht, 1992), 311.
11. See Soraya de Chadarevian, "Die 'Methode der Kurven' in der Physiologie zwischen 1850 und 1900," in *Die Experimentalisierung des Lebens. Experimentalsysteme in den biologischen Wissenschaften 1850/1950*, ed. Hans-Jörg Rheinberger and Michael Hagner (Berlin: Akademischer Verlag, 1993), 37.

12. Ibid., 38.
13. See Jacques Lacan, "Psychoanalysis and Cybernetics, or On the Nature of Language," trans. Sylvana Tomaselli, in *The Seminar of Jacques Lacan*, ed. Jacques-Alain Miller (Cambridge: Cambridge University Press, 1988), 2: 294–308.
14. Robert Steiger, *Goethes Leben von Tag zu Tag, Band 1: 1749–1775* (Zürich: Artemis, 1982), 5.
15. Ibid.
16. Hans H. Hiebel, *Kleine Medienchronik. Von den ersten Schriftzeichen zum Mikrochip* (Munich: Beck, 1997), 24.
17. Friedrich Kittler, "Am Ende der Schriftkultur," in *Der Ursprung von Literatur. Medien, Rollen, Kommunikationssituationen zwischen 1450 und 1650*, ed. Gisela Schmolka-Koerdt, Peter M. Spangenberg, and Dagmar Tillmann-Bartylla (Munich: Fink, 1988), 293.
18. Samuel Butler, *Unconscious Memory* (London: Cape, 1924), 65f.
19. Gottfried Wilhelm von Leibniz wrote to Christian Goldbach on April 17, 1712, telling him: "Music is a hidden arithmetic exercise of the soul, which does not know that it is counting."
20. Adolf Fick, "Die medizinische Physik," in *Gesammelte Schriften, Band 2* (Würzburg: Stahel, 1903), 4.
21. Bill Viola, "The Sound of One Line Scanning," in *Reasons for Knocking at an Empty House: Writings 1973–1994*, ed. Robert Violette (Cambridge, MA: MIT Press, 1998), 153–68.
22. See Friedrich Adolf Willers, *Mathematische Maschinen und Instrumente* (Berlin: Akademie-Verlag, 1951), 262ff.
23. See Lorraine Daston and Peter Galison, "The Image of Objectivity," *Representations* 37 (1992): 67–106.
24. De Chadarevian, "Die 'Methode der Kurven,'" 38.
25. Emil Du Bois-Reymond, "Der Physiologische Unterricht sonst und jetzt," in *Reden, Band 2* (Leipzig: Veit, 1887), 366.
26. De Chadarevian, "Die 'Methode der Kurven,'" 45. In reference to Étienne-Jules Marey, *Du mouvement dans les fonctions de la vie* (Paris: G. Baillière, 1868), 24.
27. Christoph Asendorf, *Ströme und Strahlen. Das langsame Verschwinden der Materie um 1900* (Gießen: Anabas, 1989), 6.
28. Étienne-Jules Marey, *Die Chronophotographie*, trans. A. von Heydebreck (Berlin: Mayer & Müller, 1893), 14.
29. Charles S. Peirce, "Of the Nature of the Continuity of Time and Space," in *The New Elements of Mathematics*, ed. Carolyn Eisele (Atlantic Highlands, NJ: Humanities Press, 1976), 3: 59.
30. Ibid., 60.
31. Mann, *Der Zauberberg*, 490.
32. Marey, *Die Chronophotographie*, 7.
33. Ibid., 10ff. Marey here refers to Janssen's "astronomical revolver."
34. Ibid., 44.
35. Ibid., 3.
36. Ibid., 24.

37. Claude Elwood Shannon, "A Symbolic Analysis of Relay and Switching Circuits," *Transactions of the American Institute of Electrical Engineers* 57 (1938): 713–23.

38. Marey, *Die Chronophotographie*, 35.

39. Michel Frizot, "Analyse und Synthese der Bewegung. Étienne-Jules Mareys Methode," in *Apparaturen bewegter Bilder*, ed. Daniel Gethmann and Christoph B. Schulz (Münster: Lit Verlag, 2006), 146.

40. Marey, *Die Chronophotographie*, 2.

41. Ibid., 88.

42. For a discussion of Eduard and Wilhelm Weber's mathematical formalization of the *Mechanics of the Human Walking Apparatus* (1836), see Hans-Christian von Herrmann, "Kinästhetik. Zur Geschichte des Studiums der menschlichen Bewegung (1836–1916)," in *Apparaturen bewegter Bilder*, ed. Daniel Gethmann and Christoph B. Schulz (Münster: Lit Verlag, 2006), 155–61.

43. Marey, *Die Chronophotographie*, 91.

44. Walter Benjamin, "Little History of Photography," trans. Edmund Jephcott and Kingsley Shorter, in *Selected Writings*, ed. Michael W. Jennings, Howard Eiland, and Gary Smith (Cambridge, MA: Belknap Press, 1999), 2: 510.

45. Ibid., 510–12.

46. Ibid., 512.

47. See Florian Rötzer, "Re: Photography," in *Photography after Photography. Memory and Representation in the Digital Age*, ed. Hubertus von Amelunxen, Stefan Iglhart, and Florian Rötzer (Amsterdam/Munich: G+B Arts, 1996), 17.

48. See Marey, *Die Chronophotographie*, 46 (Figure 23).

49. Ibid., 81f.

50. Gaston Bachelard, *Epistemologie. Ausgewählte Texte* (Frankfurt am Main/Berlin/Vienna: Ullstein, 1974), 17.

51. As described in Fritz Heider, "Ding und Medium," *Symposion* 1.2 (1927): 109–57.

52. Bachelard, *Epistemologie*, 50f.

53. Marey, *Die Chronophotographie*, 38.

54. Martha Braun, *Picturing Time: The Work of Etienne-Jules Marey (1830–1904)* (Chicago: University of Chicago Press, 1992), 277.

55. Ibid., 280.

56. See Gilles Deleuze, *Cinema 2: The Time Image*, trans. Hugh Tomlinson and Robert Galeta (Minneapolis: University of Minnesota Press, 1989).

57. Walter Benjamin, *Gesammelte Schriften, Band I: Abhandlungen*, ed. Rolf Tiedemann and Hermann Schweppenhäuser (Frankfurt am Main: Suhrkamp, 1978), 1039.

58. Georg Christoph Tholen, *Die Zäsur der Medien. Kulturphilosophische Konturen* (Frankfurt am Main: Suhrkamp, 2002), 134.

59. Henri Bergson, *Creative Evolution*, trans. Arthur Mitchell (London: Macmillan, 1922), 351.

60. Ulrich Raulff, *Der unsichtbare Augenblick. Zeitkonzepte in der Geschichte* (Göttingen: Wallstein, 1999), 67.

61. Bergson, *Creative Evolution*, 323.

62. For a discussion of punch cards as differential storage, see "Die Lochkarte als Träger des Hollerith-Verfahrens," in *Festschrift zur 25-Jahrfeier der Deutschen Hollerith Maschinen Gesellschaft* (Berlin: November, 1935), 86ff.

63. Raulff, *Der unsichtbare Augenblick*, 74. See Sigmund Freud, *Der Wahn und die Träume in W. Jensens »Gradiva«*, ed. Bernd Urban and Johannes Cremerius (Frankfurt am Main: Fischer, 1973).

64. Siegfried Kracauer, *Geschichte—vor den letzten Dingen* (Frankfurt am Main: Suhrkamp, 1971), 62.

65. Raulff, *Der unsichtbare Augenblick*, 77.

66. See Peter Galison, *Einsteins Uhren, Poincarés Karten. Die Arbeit an der Ordnung der Zeit* (Frankfurt am Main: Fischer, 2003).

67. Hermann Schmidt, *Denkschrift zur Gründung eines Institutes für Regelungstechnik* (Berlin: VDI-Druck, 1941), 9.

68. Helmut Siekmann, "Experiment und Computersimulation in der Strömungstechnik," in *Experimental Essays. Versuche zum Experiment*, ed. Michael Heidelberger and Friedrich Steinle (Baden-Baden: Nomos, 1998), 218.

69. Translator's note: the German expression "the tooth of time" (*der Zahn der Zeit*) is usually translated as "the ravages of time," but the original phrase is significant here as it also contains a reference to the toothed disc.

70. Gottfried Wilhelm von Leibniz, qtd. in Bernhard Siegert, *Passage des Digitalen. Zeichenpraktiken der neuzeitlichen Wissenschaften 1500–1900* (Berlin: Brinkmann & Bose, 2003), 183.

71. Falk Rieß, "Erkenntnis durch Wiederholung. Eine Methode zur Geschichtsschreibung des Experiments," in *Experimental Essays. Versuche zum Experiment*, ed. Michael Heidelberger and Friedrich Steinle (Baden-Baden: Nomos, 1998), 160.

72. Ibid., 167.

73. Ibid.

74. Gottfried Wilhelm von Leibniz defines these "little perceptions" in § 13 of his treatise *The Principles of Nature and Grace, Based on Reason* (1714). See Jens Gerrit Papenburg, "Hörgeräte. Zur Psychomathematik des akroamatischen Leibniz," in *Zeitkritische Medien*, ed. Axel Volmar (Berlin: Kulturverlag Kadmos, 2009), 375ff.

75. Christian Kassung and Albert Kümmel, "Synchronisationsprobleme," in *Signale der Störung*, ed. Albert Kümmel and Erhard Schüttpelz (Munich: Fink, 2003), 145.

76. Peter Mulser, "Zeitlupe Computer. Analyse von Lichtimpulsen jenseits der Meßbarkeit," in *Simulation. Computer zwischen Experiment und Theorie*, ed. Valentin Braitenberg and Inga Hosp (Reinbek bei Hamburg: Rowohlt, 1995), 101.

77. Von Helmholtz, "Über die Methoden kleinste Zeittheile zu messen," 173.

78. Berend Wilhelm Feddersen, *Entladung der Leidener Flasche, intermittierende, kontinuierliche, oszillatorische Entladung und dabei geltende Gesetze. Abhandlungen*, ed. Theodor Des Courdres (Leipzig: Engelmann, 1908), 28.

79. Ernst Mach, "Bemerkungen über wissenschaftliche Anwendung der Photographie," *Jahrbuch für Photographie und Reproductionstechnik* 2 (1888): 284–86. Rpt. in *Medientheorie 1888–1933. Texte und Kommentare*, ed. Albert Kümmel and Petra Löffler (Frankfurt am Main: Suhrkamp, 2002), 22.

80. Helmuth Plessner, "Anthropologie der Sinne," in *Gesammelte Schriften Band 3*, ed. Günter Dux et al. (Frankfurt am Main: Suhrkamp, 1980), 322.

81. Manfred Sommer, *Sammeln. Ein philosophischer Versuch* (Frankfurt am Main: Suhrkamp, 1999), 404.

82. Ibid.

83. Knut Hickethier, "Synchron. Gleichzeitigkeit, Vertaktung und Synchronisation der Medien," in *Zeit in den Medien—Medien in der Zeit*, ed. Werner Faulstich and Christian Steininger (Munich: Fink, 2002), 125.

84. See Frank Kessler, "Bilder in Bewegung. Für eine nicht-teleologische Mediengeschichtsschreibung," in *Apparaturen bewegter Bilder*, ed. Daniel Gethmann and Christoph B. Schulz (Münster: Lit Verlag, 2006), 208–20.

85. André Gaudreault, "Das Erscheinen des Kinematographen," *KINtop. Jahrbuch zur Erforschung des frühen Films* 12 (2003): 35.

86. See Christof Windgätter, *Medienwechsel. Vom Nutzen und Nachteil der Sprache für die Schrift* (Berlin: Kadmos, 2006).

87. Charles Féré, *Sensation et mouvement. Études expérimentales des psychomécanique* (Paris: Alcan, 1887), 13f.

88. For a genealogy of the timeline as a visual form, see Daniel Rosenberg and Anthony Grafton, *Cartographies of Time* (New York: Princeton Architectual Press, 2010).

89. J. Zenneck, "Eine Methode zur Demonstration und Photographie von Stromcurven," *Annalen der Physik* 305.12 (1899): 839. The author emphasizes the time-critical moment, as the repeatability of the time signal is an essential precondition for photographic registrability (see 850f).

90. Josef Czech, *Der Elektronenstrahl-Oszillograf. Aufbau, Arbeitsweise, Meßtechnik* (Berlin-Borsigwalde: Verlag für Radio-Foto-Kinotechnik, 1955), 52.

91. Translator's note: there is another play on words here in the original text, as the German term for "moment" (*Augenblick*) literally means "a glance of the eyes" (*Augen-Blick*).

92. Manfred von Ardenne, *Die Kathodenstrahlröhre und ihre Anwendung in der Schwachstromtechnik* (Berlin: Springer, 1933), 177.

93. Horst Völz, *Handbuch der Speicherung von Information* (Aachen: Shaker, 2005), 2: 431.

94. Wolfgang Hagen, "Die Entropie der Fotografie. Skizzen zur einer Genealogie der digital-elektronischen Bildaufzeichnung," in *Paradigma Fotografie. Fotokritik am Ende des fotografischen Zeitalters*, ed. Hertha Wolf (Frankfurt am Main: Suhrkamp, 2002), 1: 203.

95. Hans Schellhoss, "Die Funkortung beim Heer" (paper presented at the first conference of the "Navigation" workgroup in at the Ferdinand Braun Institute in Landsberg am Lech, March 23–24, 1944), http://www.cdvandt.org/navigation-tagung.htm.

96. Edmund Husserl, *On the Phenomenology of the Consciousness of Internal Time*, trans. John Barnett Brough (Dordrecht: Kluwer, 1991), 29.

Chapter 4

The Computer as Time-Critical Medium

CLOCKED LOGIC: THE COMPUTER

Time is a crucial criterion for media theory, as it provides insight into the being of technical media. The synchronization of signal flows, which is necessary for media to have an effect on humans, is based on a complex internal dramaturgy of time. In the digital computer, the classic cultural technique of timing becomes the literal "clocking" of chrono-logical sequences. This also applies to the networking of the computer itself; an entire hierarchy of time-critical operations is at work in the Internet (most importantly, the network time protocol for packet switching). When a defective computer produces a technological rhythm as it vainly attempts to access the hard drive and the internal synchronization ultimately fails, it becomes clear that the computer is relentlessly subject to the clock pulse of time. On the other hand, an intact computer enables signal and time axis manipulation in real time, as its operating system employs chrono-techniques like "pre-emptive multitasking" to anticipate the future and a "scheduler" to determine the optimal time window for computations. In this delicate temporal structure, informatics differentiates between "hard" and "soft" real time—a world that is oriented toward the human sense of time as well as signal processes themselves. Media-archeological analysis focuses on moments of digital time sensibility.

The concept of the computer as a time-critical medium includes key terms like stored programmability, cybernetic feedback, and recursion at the level of programming languages. Informatics speaks of the "semantic gap" between the computer and the programmer, which is simultaneously represented as a "temporal gap." Discrete cycle, computing, and dead times constitute a micro-dramaturgy that deserves to be formulated as an epistemological object of genuine media temporality. However, this analysis only succeeds with the most precise electro-technical and computational knowledge of what actually transpires: a mathematically cool, electronically hot temporality. The strict media-archeological method thus corresponds to "a theoretically cool system for melancholically hot loops over limited time resources or thermodynamically time-limited computer hardware."[1]

Time-critical symbol processing requires a much sharper focusing of the media analytical gaze than the general concept of "time-based media." Computer architectures set the basic temporal conditions of the machine; operating systems, and especially the process of "time-sharing," manage computing time for the quasi-parallel processing of incidental tasks; and signal-processing algorithms enable elaborate time axis manipulation. As the technical implementation of mathematics, the computer is not simply a symbol-processing machine; rather, its radical time-critical modes of operation also make it a complex time machine.

The so-called temporal logic of the computer serves as a model of its temporal modes in its branching sequences, transition relations, loops, and data access processes. During the development of hardware logic analyzers oscilloscopically visualize the individual computations of the computer in parallel channels. Such "monitoring" of process sequences in the computer for the purpose of optimization occurs as signal processing in real time, as a second order observation in accordance with the law of computing media. "Graphical data representation enables detailed understanding of dynamic processes on massively parallel systems," reported *Vampir*, a tool for performance analysis and "in-depth event-based analysis of parallel run-time behavior and interprocess communication" based on the parallel storage of "event traces."[2]

The digital computer does not have a problem with the processing of signals that are already symbolic and thus discretely coded, like alphabetical texts. The situation is quite different when the object of computer computation is supposed to be continuous physical time processes—in other words, the actual world with regard to the moorings of being and time. The computer only has a stake in temporal indexicality or real-world time at the level of its own time implementation. Discrete machines can hardly grasp this real-world time through digital signal processing, which is wired into the chips themselves. In the field of complex computations, "the need to treat clocked time discretely, which is associated with digitization, leads the hardware to determine the possibility or impossibility of a computation."[3] At a fundamental and thus media-archeological level, however, the "halting problem" of the digital computer still exists: no program is able to determine in advance if another program will terminate in finite time. What is required here is not an abstract algorithm, but rather the actual implementation, the being-in-the-world (*in-der-Welt-Sein*) of the mathematical operation. "Because it is only possible to show that symbol processing has come to an end through its implementation, it is not sufficient merely to imagine the Turing machine; instead, it must be allowed to run."[4] The basic ontological questions of temporality thus flare up at the macro-time-critical level of "computing."

Counting itself is already an operative implementation and therefore a temporal form; Hermann Weyl calls for the iteration of the domain of natural numbers.[5] Arithmetic initially served to define time as the measurement of movement itself (Aristotle); analysis was then able to calculate the smallest movements as Δt (Leibniz's infinitesimal calculation); this was eventually followed by cybernetic time-series analysis and "linear prediction" (Norbert Wiener) as time-critical mathematics, which was implemented in the high-speed computations of the computer. The prerequisite for this was an increase in computation speed through electron tubes, which surpassed the electro-mechanical time limits that constrained Konrad Zuse's proto-computers Z1 to Z3.

Analog and digital computers calculate with fundamentally different temporal modes. Every Turing machine must always be implemented in real analog physics in order to be effective, even though its key feature lies in the fact that this material implementation is *ideally* not crucial for the information. In contrast to the virtually instantaneous parameterization of computing tasks in analog computers through current voltage, the execution of programs in digital computers is based on discrete, incremental time; algorithms form time sequences. The symbolic formulation of the program as a list spatializes the dynamics of a process in the two-dimensional field; however, the media-technical execution of a computer program displays this space in time, thus radically dramatizing the algorithm.[6] An algorithm f must be executed in physical time in order for its result $y = f(x)$ to become real. Every computational model of the world[7] is thereby subjected to a temporalizing imperative; otherwise it remains a simple "world picture" (*Weltbild*). This is the entire difference between pure mathematics and mathematized physics. "Mathematical structures (graphs, groups, topological spaces, etc.) do not change in time whereas computer science objects (databases, machines) often do."[8] The parameter of time initially remains external to mathematical models as such, yet the Turing machine as "abstract-state-machine" already "represents an ... attempt at introducing dynamism into logic."[9]

Between numerically discrete computing time and the operative time of the analog computer, which equiprimordially acts as a simulated time process, a media-epistemological abyss opened up that has since been bridged by digital signal processing—that is, corresponding processors and algorithms.[10] In discrete signal processing, storage and time-critical moments are necessarily entangled; in discontinuous computing systems, scanning and storage are the most important operations.[11] In digital shift registers, capacitors are usually employed as buffers.

Unlike Vannevar Bush's differential analyzer from the 1930s, the Electronical Numerical Integrator And Computer (ENIAC) constructed in 1942 operated time-critically in its combination of tube technology and wiring. What is astonishing about this machine is not only its mathematical power, but also the fact "that it carried out such difficult tasks . . . in the shortest time." Galileo once had to measure the smallest fall times of bodies, and he employed an inclined plane to slow them down artificially. Electronic computing machines capable of feedback are able to measure the smallest intervals at lightning speed. Therein lies *"the special temporal relationship of this machine*: it operates in the microstructures or microprocesess of time, which cannot be utilized through human actions or thoughts"[12]—thus exceeding the human time window. This is a necessary condition for the chrono-technical effect called "real time." Media criticism must also differentiate processes in such delicate time domains.

Informatics recognizes the concept of concurrent behaviors in the computer: relative time windows *within* which an act is supposed to take place. Communication processes within computing architecture thus determine priorities and intervals. The clock pulse is not the only regulator here; rather, it involves a new kind of clockwork.[13] "The . . . distinguishing feature of the computer is its temporal creativity."[14] The computing machine here departs from the Newtonian universe, in which physical processes are reversible, and enters the realm of logical time. Indeed, the essence of the processual character of this subtle structure lies in the fact "that its being in time is not reversible but irreversible."[15]

Electronic media like radio and television and their methods of recording on magnetic tape are distinguished by the fact that their technical being is revealed in their implementation. Yet, the concept of technology only fully comes to the fore with the computer, as it represents the convergence of matter and mathematics, *techné* and *logos*. The computer is the medium in which the word becomes matter in the hierarchy of stored data from the bit to the "word." As imagined, the computer belongs to media theory; as a time-critical process implemented in the physical world, however, it belongs to media history—which, from the perspective of the computer, is inevitably time-bound and no longer necessarily historiographic. What distinguished Charles Babbage's analytical engine design in the 1830s from previous calculating machines was its time-critical element: it implied at the very least that the computational program could modify itself during the calculating process on the basis of temporarily stored results. In his description of a machine that was able to mechanically calculate everything that was mathematically calculable, Alan Turing emphasized the discrete conditions and thus the timing of the machine. John von Neumann also determined for the high-performance computer EDVAC that each step of data processing had to occur after another. The computer was thus described through its time behavior—up to asynchronous machines, the alternative to the Von Neumann architecture, which do not need to consider the slowest subcomponents.

From a time-critical point of view, the computer truly becomes accessible to media archeology through its micro-dramatic configuration. This perspective deciphers processes and events matter-of-factly as time series, while the historical discourse treats them as narrative elements. The era of the computer is written not simply as another chapter in media history, but rather as "Turing time."[16] In terms of media archeology, this requires new modes of representing temporal processes, as it makes available the entire history of technology up to the present. It is as important to write continuous and discrete conditions as they are technically signalized.

THE TIME OF THE FLIP-FLOP: 0/1 SWITCHING

The concept of chronology, which is so familiar in our culture, depends on the technical processing of binary signals. The object of switching algebra is logical and thus not time-dependent connection, but the relations between input and output signals in delay and memory circuits as well as frequency multiplication are especially relevant in their temporality.[17] The optimization of electronic data processing requires not only increasing the packet and integration density of the components on semiconductor plates, but also "reducing the delay times of an individual structure in an integrated circuit."[18] "Delay time" is the essence of binary time, as it refers to the switching time of bipolar transistors that elapses between a control signal and the point at which the collector current researches 10% of its maximum value. Digital media are fundamentally time-critical, as the basis of binary signal processing—the flip-flop circuit—divides time and thus recalls the etymological core of "time" itself: partitioning. Paradoxically, the same circuit simultaneously functions as the smallest binary storage for a "bit"; in such circuits, capacitors store the current state of both electron tubes, hence the name "decision circuit." The change of state itself is a change in current, which is triggered by a small, critical event moment.

The recoding of the notion of time as flow into a notion of time as discrete and nonlinear appears in Augustine's discussion of meter in prosody, which is critically connected to Aristotle's equation of time and counting. With the advent of the geared clock, this time-discrete aesthetic became technical at the level of measuring; in quantum physics, the world ultimately makes nothing but spontaneous time leaps. The nonlinear and nonchronological temporal modes of media are not founded on philosophy, but rather on their technical *a priori*: the discrete time of alternating current and binary signal processing.

The digital computer not only radically reformatted the concept of time *in* the twentieth century, but it also reformatted the concept of the "twentieth century" itself, as the "year 2000 problem" manifested in the last second of the last century. The "millennium bug" was a rare moment when people became aware of the time of the computer on a massive scale. The computer game "Little Computer People," which playfully dramatizes software errors, was itself affected by the Y2K bug, thus revealing the entirely unhistorical temporality of the computer.[19]

The specific temporality of the digital computer, which consists of discrete oscillatory clock pulses, transformed time itself into information. All preceding technical media were characterized by a material discursivity. For example, photography depicted a moment fixed on paper and cinematography presented an imaginary perceptual sequence through a succession of images on celluloid. On the symbolic side, vocal alphabetic writing and book printing brought about a temporally linearized cultural technique of reading.[20] However, the digital computer liquidates the material temporality of representational states in favor of a purely logical though electronically implemented code. At the operative level, the computer is constituted by radical time critique—namely, discrete scansions. At the hardware level, different successive stages of time—the linearities of past, present, and future—implode in a discrete electronic time. The computer thus proves to be a permanent media archeologist, which preserves the time-critical chronoaesthetic of World War II.

Norbert Wiener conceptualized a computing machine to calculate the trajectories of enemy aircraft; John von Neumann constructed a machine capable of performing calculations far beyond the computing capacity of humans. Computing media are at the same time systems and procedures, techno-mathematics and implementation. The results of this simultaneously logical and operative time can resolve any kind of event. All of the binary-coded operations in the digital computer proceed according to the time-discrete sequentiality of the "on/off," from which is derived all higher (program) structures right up to the narrative effects they produce through human interfaces. In such a way, the implementation of alphanumerically coded symbol sequences is able to simulate any kind of chronology and the cultural phantasms of past, present, and future better than literary narratives. If the metaphysical tradition envisages time as a continuous series of present moments, then this present becomes impossible in discrete and thus kairotic time. The relation between the moments of symbolic time in the digital computer is no longer a relation between substantial elements that can each be ontologized (as the Pythagorean mathematical aesthetic presupposes), but rather a radically different interrelatedness—the temporal essence of the binary units of information 0/1, implemented in worldly technics. Systems theory also identifies the time-critical moment that is involved in every binary circuit. The form of a statement as a border separates two sides such that only one side can be observed and the other side cannot be reached "unless the border is crossed—that is, unless time is expended."[21] The clear distinguishability of two conditions does not occur in the abstract realm of ideas, but rather necessarily in the world and thus in time, such as the flip-flop circuit as the time-critical core of binary information processing. Niklas Luhmann emphasizes "that marking must be temporalized, and therefore all calculus makes use of time.... The 'marks' ... are juxtaposed and superimposed according to definite rules.... But they only function as a sequence of operations, and thus only in time."[22]

THE CLOCK SIGNAL

According to Lewis Mumford, the clock ranks above the printing press in the list of factors that influenced the mechanization of society. "The clock, not the steam engine, is the key-machine of the modern industrial age."[23] Yet, digital culture did not begin media-epistemologically with the clock. In a genuinely media-archeological intensification of Mumford's argument, McLuhan pointed out in *Understanding Media* that clock time first became conceivable against the background of the enduring practice of alphabetic writing, which divided the flow of spoken language into smaller units that remained below the semantic threshold and whose literally elementary symbols encouraged scientific analysis. "Mumford takes no account of the phonetic alphabet as the technology that had made possible the visual and uniform fragmentation of time. Mumford, in fact, is unaware of the alphabet as the source of Western mechanism."[24] The development of clock time, time-based work organization, and time-critically implemented algorithms thus paralleled the typographic "Gutenberg Galaxy"—right up to the typewriter, which depended on the timing mechanism of the clock for the forward movement of the ribbon. In a "mechanical notation" to his analytical engine, Charles Babbage blurred the distinction between real machine and written sign by proposing a symbolic machine, which would enable not only the abstract analysis of the time behavior of the arithmetic unit in the form of a table, but also its actual reproduction as a time event—a diagrammatic simulation. "The table divides the time of a complete period of the machinery into any required number of parts; and it exhibits in a map, as it were, that which every part of the machine is doing at each moment of time."[25]

Digital computing machines deploy a calendaric of their own proper time; cultural perception is here confronted with the simulacrum of a self-referential time. The electronic computer as a contemporary condition of economic, informational, and cultural communication has no continuous time consciousness; rather, it is based on discrete hardware, pulse frequencies, and algorithmically programmed software. "Uniquitous computing" encourages an entire era to conceive of time processing in discrete steps. In the digital realm, "time axis manipulation" is spoken of differently than with classic analog media like phonograph, film, and video. The text-processing environment LibreOffice calls for the two-digit year dates in the document vault to be "interpreted" as years between 1930 and 2029, but this setting is individually variable. It is thus no longer mandatory for a computer-century to be based on the Christian calendar; rather, it transforms into an adjustable interval—a radical tempo-spatialization in computing space. Jacques Derrida defined *différance* for the culture of writing, but it has now expanded to the alphanumeric field.

The connection between time and number, which Aristotle defined theoretically, returns mechanically in the geared clock and electronically in the computer clock. However, this does not represent a recursion in a cultural-*historical* sense, but rather a superimposition. The epistemological characteristic is the quantization of time—an essential feature of the digital.

> Because microprocessors are complex structures of combinational and sequential logic, they require a clock pulse whose duration represents the smallest period for all of the processes executed in it. Through the internal time schedule controller, the *pulse generator* is thus in charge of all the processes controlled in and from the processor. In practice, quartz-controlled generators are employed. . . . For less stringent requirements . . . simple RC rectangle generators are sufficient. . . . With every cycle the microprocessor performs a partial operation step and then moves to the next condition. Consecutive operation steps constitute a *machine cycle*.[26]

Depending on the command, more machine cycles are necessary; the first is called the command read cycle ("instruction op code fetch"); a time diagram shows an entire command read cycle and diagrammatically facilitates orientation in (machine) time.

The goal of Alan Turing's symbol processing machine was to treat time discretely; it is informed not by the historical-continuous concept of time evolution, but rather by the media-theoretical (thus encompassing logic and engineering) insight into the necessity of discrete conditions T1, T2, etc. for the machinic mastery of numerical mathematics.[27] The Turing machine is a finite state machine that is always in an actual present.

> Nevertheless, a problem arises as soon as the past of the system is relevant in any way. In stateless systems no past actually exists in the system. . ., yet a

changing state exists outside the system in its output and input values. The function $sin(kt)$, for example, generates one value for every input value k to every point in time t; time is thus nevertheless located as linear progression outside the system.[28]

From a cultural-studies perspective, a long path of historical development lies between the macrotemporal *computus* used in medieval calendars and the clock signal used in computers (the inner time of the computer), yet media archeology reveals another path: the short circuit between eras and the compression of macro-historical distance to Dirac delta or impulse function intervals. The clock signal is functionally installed within the proper world of the computer and this is the very condition of its proper time, which cannot be compared to an external time. This time-critical aspect is crucial for the concept of the medium in implementation. As Charles Babbage wrote in 1855: "In order to perfectly understand a machine, it is imperative to have precise knowledge of the following: (1) The form. (2) The different parts. . . . (3) The precise moment when all of the different parts are set in motion, as well as the duration of the motion. This is the *cycle*."[29] In complex circuits, which depend largely on the synchronization of the electronic data stream, the measuring device of the logic analyzer serves not the quasi-philosophical function of verifying the validity of propositional logic, but rather the time-critical function of orchestrating virtually parallel pulse sequences. In the computer, the clock signal helps the circuits with reciprocal data synchronization in the temporal domain; therein lies the essence of the computer as a time machine. What the clockwork escapement mechanism initiated in the early modern period is evident here, too: periodic, rhythmic timing as the tuning of mathematics. Through the clock signal, the analog is articulated to the digital as the extreme compression of a dynamic process; at times, the apparently binary signal is itself composed of sine waves.

With the internal clock supply of the computer (and its antecedent, the feedback of electron tubes in the Meissner circuit), logic becomes not only mechanically operative, as in the scholastic mechanisms of Raimundus Lullus, but also time-critical. In order to construct master frequency clocks, a negation function is time-critically employed:

> Instead of drawing further conclusions from the results, as in logic, the negation leads back to the signal input, where the out-of-phase signal arrives with an infinitesimal delay due to the finite transmission rate and thus generates the inverse initial state, which reacts to the input again, and so forth ad infinitum. The negation of negation—albeit in the rather non-Hegelian domain of time—thus gives rise to a clock pulse that allows all other free-run input signals to be cut up in the microsecond rhythm.[30]

A DEFINITION OF DIGITAL MEDIA THROUGH TIME: CLOCK PULSES IN THE COMPUTER

The computer is realized through hardware, and its worldliness sets limits. The physically implemented Turing machine distinguishes itself from the ideality of its logical model through actual, not only linguistically formulated time. "Time never occurs in logic."[31] Every binary circuit built in the real world is not a purely logical function, but rather always also a time function. "Symbols are created in continuous dynamical time, and are only preserved in discrete, arbitrary structures."[32] Sampling always requires time- and value-discrete buffering; in this sense, everything that happens in the computer already represents an intermediate archival condition with respect to the world. In contrast to the purely symbolic order of the static archive, however, electrotechnically discretized relations possess a dynamic operativity and are thus temporally connected to the world.

The digital first occurs *in time-discrete implementation*, and thus as a genuine techno-mathematical operation. The basis of precise digitization is periodic timing, which introduces discreteness to computer time and relies on intermediate time, or strictly speaking the intermediate time of the clock pulse. Many digital gates operate at different speeds; the clock pulse is therefore needed to order these times and make them mathematically concurrent.

> Through the clock pulse time is microscopically quantized into computing time and rest time. For time-economic reasons, rest time is actually minimized as much as possible, but it is nevertheless indispensable to the secure exchange of information. . . . It also serves the performance-economy, such as battery runtime, because in contemporary semiconductor technology no electrical output is recorded during rest time. Looking deeper into digital circuit technology: before every time change, such as 0 -> 1, a "set up" time is defined up to which all gates must be finished—and they will be if the design is correct.[33]

This organization of time is an important criterion for stability, as it ensures that digital data processing is not only logical but also time-critical—that is, it can take place in the real world.

DOES TIME COUNT? THE DIGITAL, THE COMPUTER, AND THE SUBLATION OF MEDIA HISTORY

The time of the digital, insofar as it is based on relays and switches, permits approximation. "Therefore at any given time the circuit between any two terminals must be either open (infinite impedance) or closed (zero impedance)."[34]

This variable is explicitly "a function of time."[35] But what does time mean in this context? "Time *is* not. There is, It gives time. The giving that gives time is determined by denying and withholding nearness."[36] Shannon himself calls this 0/1 alternative the "hindrance" of the circuit. However, intermediate time itself (which Norbert Wiener defined as the "time of non-reality") does not count. "Between 0 and 1 there is no time. . . . It is the withdrawal of the real, through which the symbolic emerges."[37] Techno-mathematically discretized time only exists in the *ideal*, and this is the entire difference between calculating machines as theory and as actual technical event. An entire time world lies between theoretical and technical informatics:

> Because automaton theorys works with abstract concepts, this transition from one state to another in the theory occurs without intermediate stages. Automaton theory thus does not question how such a transition is actually technically executed by an automaton. It is only interesting, for example, that a flip-flop transitions from one stable condition to another within a certain period of time.[38]

In the digital mode, the signal is negotiated no longer as a continuous physical event, but rather as information. This paradigm shift in the concept of communication was introduced through the combination of electrotechnics and mathematics. It becomes plastic at the moment when amplifiers in communication lines (for which, in the case of the telephone, Robert von Lieben developed the electron tube), which had always also amplified interference, were replaced by the "regenerative repetitor," which was able to amplify the binary-coded signal to a large extent without errors because intermediate and boundary values were omitted. Logical time takes the place of physical transmission time.

The digital computer was more than the sum of all previous media, as it "represented a more refined and more effective intervention in time structures, as well as . . . the synthesis of various existing technologies."[39] In this respect, the temporal modes of all previous media were (in Hegelian terms) *sublated* in the computer. This was possible because it made the stream of time concretely discretized and thereby algorithmizable as software. The transmission of "streaming media" in the Internet (video, speech, and music) requires effective compression if the time window is supposed to remain close to real time. So-called "codecs" thus encode and decode signal streams.

Is the history of all previous media also sublated in the computer? On the contrary, the computer actually reconfigures the categories of media history itself. There were good reasons for distinguishing cultural techniques like counting and writing from technical apparatuses, in which processes are automatized and detached from human intervention. Technical media in the narrow sense were thus first introduced with photography, followed by phonography and cinematography. Such high-tech media completely broke away from traditional cultural techniques. Fully electronic media (which were not simply electrically augmented mechanical media) represented a new media episteme; they constituted a world in its own right, which was actually a product of cultural knowledge but which was entirely informed by the laws of logic and physics. The concept of media (besides its spiritualist derivatives) as discursively and scientifically formative was first due to this escalation in the form of electronic media, culminating in the moment when the concept of *media* was thematized outside the physical sciences. However, the definition of the computer as the combination of symbolic practices and high-tech or electrotechnical apparatuses results in a unique situation that defies the evolutionary model of media history. Apart from electronic media like radio and television, the computer practices a form of alphanumeric *symbol* processing that overturns the analog-technical world of continuous or discrete *signals*. It is thus very near to the fundamental operation of historiography itself in so far as it is based on discrete writing, which was so foreign to the time world of analog signal storage (like phonography). In view of the computer—in other words, from a media-archeological perspective—knowledge of the genealogy of Western symbol systems is as relevant as the knowledge of electrophysical technologies. The computer is the first true techno-logy, as it represents the alliance of *techné* and *logos* (as formulated by the ancient Greeks) or electrotechnics and mathematics (as formulated in media studies). While the history of the alphabet did not play a significant role for the concept of electronics, it is unexpectedly central for the *anamnesis* of the digital computer. McLuhan's expansive approach to the concept of media thus obtains a new media-theoretical relevance. On the one hand, he ushered in the concept of media with the title of his 1964 classic *Understanding Media*, thereby establishing media studies; on the other hand, he always kept the concept of the medium open to new connections and couplings that may still be yet to come. Part of the unique nature of media studies is that its object of study is constantly metamorphosing (unlike sciences with temporally closed subject areas, like ancient philology and classical archeology). The computer thus not only contains all of the temporal modes of previous media, but it also involves an emphatic time (or another macrotemporal time-relation) that is different from all previous media—the sublation of mathematical time in the computer.

COMPUTER TIME

The timing of the everyday flow of time through the geared clock and its escapement represented a mechanical precursor of the chronotechnical practice of discrete time coding. The clock pulse of the computer "enables us to introduce a discreteness into time, so that time for some purposes can be regarded as a succession of instants instead of a continuous flow. A digital machine must essentially deal with discrete objects."[40] Timing thus functions as the foundation of the *operative*, as the digital itself is implemented in real physics. Christopher Burton programmed a simulation of the Pegasus computer in the late 1950s. Time behavior separates replicas, simulations, and emulations of historical computers. "The simulation is . . . a bit-level simulation, that is, it is implemented at the logical equivalent of the molecular level of the machine. It replicates the behavior of the operating system exactly and animates the on-screen console in response."[41] The ancient Greek vocal alphabetic was a sub-semantic simplification of language; however, prosody first gave poetry its rhythm. Is this rhythm now migrating into computer algorithms? The difference between clock pulse and rhythm lies in the accuracy of time-critical *chronoi*—as defined by Aristoxenus on the basis of ancient Greek prosody[42]—which at the same time also distinguishes humans from machines as signal processing beings. Rhythm is imprecise, which is also inherent to the analog computer in contrast to the digital computer. "No analogy machine exists which will really form the product of two numbers. What it will form is this product, plus a small mechanism and the physical processes involved."[43] Accurate timing is a condition of digital precision as soon as it is performed by high-speed machines and not simply by people with pencils on graph paper. In his definition of "clocking," Turing added a note concerning his automatic computing engine (ACE): "All other digital computing machines except for human and other brains that I know of do the same."[44] In opposition to all of the reductions of neuronal signal processing to binary circuits, John von Neumann also noted that the brain communicates on the basis of chemistry. In place of accurate timing, the brain is dominated by neuronal "summation time," which corresponds to the integrator in the analog computer rather than the binary time-discreteness of the digital computer.[45] Neuronal oscillators actually differ from their technical equivalents in digital computers through internal phase shifts. Julian Bigelow emphasizes that the digital conceals the "forbidden ground in between" like an abyss in the real. "It does not seem to me enough to describe a digital process as being one in which there are two or more discrete levels in which you are only interested in saying whether you are at level A or at level B. I think it is essential to point out that this involves a forbidden ground in between and an agreement never to assign any value whatsoever to that forbidden ground."[46] In digital computers, it is understood that the value of this forbidden ground should never be written. The psychologist John Stroud thus recommended handling these transitions in practice as if they simply did not exist, and the logician Walter Pitts agreed to ignore continuity in the present.[47] However, this third exists in the domain of time—a special kind of ternary logic. Counting and computing are not simply functions of temporal operations, but rather they also make the concept possible in the first place; the consciously ignored intermediate time of binary conditions is the condition of possibility of the logical operativity of digital computers. McCulloch adds: "We have still a flavor of the continuum. When the probability of the Zwischen state is zero or negligible, we think chiefly in other terms."[48] The "forbidden ground in between" and the "Zwischen state" constitute the media-archeological level of media-technical temporal modes.

The computer is able to practice highly differentiated, discrete rhythms, whereas people physiologically perceive or cognitively assume simply continuous time—a rupture between actual signal processing and its perceptual effects. In early fully electronic computers, the timing rate was still in the audible frequency range up to 20000 Hz. In practice, as with the Z 22 of the Zuse KG, this was used by engineers as a form of conscious sonification to determine if a program was running regularly or in endless loops, thus crashing. The technicity of the medium, which otherwise remained hidden behind its contents, only manifested in moments of rebelliousness, interruption, and temporal accidents.

The time counted by the computer does not refer to actual clock time, but rather to the time that has passed since the computer was switched on; these time signals are derived from an internal piezoelectric crystal, and they thus represent an articulation of proper time. The delay time of a program constitutes a machine-sovereign temporal mode, which makes it autonomous of calendar, cultural, and historical time: the autopoiesis of media time. The circuit boards of processors now include battery-buffered real-time clocks, which continue to count after the computer is switched off; as a result, the internal counter immediately inputs the current time whenever the computer is switched on, and the computer thus conforms to global time. The time connected to the computer is always necessarily machine time, which replaces the previously merely symbolic order of calendar time. With the rise of machine time, calendar time collapses. With the positing of its own time base and the circulation of time signals as information, the system in the digital computer forms its own micro-temporal cosmos, which struggles with the entropic time of physics. "This clocking . . . must keep the pulses in step as well as prevent degeneration of the pulses over a number of cycles."[49]

THE IMPLEMENTATION OF MATHEMATICS AS MEDIA TIME

Technical knowledge relations not only pass through the people who create media but are also further developed and thus processed by them;

conversely, the proper names of thinkers and inventors almost negentropically outlive their mortal, historical biographies. Alan Turing also received the highest honor, as his name was bestowed on a logical machine—the Turing machine—which surpassed the individual person. In his 1936 essay "On Computable Numbers," Turing described how a person in the moment of computation (either mentally or on paper) is in a machine condition and thus in a transsubjective state. The Turing machine is a provocation of the historical foundation of mathematical knowledge in people. From a technical-historical or media-material perspective, the computer follows the series of electronic media like radio and television in a linear fashion. However, in a media-archeological sense—that is, deciphered as a technology in a broader sense of the word—it is not only an electrotechnical, but also a logical machine, which thus unexpectedly and nonlinearly calls for an entirely different, ahistorical state of mathematics and symbol processing. Inhuman and already predetermined evidence appears in mathematical objects (in geometry and arithmetic); they are thus also outside historical time. "Platonists are certain: every intelligent species inevitably develops the same mathematics as us, as it must be created from the same world of ideas."[50]

After World War II brought about a massive boost to innovation in the field of telecommunications, cybernetics no longer questioned individual media, but rather the systematic connection of their signal processing function. This epistemology did not evolve in the sense of the history of ideas, but rather it was the result of an inherent logic, a *logical proper time* of media technologies themselves. The waves and rhythms of media knowledge develop asynchronously in each historical context, and thus they are articulated not exclusively in the verbal written language, but rather in techno-mathematical argumentation. "When Shannon explicitly says that we have no need for a communications system for eternal truths . . . because such truths must be continuously reproducible at different times and places without technical transmission, it becomes abundantly clear how the essence of media diverges from our everyday concept of faith."[51] Media-archeological records are apparently not passed on in the mode of history, but rather in the repeated attempts of a genuinely techno-logical knowledge; roughly a century after Leibniz proposed a dual computer, for example, the engineer-captain Johann Helfreich Müller returned once again to this still unknown idea that a machine is capable of dual computations.[52]

The syllogistic mental operation as the primal scene of all logical machines—logic implemented in the real—is already played out in time through the quasi-medial necessity of an intermediate state (*medius terminus*). The conflict of logic versus time appears to be sublated. With the memory-programmable computer, logical calculus—reduced to Boolean logic statements with true/false values technified as one/off switching—is brought into the material world and thus becomes "real time." "Once we have the possibility of embodying this 0, this 1, the notation of presence and absence, in the real, embodying it in a rhythm, a fundamental scansion, something moves into the real."[53] A mathematical rather than metaphorical concept of media archeology then comes into play.[54] What is at work here is the calculating medium of the computer. "It is unique in human history that a culture should attempt to calculate and master the world with real numbers."[55]

The temporality of mathematics lies not in the realm of ideas, but rather in the worldliness of its media-technical operations—whether in the form of the "intermediate storage" of interim results in the registries of the central processing unit (CPU) of an electronic computer or in the classic form of the mathematical gesture on paper (and between them the overwriting sequences of the Turing machine on a tape loop).

> Every operation is almost completely sublated by a trace or . . . every operation virtually coincides with its grapheme-calculated trace. The most recent contribution of mathematics always also emanates a historical amnesia due to the fact that mathematics constantly overwrites and equalizes its history at the moment of its operation.[56]

The internal temporality of mathematics lies in its modes of implementation, but it is largely invariant with respect to external historical time. The temporal mode of mathematized media is thus likewise defined. As a condition of its essential *modus operandi*, circuit algebra can be functionally abstracted and raised above its concrete worldly implementation—like the concept of the virtual machine in the programming language Java. Even when an era of historical time elapses and this computer architecture is housed in a new physical form, its algebraic sense of time remains invariant.[57] In the foundational crisis of mathematics around 1900, "intuitionism placed a spatial continuum under the primacy of time. For Brouwer, therefore, mathematics had to prove itself solely through its activity, whereas for Hilbert it had to prove itself on paper."[58] Both of these modes—the abstract mathematical space of signs and machine activity—are entangled in the computer. Time lies here in the operative. While second order cybernetics later noted that every act of differentiation (what Spencer-Brown called "drawing a distinction")—in other words, the constitution of an observer difference—produces a microtemporal interval, L. E. J. Brouwer already pointed out that every mathematical operation has an irreducible temporality that precedes pure symbolic demonstration as an act of thought in the sense of operative diagrammatics. The thinking machine, the computer, exceeds this essential temporality in the field of symbolic manipulation itself—the time of the digital.

In a typewritten memorandum from April 23, 1942, on "digital computation for a. a. directors," which was written in the context of a conference on electronic fire controls in air defense, George R. Stibitz distinguished between analog and discrete-numerical computation. "Computing mechanisms have been classified as 'analog' or as 'pulse' computers. The latter term seems to me less descriptive than the term 'digital.'"[59] Stibitz added that "digital computers introduce a consideration not found in kinematic analog computers, namely the ordering of computation steps in time." He was referring here to the "number train" of 0 s and 1 s. "Digital computation is dynamic in character," and the digital computer is thus more than a traditional calculating machine; it is a genuine media-technical temporal mode born from the marriage of mathematics and logic.

From a media-archeological perspective, the computer operates in ahistorical conditions. For the nineteenth-century protocomputer designed by Babbage, the question of its proper temporalization arises: Should it be treated media-archeologically or historically? The core of the Difference Engine No. 1 was actually built (based on the techno-mathematical principle of *finite* differences) and presented at the 1862 London Exhibition; the detailed blueprint of the Difference Engine No. 2 only existed as a paper machine, however, until the arithmetic unit was finally realized in the London Science Museum on the occasion of Babbage's 200th birthday in 1991—"a modern original of an old design."[60] The trusted media-temporal concepts of museum curators and restorers of antique media here become confused. Unlike traditional kinds of technical drawings, digital computers as *per definitionem* symbolic machines represent a new type of paper machine whose temporal mode is removed from classic media history, as it can be replicated without loss at a material level (and as software). "Logical simulation as a virtual object in some respects survives the forensic test of historical utility."[61] It simulates a literary experiment as conscious anachronism: the conjecture that the computer would have already prevailed in Victorian England as the model medium. This results in asymmetries between social discourses and technical actuality.[62]

Turing's symbol-processing machine strictly requires that the time parameter be treated as *discrete*; it is based not on a historical-continuous concept of the evolution of time, but rather on a media-archeological conception of sequences of states. And yet each present of a finite state machine is the logical function of its past. "A current *state* is determined by past states of the system. As such, it can be said to record information about the past, i.e. it reflects the input changes from the system start to the present moment."[63] Turing's design of the principles of a symbol-processing machine was the answer to the problem of computable numbers, as his symbolic technology handled uncountable infinities in countable finite quantities. The clock, as

Turing wrote in 1936, introduces technically concrete "discreteness into time, so that time can for some purposes be regarded as a succession of instants instead of as a continuous flow. A digital machine must essentially deal with discrete objects."[64] This resembles Aristotle's approach to time: an apparently infinite differentiable flow is made countable through scalar measurement; time and clock go hand in hand.

In November 2007, veterans in the Nixdorf computer center in Paderborn and at the finally rebuilt "Colossus" in Bletchley Park successfully reenacted the encoding and decoding of intelligence using Enigma and Lorenz machines from World War II. For a series of operative moments, therefore, the museumized media were once again implemented, and within the framework of operative time, which sublates historical difference, they were in an equiprimordial rather than historical state. This was also true of the syllogistic machines of the early modern period. In contrast to Turing's automatic machine, which was programmed with tables, Llull's *Ars Combinatoria* was a system that depended on active reader participation and thus involved "cold" calculation (in McLuhan's sense).[65] And it required not only logical but also massive theological participation. The operation carries a historical index as vector, but the machine is also combinatorially operable without any historical context. It was also widely received—minus the discursively concrete impetus, that is—because it operates at the level of syllogistics, which was considered more or less invariantly valid from Aristotle to the twentieth century (via Boole and Shannon) and through all cultural-historical relativization, like a phase shift.

"The computer is not entirely realized in its own prehistory," Georg Christoph Tholen notes regarding the archeology of control media (rolls and chimes, calculators, punched cards, as well as calculus and logical machines).[66] The culture of nonlinear information processing in computers is not only an era of media history but also a challenge to the historical model itself. Vilém Flusser (who still wrote media histories) emphasizes that the "processual, historical . . . consciousness" is replaced by a "formal, calculatory, and analytical consciousness."[67] This was from the beginning (*en arché*) the media-archeological alternative to narrative.

THE MASTER FREQUENCY CLOCK

The more complicated the algorithms, the more rhythmic they are. The equalizing of parallel signal streams is time-critical and requires delicate synchronization. In the computer, 0 and 1, low and high, are time states or time values; the indeterminate zone cannot become operative in discrete time intervals. A logic analyzer registers each now-state and the immediate post-history of such signals. The timing mechanism in the computer does not display time, as the geared clock once did; rather, it generates a self-referential and thus idiosyncratic time world:

> The inner game of the machine is a constant clocked conversion of symbols. The rhythmic pattern of state transitions thereby follows strict formal rules. A fundamental property of computers is that future states only depend on the current state ... and the inputs. ... The *inner current state* is detached from the outer world. The computer thus only retains *the past in the present*.[68]

The distinction between clock pulse and rhythm, as thematized in ancient Greek music theory[69] and the neuronal oscillators in humans, comes into play here.

Konrad Zuse's electromechanical computer Z1 still operated with a clock pulse rate of 1 Hz, the Z2 with 3 Hz, and the Z3 already with 5–10 Hz. The fully electronic ENIAC rapidly increased the clock pulse rate to 100 Hz, and modern computers have long operated in the gigahertz range. This does not simply represent a quantitative increase, but it also changes into a new quality. Within the framework of complexity theory, time is no longer specified in absolute seconds, but rather it is relative to input size. The attention of time-optimized programming demands a chronotechnical view of process control. The concept of *algorithmic depths* thus also includes the elegant time complexity of an algorithm as a measure of information content.

John von Neumann's "First Draft of a Report on EDVAC"[70] stated that the acceleration of computing power required the analysis of time-critical processes, which was not yet possible with traditional computing times. This included above all shock wave equations in connection with hydrogen bomb explosions. "Neumann's computer concept inaugurated a paradigm shift in the concept of time. It describes a complex machine whose runtimes ideally tend towards zero but are in fact randomly minimal/maximal. Its paradigm is the atom bomb, whose momentary flash breaks all traditional concepts of time in the macrocosmos of the world."[71] Neumann's computer concept, with his principle of strict sequentiality in data processing, required an extreme minimization of the time intervals of calculation steps. As soon as the system started up it already needed the mastery frequency clock for its internal time control. It is important here to differentiate between machine clock pulses and machine cycles (for different operations). In the computer, the discretization of time through "clocking" is virtually complete: "extremely close moments, which are only separated by a few nanoseconds."[72] The sublime dissimulation of the temporality generated in the digital computer through quartz crystals lies in the fact "that it can separate a second of analog time with a divisor, which is beyond our powers of imagination."[73] Only chronotechnical media still understand such a time.

INTERRUPT: THE TIME-CRITICAL COUPLING OF COMPUTER AND WORLD

Interfaces between humans and machines, as well as machine-to-machine, constitute a coupling with time. Informatics calls this—notwithstanding the historiographic concept—an *event*. During the Cold War, US radar monitoring of airspace fed signals over telephone lines into the "Whirlwind" supercomputer. The processing of these signals in real time required a discrete dialogue between input and processing, which resulted (in the media-archeological sense) in the "foundational" (Claus Pias) introduction of an interrupt signal that was able to interrupt processing at regular intervals and sensorially perceive the environment. This engineering solution resulted in the rotating magnetic drum, which could temporarily store data until a predetermined time when processing power would again be available.

> Communication between input, computing, and output units thus raised a time-critical question concerning the issue of a collective and yet at the same time differentiated systemic rhythm. The triggering of communication through an interrupt had little to do with the clock pulse of the central computing unit; rather, it was the most economical common denominator for peripherals with different amounts of data.[74]

It also enabled a critique of uniform techno-beats.

> A system was no longer dominated by a collective rhythm, but rather a variety of rhythmic interruptions. As a result, anything that was not available to the system at a certain location and time or was not temporarily buffered did not exist. Continuities, such as the tracking of a moving target, were thus only an effect of an extremely rapid yet ineluctably discontinuous triggering.[75]

The intrusion of world time into the model medium of the present, the computer, occurs in different ways. Interrupt processing gives microprocessors the option to interrupt a running program through a control signal in order to allow another program to run first, after which the initial program continues.[76] Through such incursions of the external world (like sensors), the computer is "informed" in the sense of mathematical communication theory, especially in the case of bidirectional communication. The interrupt reacts almost instantaneously to an external event. Certain processors are microprogrammed in such a way that the code regularly calls the analog-to-digital converter if it is necessary to react to changes. As hardware, this chrono-logical responsiveness to time signals has been entirely "made flesh."

The so-called "masked interrupts" in computer programs enable interruptions that can be approved or refused by the programmer through corresponding orders. "During the execution of a time-critical program part, for example, interrupt approval is blocked."[77] An example is the weather forecast, which is continuously reevaluated on the basis of actual weather that the forecast itself cannot influence. In contrast to the thermodynamic world of the weather, the world of technological media represents a culture of the second order. The everyday success of media-technical operations is constant proof of its claim to validity. Turing's theoretical machine in 1936 defined computability as a continuous approximation of real processes. Temporal continuities called "world" are not entirely comprehensible from the machine itself, which operates with discrete symbols; informatics responded to this with algorithmic recursions as a mathematically operative time figure.

TURING'S MACHINE TIME: CLOCK PULSES AND SAMPLING

The operativity of finite state machines constitutes a linear sequence of events in time. "These events occur only at discrete 'moments'—between which nothing happens. One may imagine these moments as occurring regularly like the ticking of a clock, and we identify the moments with integers—the variable for time, t, takes on only the values 0, 1, 2, . . ."[78] Time thus becomes a property of computer numbers. A human observer "may think of the machine as operating continuously, and of our moments as corresponding to the instants at which he takes a sequence of 'snapshots' of the machine's condition"[79]—a perception that is familiar since chronophotography. The deterministic machine is always in a particular temporal state. The Turing machine operates not in a timeless space, but rather time-critically. "At any moment there is just one square . . . which is 'in the machine'"[80]— that is, on the endless tape underlying the model. This introduces a new concept of data not simply as abstract information, but rather as a temporal condition.

Being-in-the-world means being-in-time. The central theme in the algorithmic modeling of the Turing machine is the concept of finitude (the "halting problem"). A computing machine is always in a definite and thus discrete state due to the temporal clock pulse. In Turing's essay "On Computable Numbers," time also plays a technical role in the succession of processing steps in the machine clock.[81] The complex synchronization of data processing in the computer is modeled on industrial clock systems with master and slave clocks. Without such clocks, the computer would also be unable to process the available information in precise increments. The master frequency clock sets the start, end, as well as sequence and frequency of its calculation steps.[82] Yet, the computer as time machine does not actually fit into the technical-historical continuity of time devices:

Computers thoroughly differ from clocks. Thanks to their capacity, which seems capable of managing everything all at once, they make time disappear rather than calling attention to it. The symbols they use to express time are not "analog," like the continuously advancing hands of mechanical clocks, but rather alternating "digital" signals that flash on demand in chopped lines.[83]

If digitization is the translation of the environment into the drama of information processing, then the time-critical constitutes its operative core. The Nyquist–Shannon sampling theorem defines the substitutability of a constant signal through a finite number of values—computable numbers. With a sampling rate of 44.1 kHz a value in the sonic range is time-discretely calculated such that it can be faithfully reproduced without any audible deterioration. If all of the intermediate values were relevant at every point in time, this process would have to run analog; in contrast, algorithmic filtering operates discretely. This practice is implicit in Turing's symbolic machine design; accordingly, time must be viewed—in a complete realization of its Aristotelian definition—as a succession of moments instead of a continuous flow. "If data can be represented in the computer through pulse sequences per unit of time, then conversely the computer can also extract data from all real pulse sequences per unit of time . . . and thus infinitely sample what can be finitely sampled."[84]

Such high-tech operations could obviously only exist in a culture that was discursively familiar with these kinds of negotiations of timing and disciplining. Foucault's *Discipline and Punish* not only analyzes the panoptic *dispositif* but also draws attention to the genealogy of *time planning*, which was gradually refined to calculations in seconds and which occurs today in the architecture of the computer. The digital computer itself does not understand time in the cultural sense; it evades the emphatic concept of time in favor of a purely signified tempor(e)ality in the form of timing and rhythm. Mechanical clocks determined the timing of industrial society; atomic clocks do the same for the post-industrial information age. Their accuracy is the condition of possibility for the operation of highly sensitive time-critical electronic communication and navigation systems. Dynamic systems (whether technological or natural) involve the execution of temporal processes, so their state can be indicated at any given moment. This time axis proceeds mostly in one direction and is understood as a continuum. In order for the digital computer to be implemented and placed in a medial state, an operative combination of current, voltage, and logic is required—in other words, an alignment of computation and temporality, which leads (in contast to Heraclitean becoming) to a radically discrete concept of time:

> In contrast to this [continuous concept of time], it is convenient to study discrete-action devices in terms of a hypothetical discrete time. Let us imagine that the continuous time axis can be divided into an infinite number of finite

intervals, not necessarily of equal length. Moving along the axis from $t = 0$ toward $t = \infty$, we mark the points separating these intervals by characters t_0, t_1, t_2, ... These points then constitute a countable set. Let us further agree to represent the characters t_0, t_1, t_2, ... by a series of positive integers 0, 1, 2 ... ... The time instants t_0, t_1, t_2, ... now denoted by numbers 0, 1, 2 ... shall be called *discrete moments*.[85]

An essential protagonist on the computer motherboard is the timer, which counts electrical pulses. The pulses can come from any source, like an external quartz crystal that yields a certain number of pulses per second. The medium of time processing is not narrative, but rather counting.[86] The timer—in technical jargon it is occasionally called the "real time clock"—can also be addressed by the program itself, as its own form of time allocation. The timer in the Commodore 64 was thus operatively calculated in order to generate arbitrary numbers by the RND command, which resulted in (pseudo) randomness.[87]

In the early phase of the electronic-digital computer, the timer also had the additional function of refreshing the data stored in RAM, as memory chips were incapable of retaining electronic data in the time window of their existence for more than a few milliseconds. Another function of the timer in the computer was "the production of sounds."[88] To turn the argument around, this means that acoustic processes of time behavior in the computer—in other words, its essence as a time-based medium—could be perceived by the senses.

Leibniz's concept of the world as composed of monads already required timing. Monads "keep time with one another like separate clocks, so that they appeared to communicate with one another; but this appearance is merely a deceptive consequence of their synchrony."[89] This also applies to logical automata and their high-tech mode of implementation.

> In order to understand the function of a microprocessor, we must disentangle ourselves from what we previously learned in electrical engineering. There is no continuous current flow, as for example in a lightbulb. ... In the microprocessor individual current paths are always connected for very short periods of time, so the time in which the current actually flows is determined by the *timer*, which is part of every microprocessor system. It sends ... a clock pulse to the microprocessor. ... A certain combination of switches then triggers a desired activity.[90]

The difference between logical abstraction and actual implementation in hardware comes into play here. The synchronization of discrete microelectronics requires strict sequentiality and the time-critical coupling of logical instructions and the processing of singular data in the same world of bits. Sequential data processing in the Von Neumann architecture of the computer embodies the thermodynamic arrow of time (entropy) of the physical world in a time-logical sense. "There must be some irreversibility to ensure that calculations go forward (from inputs to outputs) and not in reverse."[91] Within this microcosm, however, every specific computer architecture has its own temporal properties, as in the configuration of *asynchronous* processors. The individual components do not wait for a central clock signal; rather, they work at maximum speed as soon as they are addressed. The acquired time depends on the skilful control of the flow of data in order not to impede reciprocal communication—the becoming-intelligent of chronotechnical time.

The momentariness of digital signal processing is intensified in the elementary unit 0/1, thus *materially* in the flip-flop. Binary logic as a time-critical process is operative in this circuit, and *it thus resonates with Bergson's notion of the musical sense of time in humans*. "This reduction of the spatialization of duration to nanoseconds simulates the vibrations of pure perception more precisely than analog time. The higher the frequency of this rhythm, the more it can be compared to a wave, even if this wave consists of the most discrete elements."[92] Mathematics first gains knowledge of the world in implementation through the act of time (mediamatics). As infinitesimal calculus, the cuts produced by time-discrete sampling and discrete calculation steps are able to foster the illusion of reality as continuity. This implies, however, that reality is perhaps not a becoming at all, but rather a nonlinear fabric of time.

TIME FUNCTIONS: SEQUENTIAL AUTOMATA AND MEMORY PROGRAMMABILITY

The limits and chances of the Turing machine manifest in time. It becomes comprehensible through the model of the endless paper roll (apparently inspired by the typewriter), which makes Cantor's concept of the "countably infinite" extremely concrete. The speed of computing routines in sequential architectures is a function of distance and time. Turing sought to access the stored records *quickly*.

> In general the arrangement of the memory on an infinite tape is unsatisfactory in a practical machine, because of the large amount of time which is liable to be spent in shifting up and down the tape to reach the point at which a particular piece of information required at the moment is stored. . . . One needs some form of memory with which any required entry can be reached at short notice. This difficulty presumably used to worry the Egyptians when their books were written on papyrus scrolls. It must have been slow work looking up references in them, and the present arrangement of written matter in books which can be opened at any point is greatly to be preferred. . . . Memory in book form is a good deal better.[93]

The intrusion of numbers into the alphabetic code (the page numbers in books) effectively made typographic memory addressable. This passive memory becomes active in the concept of memory programmability. John von Neumann defined its computational format in 1945. In the integrated program memory, commands and data are filed in the same format. The stored data thus potentially revises the current set of instructions; as a result, the diachronic is synchronously operative. Jacques Lacan's interpretation of the unconscious corresponds to Von Neumann's program memory and is thus inspired by media archeology (in the sense of a recursion rather than a historical causal chain).[94] Neuronal human memory is also not an emphatic archive in the sense of accumulation; rather, it is processual and dynamic in that it constantly converts incremental, impulse-like diachrony into synchronicity.

Memory programmability also requires the introduction of *dynamic* buffer storage for short-term values, which is realized in its most direct form in registers. This qualitative leap over previous computing machines was also accompanied by an exponential leap in the speed of computational processes, as the introduction of the electron tube permitted a maximum switching time of 200 operations per second.[95] For the memory programmable computer, "not only must the memory have sufficient room to store these intermediate data, but there must be provisions whereby these data later can be removed, i.e., at the end of the (t + dt) cycle, and replaced by the corresponding data from the memory."[96] This concept was dynamically realized in the early period of electronic computers in the form of delay lines for signal chains as circulating data words. A central temporal mode of high-tech media, the Δt, was actually implemented, and this chronotechnology made time itself obsolete. Contemporary storage technologies are able to write, read, and/or delete values at any time. "The typical computer program does not exemplify solely linear thinking. Programming is an activity that relies on both hemispheres of the brain. A program consists of doubling back, of loops within loops, of branching off in different directions. It is thus an apt symbol for contemporary temporality. ... There are species of time which no longer progress."[97] The Turing machine "writes" and "reads" like humans—namely, in discrete sign sequences and saccades. "The fact that a sequence proceeds step by step in time and cannot be viewed at a glance in its entire infinite expanse is an immediate consequence of our situatedness in time. The task is therefore to examine the relationship between mathematical objects and temporality, this exquisite human moment of being (*Dasein*)."[98] Current computer architecture is dominated by temporal sequentiality instead of delicate parallel data processing; the rapid central clock pulse and the time management of the operating system compensate once again for this temporal disadvantage. Every modern operating system is actually characterized by multitasking or quasi-parallelism in data processing, which connects increasingly complex synchronization problems. The computer was never as time-critical as it is today. In the concept of memory programming, the necessary connections corresponding to the commands coded in working memory are only established when needed, which results in greater flexibility in the time-critical field. "Thus all sorts of sophisticated order-systems become possible, which keep successively modifying themselves and hence also the computational processes that are likewise under their control. In this way more complex processes than mere iterations become possible."[99] A true chronopoetics thus emerges.

TIME-CRITICAL PROGRAMMING

The programming of computers always also requires time-critical consideration of data synchronization. The so-called "profiler" finds out how long the machine is needed for each operation. This unique temporal dramaturgy effectively makes the machine a world unto itself, in so far as "world" is conceived as a temporal structure. What is attractive about programming in assembly language is the fact that it enables control of the operation of the computer "up to the final ticks of the system clock."[100] At the level of media archeology, the greatest possible proximity between human and computer lies in time-critical machine programming. The programmer is also closely oriented towards the working and temporal modes of the processor through the abbreviated process of so-called "mnemonics" in assembly programming, which allows every procedural step to be predetermined. Many time-critical applications are thus still programmed in assembly despite the user-friendly standard language.[101]

Time-critical *techné* manifested in ancient Greece with the playing of music.[102] This connection becomes evident as techno-mathematical reentry for the current era in the *Lexicon of Music Electronics*, whose entry on "machine language" emphasizes the time-critical element. The time-critical is here associated with hardware.

> A program written in M. can be immediately understood and executed by the central processing unit of the computer, which normally ensures a much faster execution compared to programs in a higher (but nevertheless easier to learn) programming language. For example, it is useful to program in M. for direct sound synthesis, . . . as many computational processes must occur in the shortest time.[103]

This time form thus reveals itself sonically "in the domain of sound synthesis, in which the time structure of processes is to some extent the only object."[104]

For a long time, electronic images were also only representable as sequences of pixels in time and consequently as mathematical functions of the time axis, which virtually made them sonic events. However, current graphics memory no longer requires refresh cycles, which makes the temporal dynamization of the image nothing more than a media-archeological interlude.

In 1971, the programming language Smalltalk made it possible to rewrite a program "on the fly" during its runtime, which was previously only feasible in machine code. A relatively entangled temporal relationship emerges, especially with respect to sound objects—in other words, whenever the program refers to a process that takes place entirely in time. "For example, does an algorithm for sound synthesis refer to a sonic event or to the machine that created it?"[105] In the SuperCollider programming environment, signal processing enables the instrumentalization of the time parameter at the programming level for the composition of music. The processing of musical signals directly on the basis of their digital sampling requires the use of techno-mathematical intelligence. Signals are thereby conceived not semiotically as signs, but rather media-analytically as a time function. Pseudo-parallelism is achieved here through skillful synchronization. Complex signals result in chronotechnical complexity, and the concept of time shifts from a philosophically and culturally emphatic signifier to an operative signifier.

The MIDI standard of 1982 already implemented the temporal element in computer music by allowing the instruments to be detached from the source. This occurred even more completely in the "physical modeling" of the instruments themselves. This approach operates at the level of the acoustic real and links computing time directly to the sonic event rather than its musical-harmonic idealization in the Pythagorean sense. It is precisely the psycho-acoustic reality that stochastic calculus seeks to simulate. Concrete instruction codes in SuperCollider generate quasi-random sequences of numbers in order to produce frequentative imprecision, which is perceived by the ear as a signature of the natural rather than synthetic abstraction.[106] SuperCollider was developed for the composition of electronic music, and it therefore enables manipulation at the time-critical level. A cascade of time-critical operations structures time in the microtonal range. The "yield" command allows a smooth merging of the sound and thus a time-critical coincidence. With the "stream" command, the system overrides moments of waiting and pausing by providing a new value for every point in time. Such are the sonic temporal modes of informatics.

NOTES

1. Martin Carlé, comment from the seminar "The Computer as Time-Critical Medium," Helmholtz Center for Cultural Techniques, Humboldt University, Berlin, winter semester 2002–2003.

2. Developed by GWT-TUD GmbH, Dresden; see *Vampir 8.5*, http://www.vampir.eu.

3. Sybille Krämer, "Was haben die Medien, der Computer und die Realität miteinander zu tun?," in *Medien Computer Realität. Wirlichkeitsvorstellungen und neue Medien*, ed. Sybille Krämer (Frankfurt am Main: Suhrkamp, 1998), 19.

4. Philipp von Hilgers, *Kriegsspiele. Eine Geschichte der Ausnahmezustände und Unberechenbarkeiten* (Munich: Fink, 2008), 171. Von Hilgers here refers to Gregory J. Chaitin, *The Limits of Mathematics: A Course on Information, Theory and the Limits of Formal Reasoning* (Singapore: Springer, 1998), 11.

5. Nils Röller, *Medientheorie im epistemischen Übergang. Hermann Weyls Philosophie der Mathematik und Naturwissenschaften und Ernst Cassirers Philosophie der symbolischen Formen im Wechselverhältnis* (Weimar: Verlag und Datenbank für Geisteswissenschaften, 2002), 83.

6. See Bruno Bachimont, "Formale Zeichen und digitale Computation," in *Spektakuläre Experimente. Praktiken der Evidenzproduktion im 17. Jahrhundert*, ed. Helmar Schramm, Ludger Schwarte, and Jan Lazardzig (Berlin: Walter de Gruyter, 2006), 407.

7. See Konrad Zuse, *Rechnender Raum* (Braunschweig: Vieweg, 1969).

8. Juri Gurevich, *Logic and the Challenge of Computer Science* (New York: Computer Science Press, 1988), 4.

9. Matthew Fuller and Andrew Goffey, *Evil Media* (Cambridge, MA: MIT Press, 2012), 80.

10. See Morris Rubinoff, "Analogue vs. Digital Computers: A Comparison," *Proceedings of the Institute of Radio Engineers* 41 (1953): 1254–62.na

11. A. Kley and G. Meyer-Brötz, "Analoge Rechenelemente als Abtaster, Speicher und Laufzeitglieder," *Elektronische Rechenanlagen* 3.3 (1961): 120.

12. Max Bense, "Kybernetik oder die Metatechnik einer Maschine," *Merkur* 5 (1951): 205–18; rpt. in *Ästhetik als Programm. Max Bense/Daten und Streuungen*, ed. Barbara Büscher, Hans-Christian von Herrmann, and Christoph Hoffmann (Berlin: Vice Versa, 2004): 57f.

13. See Christian Steininger, "Zeit als kulturwissenschaftliche Schlüsselkategorie," in *Zeit in den Medien. Medien in der Zeit*, ed. Werner Faulstich and Christian Steininger (Munich: Fink, 2002), 9–44.

14. Marshall McLuhan and Eric McLuhan, *Laws of Media: The New Science* (Toronto: University of Toronto Press, 1988), 53. "While clocks are all set to the same exacting sequence, duration, and rhythm, the computer is free to manipulate all three of these temporal dimensions by merely changing the program." Ibid. McLuhan here refers to David Bolter, *Turing's Man: Western Culture in the Computer Age* (Chapel Hill: University of North Carolina Press, 1984), 38f. Bolter himself time-critically intensifies it. "In another sense, the computer processes time itself." Ibid., 102.

15. Bense, "Kybernetik," 59. Bense here refers to Wiener, *Cybernetics*, 30–44.

16. This was the title planned for the last volume of Friedrich Kittler's *Music and Mathematics* series.

17. "In practice, logical and temporal links often co-occur." Heinz Greif, *Messen, Steuern und Regeln für den Amateur* (Berlin: Deutscher Militärverlag, 1971), 197.

18. Otger Neufang, *Lexikon der Elektronik* (Braunschweig: Vieweg, 1983), v.

19. See Stefan Höltgen, "Vom Bug-on-a-chip zum House-on-a-Disc. 'Little Computer People' und die Archäologie des Computerfehlers," *Retro. Kulturmagazin für Computerspiele* 21 (Autumn 2011): 12–14.

20. See Walter Ong, *Orality and Literacy: The Technologizing of the Word* (London: Methuen, 1982).

21. George Spencer-Brown, *Laws of Form* (New York: Julian Press, 1972); qtd. in Niklas Luhmann, "Die Form der Schrift," in *Germanistik in der Mediengesellschaft*, ed. Ludwig Jäger and Bernd Switalla (Munich: Fink, 1994), 405f.

22. Niklas Luhmann, "Die Realität der Massenmedien. Niklas Luhmann im Radiogespräch mit Wolfgang Hagen," in *Warum haben Sie keinen Fernseher, Herr Luhmann? Letzte Gespräche mit Niklas Luhmann*, ed. Wolfgang Hagen (Berlin: Kulturverlag Kadmos, 2004), 68f.

23. Lewis Mumford, *Technics and Civilization* (London: Harcout, Brace & Company, 1934), 14.

24. Marshall McLuhan, *Understanding Media: The Extensions of Man* (New York: McGraw-Hill, 1964), 160.

25. Dionysius Lardner, "Babbage's Calculating Engine," *Edinburgh Review* 59 (July 1834): 314; qtd. in R. H. Babbage, "The Work of Charles Babbage," in *Proceedings of a Symposium on Large-Scale Digital Calculating Machinery*, ed. Howard Aiken (Cambridge, MA: Harvard University Press, 1948), 15.

26. Manfred Krauß, Ernst Kutschbach, and Eugen-Georg Woschni, *Handbuch Datenerfassung* (Berlin: VEB Verlag Technik, 1985), 165.

27. See also Zuse's *Rechnender Raum*, in which the computer pioneer claims that the universe is calculable as a constant mutation of discrete states—namely, as a chain of pulse transmission relays.

28. Julian Rohrhuber, "Das Rechtzeitige. Doppelte Extension und formales Experiment," in *Zeitkritische Medien*, ed. Axel Volmar (Berlin: Kulturverlag Kadmos, 2009), 208f.

29. Qtd. in Bernhard Dotzler, *Diskurs und Medium. Zur Archäologie der Computerkultur* (Munich: Fink, 2006), 182.

30. Kittler, "Real Time Analysis," 193.

31. John von Neumann, "Probabilistic Logics and the Synthesis of Reliable Organisms from Unreliable Components," in *Automata Studies*, ed. Claude E. Shannon and J. McCarthy (Princeton, NJ: Princeton University Press, 1956), 44.

32. H. H. Pattee, "Discrete and Continuous Processes in Computers and Brains," in *Physics and Mathematics of the Nervous System*, ed. M. Conrad et al. (Berlin: Springer, 1974), 129. Pattee here refers to Emil Post, "Absolutely Unsolvable Problems and Relatively Undecidable Propositions: Account of an Anticipation," in *The Undecidable: Basic Papers on Undecidable Propositions, Unsolvable Problems and Computable Functions*, ed. Martin Davis (New York: Raven Press, 1965), 420.

33. Frank Winkler, message to author.

34. Claude Elwood Shannon, "A Symbolic Analysis of Relay and Switching Circuits," in *Collected Papers*, ed. N. J. A. Sloane and Aaron D. Wyner (New York: IEEE Press, 1993), 472.

35. Ibid.
36. Martin Heidegger, *On Time and Being*, trans. Joan Stambaugh (New York: Harper & Row, 1972), 16.
37. Siegert, *Passage des Digitalen*, 9.
38. Konrad Zuse, "Rechnender Raum," *Elektronische Datenverarbeitung* 8 (1967): 343.
39. Zielinski, *Deep Time of the Media*, 31.
40. Alan M. Turing, "Lecture to the London Mathematical Society on 20 February 1947," in *A. M. Turing's ACE Report of 1946 and Other Papers*, ed. B. E. Carpenter and R. W. Doran (Cambridge, MA: MIT Press, 1986), 111.
41. Doron Swade, "Virtual Objects: Threat or Salvation?," in *Museums of Modern Science*, ed. S. Lindquist, M. Hedin, and U. Larsson (Canton, MA: Science History Publications, 2000), 144f.
42. See Lionel Pearson, introduction to *Elementa Rhythmica: The Fragment of Book II and the Additional Evidence for Aristoxenian Rhythmic Theory*, by Aristoxenus (Oxford: Clarendon Press, 1990), xxxiv. Pearson elaborates: "One of the difficulties in reading Aristoxenus is to distinguish the special or technical use of a word from its general meaning."
43. John von Neumann, "The General and Logical Theory of Automata," in *Collected Works of John von Neumann*, ed. Abraham H. Taub (Oxford: Pergamon Press, 1963), 5: 293.
44. Turing, "Lecture to the London Mathematical Society," 111.
45. John von Neumann, *The Computer and the Brain* (New Haven: Yale University Press, 1958), 43f.
46. Ralph W. Gerard, "Some of the Problems Concerning Digital Notions in the Central Nervous System," in *Cybernetics: The Macy Conferences 1946–1953*, ed. Claus Pias (Zürich: Diaphanes, 2003), 1: 187.
47. Ibid., 186f. See also Claus Pias, "Time of Non-Reality. Miszellen zum Thema Zeit und Auflösung," in *Zeitkritische Medien*, ed. Axel Volmar (Berlin: Kulturverlag Kadmos, 2009), 267–82.
48. Gerard, "Some of the Problems Concerning Digital Notions in the Central Nervous System," 197.
49. T. K. Sharpless, "Mercury Delay Lines as a Memory Unit," in *Proceedings of a Symposium on Large-Scale Digital Calculating Machinery,* ed. Howard Aiken (Cambridge, MA: Harvard University Press, 1948), 103–9.
50. Wolfgang Blum, "Ein alter Streit flammt wieder auf: Warum folgt die Welt mathematischen Regeln?," *Die Zeit* 35 (1998): 117.
51. Friedrich Kittler, *Optical Media*, trans. Anthony Enns (Cambridge: Polity Press, 2010), 44.
52. Ludolf von Mackensen, "Leibniz als Ahnherr der Kybernetik. Ein bisher unbekannter Leibnizscher Vorschlag einer 'Machine arithmeticae dyadicae,'" in *Akten des II. Internationalen Leibniz-Kongresse Hannover, 17.-22. Juli 1972* (Wiesbaden: Steiner, 1974), 2: 256.
53. Lacan, "Psychoanalysis and Cybernetics," 2: 303–4.
54. See also Martin Kusch, "Discursive Formations and Possible Worlds: A Reconstruction of Foucault's Archeology," *Science Studies* 1 (1989): 17–25.

55. Friedrich Kittler, "Die Maschinen und die Schuld, im Interview durch Gerburg Treusch-Dieter," *Freitag* 52.1 (1993): 12–13.
56. Von Hilgers, *Kriegsspiele*, 127.
57. See Daniel Hillis, *Computerlogik. So einfach arbeiten Computer* (Munich: Goldmann, 2001), 33.
58. Von Hilgers, *Kriegsspiele*, 157. Von Hilgers here refers to Hermann Weyl, "Über den Symbolismus der Mathematik und mathematischen Physik," in *Gesammelte Abhandlungen* (Berlin: Springer, 1968), 4: 529.
59. Qtd. in Robert Dennhardt, *Die Flipflop-Legende und das Digitale. Eine Vorgeschichte des Digitalcomputers vom Unterbrecherkontakt zur Röhrenelektronik 1837–1945* (Berlin: Kulturverlag Kadmos, 2009), 157.
60. Swade, "Virtual Objects," 142.
61. Ibid., 146. "Turing . . . argued that what defined a computer was not the medium of its physical implementation but the logical rules that define it."
62. William Gibson and Bruce Sterling, *The Difference Engine* (London: Gollancz, 1990).
63. "Finite-State Machine," *Wikipedia, The Free Encyclopedia*, http://en.wikipedia.org/wiki/State_machine.
64. Qtd. in Kittler, "Real Time Analysis," 193.
65. Raimundus Lullus, *Ars brevis*, ed. Alexander Fidora (Hamburg: Meiner, 1999), xxx f.
66. Georg Christoph Tholen, "Die Zäsur der Medien," in *Medientheorie und die digitalen Medien*, ed. Winfried Nöth and Karin Wenz (Kassel: Kassel University Press, 1998), 80.
67. Vilém Flusser, *Ins Universum der technischen Bilder* (Göttingen: European Photography, 1999), 206.
68. Georg Fleischmann and Ursula Damm, "Innere Zustände," in *Der telematische Raum*, ed. Frank Wagner (Berlin: Neue Gesellschaft für Bildende Kunst, 1997), 74.
69. Aristoxenus, *Elementa Rhythmica: The Fragment of Book II and the Additional Evidence for Aristoxenian Rhythmic Theory*, ed. Lionel Pearson (Oxford: Clarendon Press, 1990).
70. John von Neumann, "First Draft of a Report on EDVAC," Moore School of Electrical Engineering, University of Pennsylvania, June 30, 1945.
71. Wolfgang Hagen, "Computerpolitik," in *Computer als Medium*, ed. Norbert Bolz, Friedrich Kittler, and Georg Christoph Tholen (Munich: Fink, 1994), 143.
72. Maurizio Lazzarato, *Videophilosophie. Zeitwahrnehmung im Postfordismus* (Berlin: b-books, 2002), 110.
73. Ibid.
74. Claus Pias, *Computer Spiel Welten* (Vienna: Sonderzahl, 2002), 72.
75. Ibid., 72f.
76. Richard Böker et al., *Mikroelektronik für Einsteiger* (Düsseldorf: VDI, 1983), 32ff.
77. Ibid., 33.
78. Marvin L. Minsky, *Computation: Finite and Infinite Machines* (Englewood Cliffs, NJ: Prentice-Hall, 1967), 12.
79. Ibid.

80. Alan M. Turing, "On Computable Numbers," in *The Essential Turing*, ed. B. Jack Copeland (Oxford: Clarendon Press, 2004), 59.

81. Martin Warnke, "Synthese Mimesis Emergenz. Entlang des Zeitpfeils zwischen Berechenbarkeit und Kontingenz" (paper presented at the conference " Unschärfe. Jenseits der Berechenbarkeit [Hyperkult 13]," Lüneburg University, July 22–24, 2004).

82. See Dirk Baecker, *Wozu Systeme?* (Berlin:Kulturverlag Kadmos, 2002), 27.

83. Arno Borst, *Computus. Zeit und Zahl in der Geschichte Europas* (Munich: dtv, 1999), 104f.

84. Hagen, "Computerpolitik," 144.

85. Mark Aronowitsch Aiserman et al., *Logik, Automaten, Algorithmen* (Munich: Oldenbourg, 1967), 59f.

86. This is Paul Ricoeur's understanding in his three-volume work *Zeit und Erzählung* (Munich: Fink, 1989).

87. Neil Boyle, "Random Numbers in Machine Language for Commodore 64," *Compute!* 72 (May 1986): 77ff. He concludes that "due to its questionable randomness, the timer/clock method is not recommended."

88. Thomas Little, *Das PC-Buch. Die Hardware und ihre Programmierung* (Munich: System, 1990), 111.

89. Wiener, "Time, Communication, and the Nervous System," 207.

90. Hans-Joachim Sacht, *Mikroprozessoren. Kleincomputer für alle* (Munich: Humboldt-Taschenbuchverlag, 1978), 33f.

91. Neil Gershenfeld, *The Physics of Informatic Technology* (Cambridge: Cambridge University Press, 2000), 1. Fundamental to this issue is Leo Szilard, "Über die Entropieverminderung in einem themodynamischen System bei Eingriffen intelligenter Wesen," *Zeitschrift für Physik* 53.11/12 (1929): 840–56.

92. Lazzarato, *Videophilosophie*, 110f.

93. Turing, "Lecture on the Automatic Computing Engine," in *The Essential Turing*, ed. B. Jack Copeland (Oxford: Clarendon Press, 2004), 379.

94. See also Annette *Bitsch, Diskrete Gespenster. Die Genealogie des Unbewussten aus der Medientheorie und Philosophie der Zeit* (Bielefeld: transcript, 2009).

95. B. Randell, ed., *The Origins of Digital Computers* (Heidelberg: Springer, 1973), 350.

96. A. W. Burks, H. H. Goldstine, and John von Neumann, "Preliminary Discussion of the Logical Design of an Electronic Computing Instrument," in *Collected Works of John von Neumann*, ed. Abraham H. Taub (Oxford: Pergamon Press, 1963), 5: 34–79.

97. Krämer, "Was haben die Medien, der Computer und die Realität miteinander zu tun?," 13.

98. Oskar Becker, *Mathematische Existenz. Untersuchungen zur Logik und Ontologie mathematischer Phänomene* (Tübingen: Niemeyer, 1973), 197.

99. Von Neumann, *The Computer and the Brain*, 20.

100. Borland, *Turbo Assembler. Benutzerhandbuch* (Langen: Borland, 1992).

101. Alfred Görgens, *Einführung in die EDV. Ein Wegweiser in die Welt der Computer* (Cologne: Buch und Zeit, 1987), 75.

102. Martin Carlé, "Augmented Phenomenology" (PhD diss, Humboldt University, pending completion).

103. Bernd Enders, *Lexikon Musikelektronik* (Mainz: Schott, 1997), 170.

104. Rohrhuber, "Das Rechtzeitige," 195–211. See also Nick Collins, Alex McLean, Julian Rohrhuber, and Adrian Ward, "Live Coding in Laptop Performance," *Organised Sound* 8.3 (2003): 321–30.

105. Ibid., 210.

106. The musical ear has long suffered from the pure but false tones of synthesizers.

Part II

MEDIA-INDUCED DISRUPTIONS OF THE HUMAN PERCEPTION OF TIME

Chapter 5

Experiencing Time as Sound
Recorded Voices, Magnetic Tapes

TIME SHIFT: THE ANSWERING MACHINE

Media archeology defines telephony first and foremost as a technology that transmits speech by electrical means. The technological coupling of the telephone and the answering machine—the first practical application of magnetophony—thus fundamentally unsettled the logocentrism of the Western culture. The first such device was the "telegraphone," which was developed by Valdemar Poulsen, an engineer who worked for the Copenhagen Telephone Company. This device, which received a prize at the Paris World's Fair of 1900, was able to record conversations through electromagnetic induction on an unwinding wire spool. The patent application is unambiguous in its emphasis on temporary storage rather than archiving, as it clearly describes the device as a *process for receiving and temporarily storing messages, signals, and the like*.[1] The primarily media-technical and thus inhuman function of this apparatus indicates that it is entirely indifferent with respect to the cultural meaning of the vocal or other signals that it memorizes. The first written commentaries already emphasize that the telephonic messages can be retrieved "at any time."[2] This temporal delay is the technological alternative to emphatic permanent storage. The "telegraphone" was thus the ideal storage medium for electrical transmissions; it constitutes so to speak a continuum of both processes in the electromagnetic field. In Poulsen's patent application, the technical bridging of distance is also understood not only simply in a geographical sense but also in a temporal sense. "The telegraphone can be used locally but is especially useful for distance phonography, that is recording, storing and replaying speech as delivered from a standard telephone in the absence of the called subscriber."[3] This explains the miracle that an electromagnetically transmitted speech, articulated 100 years ago, can still be heard today *through the same medium*. Technical memory was able to store calls, which in extreme cases could still be retrieved after the death of the caller without creating the impression of historical discontinuity. The time of the conversation is thus simultaneously shifted and sublated by a temporal interval (Δt). This is the technical meaning of Jacques Derrida's neologism *différance*.[4] Like a dictation device, the answering machine constitutes an everyday correlate to the microelectronic delay line.

The establishment of mailboxes in the postal system already made the delivery of symbolic messages independent of the presence of the receiver.[5] Phonographically stored messages also suspend the actual voice of the present in the real. The shifting of presence on the time axis is originally inscribed in sound recording: the temporally invariant recording of a voice on the phonograph complementarily joins the space-bridging concept of telephony. When Charles Cros presented a technical process of sound wave recording to the Academy of Sciences in Paris in 1877, he called the apparatus the "paléophone." The purpose of the medium—to let "the voice of the past" speak again—was thus inscribed in its very name. In the replay, the material-diagrammatic registration of vocal frequencies on a kymograph, which was developed from Edouard-Léon Scott de Martinville's phonautograph for the scientific analysis of speech, became a time machine, as the stored wave forms represented a diagram in its sonic implementation—an operative medium. The telegraph constituted an intermediate medium of typographic *dispositif* and telephonic sender, and Edison later captured its noise as the primal scene of sound recording.

The answering machine transforms the unique moment of the physical call into its time-technical reproduction. This results in the effect of a past that first manifests through repetition; G. W. F. Hegel philosophically anticipated this game of identity and difference, yet the actual repetition of an event is accomplished not in the symbolic order (i.e., in text-based archives), but only signal-technically in the real. Recording media follow a strange path, as they enable a temporal phase shift that is characterized in English as "untimely" and in German as "out-of-time" (*unzeitgemäß*). The answering machine robs the human voice of its phonocentric uniqueness, as articulation is always already fleeting. In the list of possible applications of the phonograph that Thomas Alva Edison published in the *North American Review* in June 1888, the final entry reads: "10. Connection with the telephone, so as to make that invention an auxiliary in transmission of permanent and invaluable records, instead of being the recipient of momentary and fleeting communications"[6]—like a sonic monument. Emphatic time is thus incorporated into the technical work.

AUDIOVISUAL SIGNALS FROM THE BEYOND

While phonographic recordings are subject to deterioration through progressive use, symbolic codes, and symbol processing (whether as text or telegraphy) are largely invariant with respect to the physics of their embodiment. The challenge of modernity was to develop a concept of information that was essentially beyond materiality and energy, even though the concrete transmission of signals still remained irreducibly dependent on material and energy-powered vehicles. Paul Nipow's patent application from 1884 defined the technical purpose of the "electric telescope" as follows: "to make an object situated at location A visible at any other location B."[7] It is a matter not of material transport, but rather of converting current into information—an immaterial visualization. The medium of this event is signal transmission, which is initially electric but later based on vacuum tubes that enable electronic amplification. That which is electrically transmitted can also be electrically stored—possibly even in the same channel, such as the line or wire.

In 1898, Poulsen's patented telegraphone for the magnetic recording of electroacoustic vibrations was still primarily based on piano wire, which could thus be both the subject and object of sound. In 1900, Otto von Bronk also proposed using selenium cells to convert light into electric voltage, so that "images could be transmitted or reproduced at any time."[8] Immediate audiovisual perception was thus replaced by electro-temporal latency.

The shocking experience of transcending space through disembodied telecommunication led to the inverse hallucination of transcending time through communication with the dead, which was illustrated in ghost stories of telegraphy. When Rudyard Kipling became familiar with Marconi's wireless telegraph in 1898, he wrote the story "Wireless," in which ghostly signals received with antennas produce a fragment of a poem by John Keats. The experience of literally receiving transmissions from the beyond in a spatial sense thus epistemologically *induced* the hallucination of receiving transmissions from the beyond in a temporal sense—like resonant electromagnetic vibrations. The sense of presence engendered by the telephone is always uncanny. At first glance, it seems close to the spiritualist concept of media, which was also involved in the emergence of a corresponding technology in the nineteenth century—namely, the electric telephone of Reis and Bell. Marcel Proust succinctly expresses this parallel. In a scene from his epic novel *In Search of Lost Time*, the narrator finds himself in the city of Doncières, where he receives a telephone call from his grandmother in the post office. Proust describes the effect of the telephone, "by whose intervention the absent rise up at our side, without our being permitted to set eyes on them."[9] This presence effect creates the impression of not only spatial but also temporal distance.

[It was] a premonition also of an eternal separation! Many are the times, as I listened thus without seeing her who spoke to me from so far away, when it has seemed to me that the voice was crying to me from the depths out of which one does not rise again, and I have felt the anxiety that was one day to wring my heart when a voice would thus return (alone and attached no longer to a body which I was never to see again), to murmur in my ear words I longed to kiss and they issued from lips for ever turned to dust.[10]

It thus also anticipates imminent separation. Romantic literature (such as Chateaubriand's *Memoirs from Beyond the Tomb*) preferred to focus on the voice from the grave; in the twilight of the nineteenth century, this despotic signifier became technical.[11] The voice that has been puzzling the dog Nipper on the logo of the HMV ("His Master's Voice") record label since 1900 was analytically mastered by humans with the help of the sonagraph—a measuring medium that visualizes speech as a special case of frequency distribution *in time*.

In contrast to the voices emanating from the phonograph, the reception of classical radio voices has a direct indexical time reference to the acoustic event of sending in an electrophysical sense. "The basic characteristic of the relation between radio and time is the time-coincidence of the 'phenomenon' to which we are listening and the broadcast performances."[12] In calculating space, this precise time reference is cut. The temporal sense, which in human perception is connected with vocal articulation (fed back to the ear) and yields corresponding presence affects, is disrupted by the timeless space of computer voices, which remain invariant with respect to time due to the fact that discrete buffering is a necessary condition of their transmission.

Computer-synthesized speech is now able to produce presence (i.e., "live") affects that are in no way inferior to the living presence of natural speech. The Turing test could thus be time-critically modified to determine whether the received message comes from the present or the past.

(DE-)DISTANCIATION: THE APPEARANCE OF "LIVENESS" (RADIO AND TELEVISION)

Electronic telecommunication media initially constituted a memoryless technical enframing (*Gestell*); their analog temporal mode was defined by virtually instantaneous electromagnetic signal transmission. In order for them to operate, only electrical energy had to be stored, yet the dependence of electronic presence on stored electricity was not apparent until the moment when the power supply of telegraphs, telephones, and radios broke down. Every army signal corps and broadcasting organization thus has a reserve of "storage batteries" at its disposal "in order to maintain communication."[13]

The media for storing real physical events (photography, phonography, and cinematography) were joined in the twentieth century by the media of pure transmission, the most prominent of which was initially radio.

> It converts and transmits sounds, but it does not store them. Electromagnetic waves would be the memory, but they are . . . only relativistic effects of electrical energy. An important phenomenological feature corresponds to them—namely, radio is in a strictly formal sense always only now, this moment, unstorable, bound to the present moment of the radio event on a continuous time axis. That's why McLuhan says the medium is "hot."[14]

McLuhan's systematic distinction between "hot" and "cold" media can thus be extended to encompass the media-technical modes of creating presence and disrupting temporal perception—time-intensive and time-extensive. As Theodor W. Adorno wrote in a typescript from 1940 that not coincidentally bears the plain text title *Current of Music*, "The 'radio voice' creates a strong feeling of immediate presence. It may make the radio event appear even more present than the live event"[15]—a form of hyperpresence, in which the concept of radio transmission is ultimately transferred metonymically to the musical event itself. The temporal characteristics of electrotechnical "liveness" correspond to the perceptual impression of "real time." "This feeling of presence necessarily means a feeling of immediacy, too. There is no gap and no mediation between the time something is going on and the time at which you are listening to it."[16]

Adorno thus identifies the fundamental relationship between radio and time: "the time-coincidence of the 'phenomenon' to which we are listening and the broadcast performances."[17] At the limits of the light speed of electromagnetic transmission, *delta-t* tends toward zero: "This time difference is so infinitesimal that it may safely be overlooked."[18] The virtually real temporal indexicality[19] of so-called live broadcast media is the secret of their effectivity. "Radio always tends to make us forget that is gives us in other respects a mediated phenomenon."[20]

The time relation of telecommunications, which is denoted by the term "live," not only applies to sound transmission. In Albert Robida's futuristic novel *The Twentieth Century* (1883) telephony is joined by the option of transmitting images over distance. "The performance is truly witnessed with the eyes and ears. The illusion is complete, absolute!"[21]

With the development of magnetic videotape, electronic television—until then primarily a technology of direct transmission—entered the virtual world of time-shifted broadcasting ("deferred time").[22] This time delay had previously only been common in the domain of the symbolic order through archival latency (as the task of the archive is to preserve records for future reference). From the beginning of radio drama, the key technoaesthetic features of radio—namely, the simultaneity of production, transmission, and reception—were threatened by recording media,[23] and this situation was once again repeated with the arrival of magnetic videotape, as television drama faced a similar threat.

From this time on, broadcasting occurred in a time horizon that encompassed storage and transmission. Audiences could now hear a radio broadcast whether actually "live" or a play from the archive. From a signal-technical perspective, both cases strictly involve a "live" event (regardless of content), yet digital coding adds the first element of ultra-short intermediate storage (which is audible as a time delay with respect to the same broadcast on analog radio). Radio is a constant mixture of "live" transmission and recording from the electronic sound archives; as transmission from magnetic tape, "live" and archive represent two extreme values of what is always already differentially entangled in the electromagnetism of the technical medium—in terms of media-archeological categories, they are thus equiprimordial. In between is the concept and practice of "live on tape"—a term used in television production to refer to the indistinguishability of recorded image signals and their "live" camera transmission. In the "live" broadcast, the sensation of simultaneity is not an illusion; rather, it is real in terms of transmission technology—apart from infinitesimally short *propagation delays* caused by the electronics. Prior to the development of photography and phonography, the perception of the present was always already categorically separated from the past, as it was not possible to record fleeting images and sounds, and the time index of the past was thus a function of its irretrievability at the signal level. This culturally self-evident concept of the past was disrupted by technical recording media. In 1941, Gerd Eckert wrote: "In radio broadcasting . . . the event speaks to us in its own unadulterated speech, even when it is prerecorded."[24] No individual "I" is speaking here, but rather the "it" of the apparatus. Radio broadcasting thus becomes a means of time management, and it creates the impression of real presence even when no body is actually present. The separation of voice and sound from the body through technical recording and transmission set new conditions for the Western temporal economy, as the time that had previously dominated cultural techniques became media-technically available.

Unlike the sheer physics of mechanical media, like gramophone and film, a theory of electronic storage media necessarily points to dilatory time. There exists here no abrupt separation between the past (recording) and the present (replay), but rather a transitive shift from the technical signal character of "live" transmission to the recorded signal, as the process is basically the same: electromagnetic induction, which is itself a model of a relatively entangled processuality of time and event. Media storage is thus less an archive of the transmission than a form of decelerated, temporally extended present.

When transmission is described as the linking of systems (which can thereby change their form), this interconnection can open up a temporal gap that varies in the moment of implementation from an *ideally* pulse-like zero time difference to emphatic dilation. The time delay is a horizon that potentially stretches from immediate to long term, while still retaining its transmission-symmetrical quality.

At the radio speaker or television screen, human perception was unable to determine whether a transmission involved spatial or temporal distance—in other words, whether the transmission was recorded on tape. Because the auditory and optical senses physiologically deem it present, a television broadcast must explicitly announce this at the symbolic level as meta-information: "This program was recorded." The time-shifted transmission of a recording involves both sublated and deferred time—*différance* as differential spatio-temporalization, to use the more precise terminology of differential calculus. Recorded and transmitted human articulations are not simply allegories of time or life, as in painting and sculpture,[25] but rather moments of lived time that are actually *sublated* in the electromagnetic field.

All that remains is to consider how time figures that continually appear as phenomena in the field of electrophysics influence the model of time through which they are cognitively understood. It actually already began with photography: while it initially stored moments that were quite prolonged, it suddenly shifted to moments that were infinitesimally short—depending on the shutter speed of the camera and the sensitivity of the chemical storage medium. The recording of a "live" radio broadcast *for the purpose of* its permanent retrievability is not the antithesis but rather another form of transmission; technical storage and transmission allow something to be *simultaneously* historical and actual. This relationship is also reflected in the sonic event itself.

> As Jean Paul and many others have said, sounds have no "present," no objective reality separate from or prior to their resonances . . . in (parts of) ears and bodies. The "present" in which their oscillations are presentlessly realized is their resonance or reverberation. As an object of acoustics (of countability and calculability), the reality of a sound is also a figure and thus subsequent to that which actually not *is* but rather ("eternally") *becomes*.[26]

It is therefore a presence that does not exist at a precise point in time, but rather only reveals itself in the extension of protention and retention. Sound, which is mathematically addressable through Fourier analysis (like any oscillation event), thus shares an essential characteristic with time itself in so far as it (following Aristotle) is conceived as a counting movement.

At the level of acoustics, an echo is an indicator of a time-critical transition from the present to the past. Compared with this, the transition in the electromagnetic field is virtually instantaneous; however, an infinitesimal time delay also occurs here, which always already represents a differential between present and past. The echo is an acoustic phenomenon whose uncanniness lies in the fact that the sound in the present and the sound from the past blur together. In principle, all oscillatory processes create not only reverberations, and thus a transition from the present into the past, but also pulses; they cannot be reduced to two- or three-dimensions (in terms of the human ear or the sense of sight), but rather they always also extend into the temporal dimension. The human perception of reality physiologically encompasses each physical presence in space and time, but this immediate perception is not selectively oriented toward the present moment; rather, as Bergson emphasizes, it is also rapidly converted into memories and associations (which happens completely independently as dreams). Telegraphic image transmission already produced the sensation of "liveness," although at first it was unavoidably the function of a storage medium and was written on a storage medium (such as paper strips). Film was also able to affect perception in its immediate presentness, although it was dependent on celluloid. Television was the first medium able to broadcast "live" in a technical sense; this mode is linked to actual and perceived synchronicity, and it is thus radically temporalized. The localization of perception is sublated, but the temporalization remains authentic. Through the impression of synchronicity, "live" television was also able to create the impression of being-there (*Dabeisein*) across spatial distance—a creation of presence according to the power of electronic transmission media. But how real is "liveness"?

> One can no longer distinguish, visually or aurally, between that which is reproduced and its reproduction . . . not even discern *that* or *when* reproduction or repetition, in the manifest sense of recording or replaying, is taking place. We must be informed whether or not what we are seeing is "live". . . . We cannot distinguish through our senses alone between what we take to be simply "alive" and what as reproduction, separated from its origin, is structurally posthumous . . . what Derrida called the irreducible "iterability" of the mark.[27]

In principle, this extensive indistinguishability also applies to the letter (in mailboxes or archives)—that is, literal writing; at the moment of reading, letters are just as immediate to the sense of sight as electromagnetic waves are to the ear. In terms of frequencies, however, it is blatantly obvious that the worldliness of oscillations (the superimposition of waves to form sound or light) mediates itself—and, as a recording, this applies to the worldliness of the past or distance. With respect to information, the difference between remoteness as past and remoteness as distance is entropic: a transmission from the remote present has the potential for veto or intervention, such as when the message follows a gunshot—which does not apply to messages archived on magnetic tape. The domain of the symbolic (writing, letters) was for the longest period of time not worldly, but this changed with the development of binary code, which—implemented in digital media—was not only able to control sounds but also synthetically generate them.[28]

MAGNETIC TAPE RECORDING: STORAGE OR PRESENCE

The same tape head that induces tonal alternating currents when in contact with the passing magnetic tape is also able to process signals that come directly from a transmission line, yet it is conversely in the same situation as a radio broadcast, as it makes no audible difference whether the sound signals are sent "live" or from tape. From the perspective of media archeology, the objective fact of the radio broadcast is in a technical sense constantly "live." The same technical differential results in the entanglement of the past (storage) and the present (broadcast). The sonic (techno-acoustic) event is invariant with respect to the time axis; it constantly yields the same phases, progressively deferred with progressing time, as the reproduction of a present moment. The sound event recorded magnetically on tape is for a long time resistant to the progress of historical time, as oscillations that seem perpetually dormant are able to emerge once again during replay. Tape recordings of radio broadcasts perpetuate something that was in most cases already recorded on tape by the broadcaster; these recordings thus involve time shifting. Radio as broadcast here means the multiplication of the media archive, which is closer to the model of memetics than technically pure radio, whose signals vanish as soon as they are broadcast.

During a radio broadcast, "live" production and recording are entangled with one another. In this respect, cinematography already distinguished itself from theater in a media-dramatic way. According to Hugo Münsterberg, theater depends on the interpolation of past scenes from memory in order for them to be part of the current dramaturgy, yet the photoplay does not rely on memory. "Suddenly there flashes upon the screen a picture of the past,"[29] and "We have really an objectivation of our memory function."[30] Sigmund Freud wrote shortly before about the "psychic apparatus"; from this time on, human memory was modeled after the technical medium. The language that von Münsterberg chose to describe such time-dramaturgical effects was itself already technical. "The modern photoartist makes use of this technical device in an abundance of forms. . . . Any going back to an earlier scene is called a 'cut-back.'"[31]

The essence of the invariant shifting of recorded time processes reveals itself on tape precisely when the recording of time-based speech and music is converted into the recording of discrete pulses and the sound or video recording is thus transformed into a data storage technology. The fleeting memory in early computers (and their external "datasettes") required such external magnetic tape recordings. A sonic time figure continues here as sound—namely, as the temporal mode of the sonic beyond actual sound.

Electromagnetically recorded time processes deconstruct the time arrow of physical entropy.

> The stability of a tape recording does not depend on the storage time. The extent to which the playback quality changes over time is only a question of the frequency of use. . . . Furthermore, every remanence (residual magnetism, no matter how slight) of the magnetic heads and tape guides of the devices must be avoided in order to safeguard the playback quality of the tapes.[32]

The alphabet as symbolic notation facilitated the ahistorical transfer of knowledge at the level of coded communication; however, magnetic tape recording enabled the sensory reproduction of presence. Audiovisual presence can be retrieved from magnetic tape reels; the electrical technology of the tape is prepared to convert the latent signals stored on the reels back into audible signals at any time. These signals do not exist as character strings in the sense of printed texts or notes, but rather as a magnetic latency that only takes place in the inductive moment of the passing machine, the unwinding, when it is converted into a signal: a genuine act of time in the processing of sound and image. A privilege of human perception thus migrates into the electronic technology itself.

Magnetic tape recording records the real from signals. The phonograph already performed the same function mechanically. Both cases involve the recording of oscillation processes—the latter kinematically and the former electromagnetically. This already points to a conceptual difference: the distinction between mechanical "grooves" (incisions in phonograph records) and electromagnetic "tracks" (traces in an electrodynamic *and* temporal sense). The temporal development of the tape signal is a function of the unwinding reel; the object of movement (the time signal) is thus accompanied by a self-moving storage medium. The recorded time event is made possible through the event time of the technological medium. Both times are integrally related. However, the static memory modules on microchips contain resident fixed values; a successively stored, digitized event here becomes "visual" in the sense of data that can be addressed as function $f(x, y)$ and described as a matrix. The readout of data takes place either sequentially or in blocks, so that the processor itself is a time machine.

Electromagnetic induction itself, as it has been mathematically calculated since Maxwell, represents a time function of movement on the microphysical level. In a technically controlled form, it occurs at the ring head of the tape recorder, such as when the microphone records human singing in order to preserve it for future playback in literally electromagnetic latency. This is how the times of culture correspond to the times of electronics. Electromagnetic sampling as sound pickup also occurs in time—a temporally repeatable mode of existence. "Once an event has been recorded the recording is, mysteriously, felt to have more reality than the original. This . . . must be the effect of the instant replay."[33]

When it is played back, a video recording on magnetic tape becomes the present for human perception. Even if the video images are visually noisy, they are perceptually updated at the physiotechnical level of the event—just as early 30-line television image, which were objectively detail-poor as optical signals, were perceived by human eyes as moving image sequences of higher quality. Like a transformer, this dynamic represents a neuronal projection of perception that cognitively overrides defects in the model and thus also the entropic traces of historicity.

Once it is turned on, the tape recorder awaits the input of latent presence, like a radio apparatus set for reception. For electrotechnical systems, it does not matter whether the electromagnetic frequencies are transmitted from the present or played from a tape; in their operative implementation, the past can have no meaning for the technical system. This dynamic proper time behaves anachronistically toward the external time that entropically decays in the objective world.

Is it the noise of early sound and image carriers that constituted the index of the historicity of their media processes? When an antique tube radio is brought back into operation in the present, it receives not music and speech from the time it was built (such as Goebbels' voice as heard on "People's Radio" during the National Socialist period), but rather contemporary medium-wave broadcasts. And when an early television set from 1953 is finally brought back into operation, it displays not the final image it received the moment it was switched off,[34] but rather contemporary programs (provided that the available channels of the apparatus can still be tuned to them).

Electronic storage media *create* presence, as their updated signals are able to address human sensory nerves. The past is thus operatively sublated—another, nonhistorical form of event. Electromagnetic transmission and recording media operate in a transitive, flat time—at the level of their technology (as the sampling of signals in the micromillimeter range) as well as their sensory address to the sense of time. The situation is different with symbolically coded conditions. A city building (provided that it does not emanate an explicitly historical aura) is hardly perceived in its temporal concreteness, even though its duration spans time and it is not a sudden, eventful presence. This durability differs from the sound of a violin in so far as the latter must be newly created in each present moment. This is a quality that oscillations share with all electronic processes, which are likewise based on oscillators. A continuously standing object is admittedly also a function of subatomic particle waves, yet human sensory thresholds make an absolute distinction between sound and stone.

Sometimes, however, media-technically induced time perception is melancholic. Just as Hamlet stares at the skull of Yorick in Shakespeare's eponymous drama, so too does the media-archeological gaze stare at the reel of a wire recorder from the mid-twentieth century. Human eyes are unable to discern whether the coiled wire stores voices or music, as this is something only the playback machine is able to reveal. The voice of the previous owner might unexpectedly be heard coming from the wire reel during replay. The technical test of the medium thus becomes the time cut(-out). Technology and communication temporally converge here—unlike the case of an antique radio put into operation in the present, where there is a wide temporal gap between the received broadcast and the technical-historical index of the apparatus. The magnetic wire reel sublates the past moment—the latency of storage. Written documents from the Middle Ages also embody a latency, but the decoding and processing of this information is strictly based on the reading ability of humans. In magnetic tape, however, a technological sampling emerges instead, which occurs without humans and which does not read signals hermeneutically, but rather listens to the noise as well. The corresponding technical infrastructure has outlasted all political-historical upheavals for nearly a century—an *epoché* of stable technological conditions of possibility for sound events.

Lacan emphasizes that it is of secondary importance for museums and mausoleums whether their objects are dead or living[35]; what is most important is the presence of a piece of the real. Like the creature composed of body parts in Mary Shelley's novel *Frankenstein*, media signals can be instantaneously brought back to life. Technological apparatuses are never dead, but rather they remain in latency; in order to be media, they need to be charged with electricity and set into operation. From the perspective of a magnetic tape, the playback of a recording is just as authentic today as it would be fifty years ago (provided that the technology remains autonomous)—a time interval, but not "historical" distance. Tape recorded singing represents another sublation of the time of articulation, as it represents written archives or musical scores. In his analysis of technologies of the word, Walter Ong emphasizes that Homer can no longer be asked about ancient singing;[36] it is only possible to understand the mechanism of oral poetry by analogy. However, electrical sound technologies, inspired by telephony, actually allow playback as "re-call"—a conversation with the dead. This culminates technologically in the moment when a voice recorder was placed next to the newly introduced flight data recorder in 1957; it preserves movement data as well as the last words of pilots before a plane crash.[37] The conversation partners in technical systems are themselves technological.

Magnetic tape requires a theory of time that is appropriate to the medium. A human voice stored on tape (as a further development of the functional principle of electrical telephony) remains electromagnetically latent—like the process of electrostatic photocopying, where a negative charge image emerges for a brief moment. This voice remains silent until it is electrotechnically recalled in the sense of Grimmelshausen's anecdote about Baron von Münchhausen, who froze the sounds of a trumpet in the winter and thawed them out in the spring. Latency time represents a techno-archivistic, unhistorical condition. At the moment of playback, all techno-sonic sounds from the past are pure presence. From the perspective of the electromagnetic pickup, as well as all electron tubes and transistors, it makes no difference whether the signals come from the past (tape) or the present (radio). The moment of their electro-technical actualization is dominated by the present. In both cases, what is taking place technologically is a spatio-temporal de-distanciation; the sound pickup is the reified time window, which is intensified in the gap of the electromagnetic field. What reveals itself here in astonishing clarity is a technological time capsule—electromagnetically sublated time, a media temporality that transcends history.

TIME (ZONE) SHIFT THROUGH MAGNETIC TAPE: TIME TRACKS AND PLAYBACK

The title of Prince's song "Sign 'O' the Times" gets right to the point: studio recordings can be heard and understood as the operative multiplication of temporality with respect to the purely philosophical concept of emphatic, totalizing time. They experiment constantly with new modifications of temporal modes through techno-musical rhythmics. "Time tracks" and "multi-track recording" provide the foundation of possibility for a new time aesthetic, yet at the same time, they also disrupt the perception of the living as soon as a singer is able to sing a duet with himself.[38]

Magnetic tape enables new ways of synchronizing and automating temporalities—particularly in the context of the recording studios of the radio and record industries. This time culture evolves in a literal time zone. The problem of geographical time difference at the macro-temporal level is a counterpart to the phase-shifted echo at the micro-temporal level of magnetic tape. In 1946, entertainer Bing Crosby announced that he wanted to record his radio show. This was a solution to the specifically American problem that programs had to be repeated twice in order to bridge the time zone difference between the East and West coasts—"broadcast delayed in time." In this context, direct recording on disc proved to be an awkward intermediate storage medium with limited audio quality. This abruptly changed in 1947, when Officer Mullin brought two tape recorders from Germany that significantly increased the sound dynamics of the recordings through the high-frequency premagnetization of the tape. Not only were tape recorders usable for radio purposes, but this technical escalation also introduced a new option for the time aesthetic of media. "During the second season of Crosby's Philco Radio Time, Mullin used the tape not only for recording and playback but he put the process of editing into the interval between both. The methods of editing were simple: cutting and assembling."[39] The "reentry" of geographically conditioned time zone differences was thus incorporated into the present time window of production itself.

Acceleration (time compression) and deceleration (replay and time delay) are options for editing recorded sounds in tape-based studios. The record of piano concerts 1 and 2 by Franz Liszt with the London Symphony Orchestra under Kyrill Kondraschin and the pianist Svjatoslav Richter has a playing time of forty-five minutes; on tape, however, it is nine and a half hours long including all of the repeated partial recordings from which the final recording was assembled. "[The tapes] were repeatedly heard, stopped, discussed."[40] The recording that was ultimately published was the only apparently continuous work of a time-discrete production.

THE MEDIA MEMORY OF ACOUSTIC PRESENCE (REVERBERATION, ECHO)

For Husserl, like Bergson, the human perception of melodies is the perfect phenomenon of an interiority that extends beyond the acoustic time window of the present in the form of protention and retention.[41] For a long time, however, this cognitive time form has become technical. Signal-technical sound recording with a phonograph or tape recorder not only made the singular acoustic event identically repeatable and thus largely invariant with respect to transformations in historical time, but it also created an independent time base. A tape echo generator is a veritable time machine, and its operational base is the use of a signal-delaying tape as an endless loop. An inverse time function was media-archeologically arranged in the electronic assembly of the Magnetophon K4 tape recorder, which was developed by the German electronics company AEG (*Allgemeine Elektricitäts-Gesellschaft*) in 1938/1939; with its three toroidal coils, it permitted "listening after writing." The designer, Eduard Schüller, found a way to listen to the recording (from the playback head) at virtually the same time that the recording was being made (from the recording head). This read-after-write control made it possible to verify the quality of the recording without any delay, and it thus introduced acoustic "monitoring" of the present in real time.[42]

In contrast to the echo, reverberation relies on delay times that can no longer be time-critically discerned by the human ear and thus represent what brain research defines as the present time window—an extended moment of the now. The difference between the mechanical generation of reverberation (spring reverb) and the switching delay in echo magnetic tape (loop) is not purely technical, but rather essentially based on time behavior: the feedback of a sound in the electronic circuit, which is also immediately neutralized again by the erasing head. When there is no acoustic pulse on the endless loop, it continues as an inaudible time process: empty time. And it is precisely when it is inaudible that the actual message of the sound signal manifests itself in this electronic system: its temporal essence as an oscillation event.

TIME LOOPS: *KRAPP'S LAST TAPE*

In the coupling of electrotechnical audiovisual media and the human perceptual apparatus, the distinction between temporal modes that signify and the emphatic time that is signified is not plausible for processes of either remembering or forgetting. An analysis of Samuel Beckett's media-dramaturgical answer to Marcel Proust's epic *In Search of Lost Time* states that "Krapp's monologue in the face of the technical medium appears to be a delirious swan song to all signifieds."[43] Beckett's one-act play *Krapp's Last Tape* stages the hybrid communication between a human actor and an inhuman protagonist embodied in the real presence[44] of a tape recorder. The rotating tape reel induces autobiographical streams of voices, some of which were imprinted on low-frequency magnetic tape decades before. The essence of media-induced time is the invariance of its temporal segments (the time shift interval Δt) with respect to the irreversible transformations of "historical" time. On his 69th birthday, at the time when Beckett's piece was performed in 1958, Krapp listens to a journal entry from the past that he recorded on tape when he was 39. On this tape, he comments in turn on an earlier recording that he made in his late twenties. Beckett thus dates the existence of the home tape recorder back to a time when such a device was not yet known—an anachronism from the perspective of the history of technology, but not in the sense of true media dramaturgy, which here constructs its own time field. For the staging, it very well makes a difference whether Krapp listens to the recordings of his voices from a tape recorder or a phonograph, as physical sound waves are mechanically registered (or electroacoustically converted) on the gramophone soundtrack, but this mechanical facilitation of memory is replaced with magnetic latency on tape.

Physical speech signals—functions of time—are here inductively transformed into spatial storage on tape. However, the movement of this storage (which distinguishes it from the classical archive) is able to elicit signals as functions of time (and tape speed) from this spatial localization.[45] The technological act of anamnesis is thus dynamic—in contrast to static fixing of notices in Krapp's register. Memory occurs not in a spatial topography, but rather in a techno-temporal dynamic that is closer to the human neuronal memory apparatus than the cultural technique of writing has ever been. This one-act play emerged at a time when television was not yet ubiquitous in private living rooms. If Beckett had written the play decades later, in the era of early home video recorders, would he have staged the piece with videocassettes instead of tape reels (as in a recent staging in the Schinkel Church of Neuhardenberg Castle)?[46] It is a media-archeological quirk of history, however, that video recording was itself an offspring of magnetic tape technology. In the media-theoretical "grounding" of Jacques Lacan's trio of psychoanalytic concepts, the phonographic recording of the voice is associated with the indexical real (signal), the written recording with the symbolic (archive), and the visual projection with the imaginary (visual media)[47]—three tempor(e)alities induced by technical media.

Krapp experiences the voice of his own youth on tape with auratic ambivalence—namely, as asynchronicity.[48] Walter Benjamin defined the aura with regard to early photography as "a strange weave of space and time: the unique appearance or semblance of distance, no matter how close it may be."[49] However, a specific *tempaurality* appertains to microphonic tape recordings. In his 1984 media installation *Distant Trains*, Bill Fontana made the past present again by recording the noises in a Cologne train station on tape and replaying these recordings through a cascade of loudspeakers on the grounds of the former Anhalter train station in Berlin.[50] The creation of presence is actually most successful in the signal-acoustic channel.[51] In Beckett's one-act play, Krapp's voice on tape is a time-shifted person; as an electroacoustic media event, the false etymology of *persona*[52] suddenly makes sense. Signal-technical recording, sublated in the electromagnetic field, makes real presence relocatable on the time axis—a fundamentally new chronopoetics with respect to the culturally familiar historical ordering of time. The previously common cultural triad of past, present, and future has been "surreptitiously replaced by two tenses, *real time* and *delayed time*,"[53] and "this *relative difference* between them reconstitutes a new real generation"[54]—thus a new temporal reality (*Zeitreal*) based on the microtemporal logic of electronic circuits. A microcosm of epistemologically delicate temporal relations presents itself on the media-archeological level of time-critical media processes, to which the early composers of electronic music (since Pierre Schaeffer in Paris) programmatically responded. It was precisely the post-historical mood of early postwar music that emancipated the technical sound from music-historical understanding; the diverse parameters of sound were identified by media-archeological hearing as a temporal essence or mode. For Karlheinz Stockhausen, the phase shifting of oscillations became a basic element of composition,[55] and its media-material requirement was magnetic tape, on which the phonetic syllables of the composition *Song of the Youths* (*Gesang der Jünglinge*, 1955–1956) actually became snippets of time. In current digital sound editing software, such time-shifting tools have long been taken for granted. If technology is recognized as "a way of revealing,"[56] then this particularly applies to media-induced processes, which show as well as produce the temporal essence or mode of signals and data.

What Krapp experiences as the macrotemporal time shift of his own voice has its corollary in the microtemporal phase shifts that first became possible with magnetic tape. A specific effect in Elvis Presley's vocal technique was due to the possibility of the echo effect; Sam Phillips, who was trained as a radio technician, employed two tape machines to produce this "slapback echo" in the studio of Sun Records in Memphis, Tennessee. In digital time processing, however, the symbolic order returns in a temporalized form—namely, as algorithm—and thus also the time figure of the "if-then" statement, which in the tape reel is still literally a "loop." A digital video recorder for television programs also incorporates this genuine media-temporal object: the time-shift function. Broadcasts that are currently running can simultaneously be recorded and viewed with a delay; they are also available "on demand." Unlike the classical functions of fast-forward and rewind, which were originally bound to the materiality of the tape reel and have now become only metaphorical, data streams recorded in this way permit the nonlinear addressability of time events.

The ferric oxide tape is the material correlate to Krapp's memorial recursions. "The life from the tape is caught in a memory loop, as machine memory increasingly replaces the fading body memory."[57] These loops can also be considered an early form of computer memory. The exposed operations of magnetic tape storage vividly demonstrate the entanglement of two temporal modes—the constant and the discrete—which was part of Von Neumann computer architectures but later disappeared in the hard drives of IBM computers, which were partially serial and partially non-linear with a configuration of rotational and discrete access moments. This becomes evident in Beckett's media-play with a striking observation: "The electronic magnetic track is actually still linear, and its direction is reversible through fast-forward and rewind. At the same time, its rotation reflects the monotonous return of the same"[58]—just as historical time (development)[59] and mythological time (the cycle) are entangled in this storage and playback medium.

An entire aesthetics of memory depends on the materiality of electromagnetic storage media. Krapp still has the corporeality of the reels at his disposal, like media-archival monuments: "Revelled in the word spool. Spooool!" The reel here becomes sensually tangible in its analog form as an acoustic signal carrier, but when it stores digital signals the tape marks the "transition to the desensualization of the data carrier."[60] Ever since the computer has been able to recognize data from magnetic tape in the form of acoustic-couplers, which call forth texts, images, and operations, alphanumeric code has been linked to phonocentrism to produce an uncanny presence in calculating space.

Beckett's one-act play stages a split personality at the temporal level—namely, the technical time shift of a voice with respect to itself. The human-machine feedback of memory and presence results in a disruption of the consciousness of time. For centuries, the human voice (especially one's own) was considered irretrievably historicized as soon as it was spoken—it was as fleeting as time itself (*tempus fugit*). The symbolic notation of Homer's vocality in the alphabet preserved the musical character of poetry, but not Homer's voice. The situation has changed since Edison's phonograph, as the individuality of the voice is now repeatable. The more time elapses, however, the greater the discrepancy between the media-technically repeated voice and its current articulation. Thomas Alva Edison himself provided a paradoxical example of this in that he media-theatrically repeated the primal scene of his first voice recording by filming himself as an old man singing the children's song "Mary Had a Little Lamb" into the horn of a phonograph. Krapp lives in a twilight field of archival simultaneity in the present, as the tapes sublate all of his voices from the past; on the other hand, he is also forced to confront the traces of his own physical deterioration. The stage truly becomes a media theater, as the tape recorder is the actual protagonist. And the conflict lies in the opposition between cultural techniques and media technologies of memory as embodied in the inventory of the tapes (written archive) and the tapes themselves (electronic memory). The reading of a journal differs drastically from technologically stored memory, in which the real of the voice disrupts all symbolic and thus historicizing distance.

Krapp does not stop each of the recorded tapes in accordance with the conventions of literary narration—that is, after the end of each sentence. "The empty spaces, the gaps, contain the real, concealed personal memories. Over the course of the play he no longer goes back to these unheard tape segments."[61] Unlike written documents, a magnetic tape also articulates silence; in contrast to the typographic *spatium* in printing, absence itself here becomes a time event. Beckett's final stage direction reads: "Tape runs on in silence." Tape technology has developed automatic switching using a piece of iron tape to induce a signal that tells the device to switch itself off; however, this is not the end of a history, but rather a chronotechnical slice of time.

Vilém Flusser defines the essence of culture as the unnatural ("negentropic") preservation of information through storage,[62] which enables the transmission of improbable messages by means of coded symbols. Is culture thus already historical or not simply technical? "Historical" time is a hybrid of entropic and symbolic time, which is at work in the archival register, whereas media-induced time is technical real time. Krapp is only able to listen to his alter ego in a form of sonic self-identification, albeit in reverse, on the basis of tape-recorded signals. This possibility depends on the media-archeological "high fidelity" of his voice recordings—the technical-critical point that separates prewar and postwar tapes. High-frequency premagnetization qualitatively pushed the sound dynamics of tape recordings from noisy to virtually crystal clear.[63] Oscillations that were previously in the inaudible ultrasonic range thus subsequently enabled the effect of the pure sound of voice and music. As a result, the formerly audible distinction between a "live" transmission and a sound recording ceased to exist. These dynamic tape recorders were first employed in the German broadcasting company of the Reich around 1940, and they led allied listening posts to falsely assume that orchestras also played there at night. This *dissimulatio artis*—the complete concealment of artistry—is characteristic of all technological media, and it is precisely what allows them to have such an effect on the human perception of time.

HISTORICAL PERFORMANCE PRACTICE IN THE ELECTROACOUSTIC STUDIO

The key word of operative media archeology is not "reproduction," but rather "reenactment." In contrast to sculptures and paintings, media art is operative and therefore a time event. Musicology employs the term "historically informed performance" in connection with the imperative to play instruments that are specific to particular historical periods. This aesthetic depends on the concept of early music, which ironically refers to works composed before 1830—it thus ends at the time Faraday discovered electromagnetic induction, which was to become the condition of possibility for electroacoustic music.

The German Museum (*Deutsches Museum*) in Munich houses the Siemens Studio for Electronic Music (1959–1969). Do faithful performances of early electronic compositions—analogous to the concept of "historical performance practice"—require them to be replayed in this museum studio? To what extent does electronic music depend on the analogous instruments, and to what extent can these instruments be emulated as digital software? The historicity of early electronic music does not depend on an (irretrievable) performance, like a piano piece from Mozart's era. Electronic music is not to be found in the studio production as progressive implementation, but rather in the final recording and composition on multi-track tape, as this is how it was first performed. If these tapes are available today, then a techno-historical equiprimordiality of performance takes place—an equiprimordial (re)enactment. One way of clearing a path to the past is the symbolic approach of writing-mediated historiography, but the other is the chronotechnical approach, which is "grounded" in material remains. The latter approach includes the media archeology of the synthesizer. At the media art festival Ars Electronica in Linz in 2009, Elisabeth Schimana performed her composition "Hell Machine" (*Höllenmaschine*) on the same synthesizer that Robert Moog once constructed for composer Max Brand. This reactivation initially required hard work on the electrotechnical material—thus media archeology in a manifest sense. While traditional historical instruments, like a grand piano from Beethoven's time, are mostly preserved in order to perform a corresponding historical composition, the message of an electroacoustic artifact from the past lies in the primacy of the present. Schimana thus created her current composition specifically for the antique synthesizer. "What is in such a machine that has not yet been experienced?" Schimana asks. [64] Electronic apparatuses of the past, conceived as media, are not in a historical condition, but rather in the mode of latent presence. Media archeology seeks to unleash the potential sublated therein.

NOTES

1. Poulsen realized a suggestion made by Oberlin Smith, who proposed electromagnetic recording by means of alternating current flowing through a coil. See Oberlin Smith, "Some Possible Forms of Phonograph," *The Electrical World* (September 1888): 116.

2. August Foerster, "Das Telegraphon," in *Die Pariser Weltausstellung in Wort und Bild*, ed. Georg Malkowsky (Berlin: Kirchhoff, 1900), 398.

3. Mark Clark and Henry Nielsen, "Crossed Wires and Missing Connections: Valdemar Poulsen, The American Telegraphone Company, and the Failure to Commercialize Magnetic Recording," *Business History Review* 69 (1995): 7.

4. Jacques Derrida, "Die différance," in *Randgänge der Philosophie* (Vienna: Passagen, 1988), 29–52.

5. Bernhard Siegert, *Relays: Literature as an Epoch of the Postal System*, trans. Kevin Repp (Stanford: Stanford University Press, 1999), 115.

6. Qtd. in Günter Große, *Von der Edisonwalze zur Stereoplatte. Eine Geschichte der Schallplatte* (Berlin: Lieder der Zeit, 1989), 15.

7. Paul Nipkow, "Elektrisches Teleskop," German Patent No. 30105 filed January 6, 1884 and issued January 15, 1885.

8. Qtd. in G. Goebel, "Das Fernsehen in Deutschland bis zum Jahre 1945," *Archiv für das Post- und Fernmeldewesen* 5 (1953): 366.

9. Marcel Proust, *The Guermantes Way*, trans. C. K. Scott Moncrieff and Terence Kilmartin (London: Vintage, 1996), 147.

10. Ibid., 148.

11. See also Mladen Dolar, *His Master's Voice. Eine Theorie der Stimme* (Frankfurt am Main: Suhrkamp, 2007), 102f. Walter Rathenau described a telephonic application of this idea. See W. Hartenau [Walter Rathenau], "Die Resurrection Co.," *Die Zukunft* 6 (9 July 1918): 72–78.

12. Theodor W. Adorno, *Current of Music: Elements of a Radio Theory*, ed. Robert Hullot-Kentor (Frankfurt am Main: Suhrkamp, 2006): 120.

13. C. Talley, "The Army's Amateur," *Radio News* 12 (April 1931): 925.

14. Wolfang Hagen, "Theorien des Radios. Ästhetik und Äther" (lecture presented at Humboldt University, Berlin, winter semester 1995–1996), http://www.hagen.de/seminare/AETHER/Aether3.htm.

15. Adorno, *Current of Music*, 120.

16. Ibid.

17. Ibid.

18. Ibid.

19. See Thomas Levin, "Rhetoric of the Temporal Index: Surveillant Narration and the Cinema of 'Real Time,'" in *CTRL[SPACE]: Rhetorics of Surveillance from Bentham to Big Brother*, ed. Ursula Frohne and Peter Weibel (Cambridge, MA: MIT Press, 2002), 578–93.

20. Ibid.

21. Qtd. in Monika Elsner, Thomas Müller, and Peter Michael Spangenberg, "Der lange Weg eines schnellen Mediums. Zur Frühgeschichte des deutschen Fernsehens," in *Die Anfänge des deutschen Fernsehens. Kritische Annäherungen an die Entwicklung bis 1945*, ed. William Uricchio (Tübingen: Niemeyer, 1991), 160.

22. Bernard Stiegler, *Technics and Time,* trans. Stephen Barker (Stanford: Stanford University Press, 2009), 2: 126.

23. See Richard Kolb, *Das Horoskop des Hörspiels* (Berlin: Max Hesses, 1932).

24. Qtd. in Friedrich Knilli, *Das Hörspiel. Mittel und Möglichkeiten eines totalen Schallspiels* (Stuttgart: Kohlhammer, 1961), 64. Knilli confirmed this statement against the background of his own sonic and psychological radio play research.

25. See Ernst Gombrich, "Der fruchtbare Moment. Zum Zeitmoment in der bildenden Kunst," in *Bild und Auge. Neue Studien zur Psychologie der bildlichen Darstellung* (Stuttgart: Klett Cotta, 1984), 40–62.

26. Bettine Menke, "Adressiert in der Abwesenheit. Zur romantischen Poetik und Akustik der Töne," in *Die Adresse des Mediums*, ed. Stefan Andriopoulos, Gabriele Schabacher, and Eckhard Schumacher (Cologne: DuMont, 2001), 116.

27. Samuel Weber, *Mass Mediauras: Form, Technics, Media* (Stanford: Stanford University Press, 1996), 121.

28. The "ILLIAC Suite," which was generated by an electronic computer of the same name, was actually algorithmically programmed by composer Lejaren Hiller,

and it required a classical string ensemble to be performed for the first time as a sound event in 1956.

29. Hugo Münsterberg, *The Photoplay: A Psychological Study* (New York: Appleton, 1916), 94.

30. Ibid., 95.

31. Ibid.

32. "Aufbewahrung von Tonaufzeichnungen," *Funk-Technik* 14 (1969): 538.

33. Marshall McLuhan to Jane Bret, January 1973, in *Letters of Marshall McLuhan*, ed. Matie Molinaro, Corinne McLuhan, and William Toye (Oxford: Oxford University Press, 1987), 460.

34. In an earlier era of scientific photography, forensic criminology sought the remaining retinal image in the eye of the victim.

35. Jacques Lacan, *Radiophonie/Television*, trans. Hans-Joachim Metzger, Jutta Prasse, and Hinrich Lühmann (Berlin: Quadriga, 1988), 13.

36. Ong, *Orality and Literacy*, 58.

37. The first flight recorder kept a record of the altitude, course, speed, vertical acceleration, and time, which were inscribed on a continuously moving metal foil, much like a kymograph. This technography is also close to voice recording.

38. Jan Hein Hoogstad, "Time Tracks" (PhD diss., Utrecht University, 2005).

39. Jens Gerrit Papenburg, "Transatlantic Echoes: Elvis Presley's Voice as a Product of German Magnetic Tape Machines and its Function in Americanisation of Postwar Germany" (paper presented at the "Cultures of Recording" conference at the Centre for the History and Analysis of Recorded Music, Royal Holloway, University of London, April 10, 2008).

40. Philips LP 82850 (explanatory notes).

41. See Wolfgang Ernst, "Im Reich von Δt. Medienprozesse als Spielfeld sonischer Zeit," in *Sound Studies. Traditionen—Methoden—Desiderate. Eine Einführung*, ed. Holger Schulze (Bielefeld: transcript, 2008), 125–42.

42. See Friedrich Engel, "Iron Oxide γ-Fe_2O_3," in *Zauberhafte Klangmaschinen. Von der Sprechmaschine bis zur Soundkarte*, ed. Florian Cramer and Elfriede Jelinek (Mainz: Schott, 2008), 114f.

43. Carl Wiemer, "Im Rauschen des Realen. 'La dernière bande'—Becketts medientechnologische Antwort auf Prousts *Recherche*," *Romanistische Zeitschrift für Literaturgeschichte* 25.1-2 (2001): 173. For a critique of this theory, see Michael Lommel, "Synästhesie der Erinnerung: Becketts *Krapp's Last Tape*," in *Medium und Gedächtnis. Von der Überbietung der Grenze(n)*, ed. Franziska Sick and Beate Ochsner (Frankfurt am Main: Peter Lang, 2004), 255–64. *Krapp's Last Tape* premiered on October 28, 1958 at the Royal Court Theatre in London.

44. This was initially a liturgical concept. See George Steiner, *Von realer Gegenwart. Hat unser Sprechen Inhalt? Mit einem Nachwort von Botho Strauß* (Munich: Hanser, 1990).

45. See Horst Völz, *Handbuch der Speicherung von Information, Band 3*, (Aachen: Shaker, 2007).

46. This production, which premiered on June 1, 2007, was directed by B. K. Tragelehn and starred Josef Bierbichler.

47. See Friedrich Kittler, *Gramophone, Film, Typewriter*, trans. Geoffrey Winthrop-Young and Michael Wutz (Stanford: Stanford University Press, 1999).

48. For more on the connection between the formation of identity and vocal sound, see Daniel Gethmann, *Die Übertragung der Stimme. Vor- und Frühgeschichte des Sprechens im Radio* (Zürich: diaphanes, 2006), 8.

49. Benjamin, "Little History of Photography," 518.

50. Translator's note: There is a play on words here in the original text, as the German word for "replaying" (*wiedergeben*) literally means "giving again." Through the process of replaying these tapes, Fontana is thus literally giving the past once again to the present.

51. See Martin Supper, "Klänge aus Lautsprechern. Klang in der Geschichte der Elektroakustischen Musik," in *Sound Studies. Traditionen—Methoden—Desiderate. Eine Einführung*, ed. Holger Schulze (Bielefeld: transcript, 2008), 27.

52. From *per/sonare* as the sound of the voice coming through the open mouth of an actor's mask.

53. Paul Virilio, *The Vision Machine*, trans. Julie Rose (London: British Film Institute, 1994), 66

54. Ibid., 72.

55. Karlheinz Stockhausen, ". . . wie die Zeit vergeht . . .," *Die Reihe. Information über serielle Musik* 3 (1957): 13–42.

56. Martin Heidegger, "The Question Concerning Technology," in *The Question Concerning Technology and Other Essays*, trans. William Lovitt (New York: Garland, 1977), 12.

57. Michael Lommel and Jürgen Schäfer, "Von Band zum Netz. Gedächtnismedien," *Navigationen. Siegener Beiträge zur Medien- und Kulturwissenschaft* 2 (2002): 47.

58. Lommel, "Synästhesie der Erinnerung," 260.

59. Translator's note: there is another play on words here in the original text, as the German word for "development" (*Entwicklung*) literally means "unwinding." The concept of time as development thus refers to the literal unwinding of the tape reel.

60. Ibid.

61. Ibid., 255.

62. Vilém Flusser, *Kommunikologie* (Frankfurt am Main: Fischer, 1998).

63. See Friedrich Engel, "The Introduction of the Magnetophon," in *Magnetic Recording: The First 100 Years*, ed. Eric D. Daniel, C. Denis Mee, and Mark H. Clark (New York: IEEE Press, 1999), 47–71.

64. Elisabeth Schimana posed this question during her presentation in the colloquium "Medien, die wir meinen" at Humboldt University, Berlin, February 10, 2010.

Chapter 6

A Close Reading of the Electronic "Time Image"

FROM THE PHOTOGRAPHIC TO THE TELEVISUAL *PUNCTUM*

Ever since the development of photography and phonography, audiovisual media have appealed directly to human sensory perception, unlike the primary cultural technique of the symbolic (written texts and musical notations). Historical distance thus seemed to be undermined and brought into close proximity. High-tech telecommunications produce simultaneities, as an undifferentiated present is replaced by the intensified actualization of the signal and the most widely varied temporalities are combined. Such a media time no longer corresponds to the linear consciousness of history; rather, it is directly connected to neuronal inner temporality.

Every medium that succeeds in reshaping the present with its own temporality is a chronotechnology. Media constitute an enframing of time (*Zeit-Gestell*); in this sense, Marshall McLuhan identified the business world as a servomechanism of clocks.[1] Yet, the temporality of television does not lie simply in the seriality of its programs, as the mode of electronic line scanning yields images on the signal-technical level itself.

What was once continuous frequency modulation in the domain of concrete radio waves and culminates today in the compression of streaming media, which no longer transmit complete image sequences but only degrees of change, is a transformation of linear media time through radical discretization. In classical photography, it was the duration of the photochemical exposure that allowed time to inscribe itself as an image; compressed time intervals in the two-dimensional image reduced the duration to virtually nothing. In a real media process, however, there is no such thing as an extensionless moment. With electrotechnical phototelegraphy, the time interval that was previously necessary to bridge vast distances was reduced to moments of electrical signals. With serially dispersed image transmission, time penetrates into the image itself. The exact synchronization of the sending and receiving apparatuses, such as by means of an isochronous pendulum, was chronotechnically delicate. The fully electronic transmission of moving images represented a time-critical escalation, as not only did the cinematographic frequency of the images have to be retained in order for them to be perceived by human eyes as living movement, but every individual image also had to be divided into lines. "Von Braun's cathode ray tube inscribed images dot by dot and line by line. In the electronic camera, a microelement of the image became a unit of time, which in turn could be manipulated. In electromagnetic recordings of image and sound elements, what can be seen and heard can be stored or processed in the smallest particles or in large packages."[2] Sound and image signals stored in the apparatus can be not only repeated at will, but also "layered, expanded, or speeded-up."[3] Cultural techniques that have long been familiar (cutting, splicing, and replacing) thus escalated into experimentation with time by highly technical means.

With lightning-fast line-writing in the cathode rays of the picture tube, the unique moment of image fixation in photography became an electronic event on the time axis itself. Barthes still looked for the essence of photography in the physical medium.

> It is often said that it was the painters who invented Photography (by bequeathing it their framing, the Albertian perspective, and the optic of the *camera obscura*). I say: no, it was the chemists. For the *noeme* "That-has-been" was possible only on the day when a scientific circumstance (the discovery that silver halogens were sensitive to light) made it possible to recover and print directly the luminous rays emitted by a variously lighted object.[4]

In the photochemical fixation of photography and film, such light rays represent the passive traces of a former referent. However, the broadcast of an early film with Audrey Hepburn on current television is more than simply the emanation of a long since extinguished star conveying information from the past; in the electronic broadcast, this film becomes a time event of an entirely different kind. The media-induced short circuit between photographic time and observer time in the *punctum* fundamentally differs from the mutual inhabitation of a temporal field. The media-technical distinction lies primarily in the fact that television does not project static images of individual moments, but rather it is a form of electronic light-writing that is permanently in motion. John Logie Baird, who developed an early method of electromechanically transmitting moving images, described this distinction by comparing his own images to photographs taken of the screen. "The photographs . . . do not do the results justice. Much is lost in the photograph and much more is lost owing to the fact that photographs can only show a still picture, whereas the movements continually presented new aspects and the effect was obtained of much greater detail than the photographs can show."[5] The photographic *punctum* of the indexical image is here replaced with a temporal index. What divides film and television is sheer electronics. One of the relics in the National Media Museum in Bradford is Baird's Televisor, which represents not a fictional surveillance medium, as in George Orwell's futuristic novel *1984*, but rather the first television receiver mass-produced in England. Baird's invention illustrates the first principle (thus the medial *arché*) of television as a time-critical process. The core idea is the successive dissolution of an image into time-variable brightness values, which are then transmitted as electrical pulses and reconstituted as an image at the other end, albeit in reverse. The radical temporality of the technical medium produces an entirely non-iconological view of the image. In order to measure delay times in the nerves, nineteenth-century physiology already developed technical kymographs ("wave writers"), whose graphic curves represented time diagrams; from the perspective of media, the relationship between time and image is reciprocal.

One method of understanding the specific temporality of media is to reconstruct them and then put them into operation; this is an operative model for the media archeology of television. Middlesex University developed an electromechanical build-it-yourself kit as a teaching aid to reenact Baird's Televisor. Before "it" happens (the television image medium is not a matter of ontology), it is worth considering its conditions of possibility—in other words, the kinematics of image scanning and playback. A Nipkow disk is punctured by holes that form a spiral shape. As the disk rotates at a high velocity, linear (actually slightly curved) light from an intense source shines through these holes onto the object to be scanned. Each point of light is registered by a light-sensitive photocell and converted into analog voltage, which is then transmitted over a wire or wireless as a time signal. At the receiving end, the voltage signals are converted back into light by a glow lamp, and another Nipkow disk rotating at precisely the same speed reassembles the points of light into a 30-line image. A central characteristic of technical media once again becomes manifest: the actual media event only occurs at the moment of successful synchronization. The holes in the rotating receiving disk must be in the same momentary position as the holes in the sending disk; otherwise, the image "wobbles." In addition to the optical impression, the observer also has a tactile relation to the image, as he must constantly readjust the rotational speed by hand. The observer is thus directly connected to the time behavior of the apparatus—a cybernetic coupling of human and machine in the temporal domain.

In the sonic technospheres, such as telephony, the virtually instantaneous communication of continuous signals was initially successful. The sonification of the electronic image also functioned as a form of acoustic monitoring in Baird's early television experiments.

> In testing out the amplifiers I used to use headphones and listened to the noise of the vision signal made. I became very expert in this and could even tell roughly what was being televised by the sound it made. I knew . . . whether is was the dummy's head or a human face. I could tell when the person moved. . ., and even when two or three people had different appearances I could even tell one from the other by the sound of their faces. I got a gramophone record made of these sounds and found that by laying this with an electrical pick-up, and feeding the signal back to a television receiver I could reproduce the original scene. . . . If the cinema had never been invented the "Phonovisor," as I christened the device, might have been worth developing; it was certainly an intriguing process. Vision into sound and sound back into vision.[6]

An image can be recognized as sound as soon as its electronic existence (in contrast to its cultural history as painting) becomes a function of spatial coordinates as well as time, and it thus shares with sonic events their essential parameter t.

However, the noisy signals of Baird's gramophonically recorded television images can only be retrieved by algorithmized media—the true archaeologists of the high-tech age—through the use of special filtering software. Is this a matter of restoring or reconstructing the image? The digital analysis of such images actually represents their "reenactment"[7] in both a performative and media-operative sense. In contrast to the symbolic techniques of the cultural tradition, signal recording facilitates a truly equiprimordial reenactment. The concept of emulation is evidence of this: the British Narrow Bandwidth Television Association "continues . . . the spanning of the Atlantic in January and February 2003 in emulation of J. L. Baird's 1928 exploit."

The brochure accompanying the "Televisor Kit" conveys the idea that media time consists of functional equivalences; the functional reenactment of high-tech events (to use Collingwood's concept)[8] is actually closer to the temporal relations of a natural science experiment than those of historicism. "The *televisor* . . . works in exactly the same way as the original, but uses modern components such as an LED instead of a neon lamp for puncture illumination." Media archeology raises a crucial question here: what differences does it make when the focus of analysis becomes the concrete physics of the media components in their current form instead of merely their function? Is the transistor as electronic semiconductor simply a functional equivalent of the vacuum electron tubes of early radios, televisions, and computers? The transistorized replica "is about one third of the size of the commercial televisor—but the performance is as good." The central criterion of media implementation remains ahistorically intact: its operativity as equiprimordial reenactment.

Technological structures are usually still evident in their original and thus media-archaeological incubation phase (they only become invisible with the rise of mass media). "It is the beginnings of invented things, which appeal to me," wrote Lance Sieveking, the author of a genuine television drama that was naturally transmitted "live" by the BBC during the experimental phase of the Baird television system. Sieveking added: "For it is at their beginnings, that we may detect their true nature"—in other words, their epistemological essentials. Sieveking's quotation constitutes the motto of the updated edition of *Television and Me: The Memoirs of John Logie Baird*—a work that offers quite idiosyncratic but also highly archeological insight into the first realization of the electromechanical television apparatus. Mathematically formulated, such a beginning, such an *arché*,[9] is a vector that conceptually unites both of the criteria of a veritable techno-logic: "the idea of magnitude (a core element of mathematics) with that of direction (a core element of physics)."[10] Conceived as a vector *field*, it provides an alternative way of formulating the history of media.

The early sense of television was not aesthetic, but rather—in media studies terms—aisthetic as sensory experience. At this level, the medium actually performed a massage[11] of the human sense of time. "Using the new *televisor* is a unique experience because you will be seeing images similar to those seen by the earliest TV reviewers some 80 year ago." What is meant here is not the iconological content of early television programs, but rather the event of the successful image itself—in other words, an awareness *that* it is happening. This is the basis of the media-archeological sublime.[12] The experience of media time not only occurs on the physiological microlevel, but also affects the emphatic macrolevel—"something quite close to time travel!" The historical-temporal distance is not actually eliminated, but rather compressed into contemporaneity through an act of induction. The current media event of the Televisor forms a quasi-electromagnetic time field around itself, which is able to induce temporal affects. The reverse temporal engineering of actual signal processing illustrates the entire difference between technical and pre-technical archives, which store time events in symbolic codes until they are reactualized in the present through reading.

THE UNDERLYING TIME OF THE IMAGE: TELEVISION

The perceptual impression of moving images first emerged with the intermittent mechanism—in other words, the temporal interval between the individual images in the cinematographic apparatus. However, an electronic image consists of continuous movement, which is based on time and which also represents time. The television image does not actually exist in its entirety at any given moment; it only consists of individual moments of light, which are transmitted and observed at particular points in time—a radical temporalization of the image. During its implementation, television represents to a large extent a self-referential chronocosm—a proper time with respect to time outside the electronic apparatus. If the capacitors wear out or sparks begin to flash between the transformer windings, the thermodynamic time of physics shakes the fragile temporal structure of the television, but as a system it still remains intact. The condition of possibility of this electrotechnical autopoiesis is an infrastructure that has remained stable for centuries (the power grid and the technical standard of electromagnetic wave transmission). This phase constitutes an era with its own media-technical logic, which is asymmetrical to other histories of events.

As the facilitation of electronics, television represents an escalation of temporality with respect to cinematography. The behavior of time becomes *critical* in the sense that it is *crucial* for the realization of the image, as the individual image lines must succeed one another so quickly that the entire image surface is recorded in fractions of a second. In contrast to the projection of stored film images, the genuinely "live" television system requires a tightly coupled time relation between the sender and the receiver of the image. The synchronization of the disassembling and reassembling devices is ensured through synchronizing signals that are transmitted at the same time as the image signals. These pulse signals are not unique to the image, but rather the chronotechnical condition of their successful transmission—a modification in the diagram of telecommunications; Claude Shannon's transmission model of communication is time-critically intensified. Viewing television from a media-archeological perspective means paying attention to the transmission mechanisms of the images, which only actually exist in their temporality. This electromechanical image analysis was initially based on absences in image scanning; in 1884, Paul Nipkow filed a patent application for the "punched disk"—an analog sampling. The disassembly of the image through scanning and its reassembly on the receiver side was thoroughly sufficient for crude image reproduction, but "there persists a desire to further refine the scanner for content-rich images."[13] The so-called "content" mentioned here is not a question of emphatic semantics, but rather a measure of the information that is logarithmically calculable, as the number of image elements increases by the square of the number of lines. This close-up view of the television image also became a political aesthetic in Germany under National Socialism. "When the *Führer* speaks at the Nuremberg Rally, for example, audiences at the television receivers in Berlin not only hear his speech, but they also see his image, all of his movements, and his surroundings at the same time"[14]—but not entirely simultaneously due to the delay of the intermediate film system. "At the same time" here articulates the media-culturally signifying metonymy of human and technical *receivers*.

A media archeology of this technical form goes one step further back. The ancient Greek *analysis* of spoken language in the form of the vocal alphabet and its communication-technical escalation in the form of the telegraph constituted the cultural techniques through which the familiar process of image transmission first became conceivable. In electromagnetic phototelegraphy and in the fully electronic television, the image is literally disassembled and thus discretized. In the electrotechnical intermediate stage (the stored image in the iconoscope), a continuous model is transformed into a discrete mosaic whose elements are recorded and transmitted successively—that is, in the time channel—at lightning speed. Instead of image *points* (as they are described in most textbooks), it would be more accurate to call them image *moments* in the sense of the functional values of continuous time signals; otherwise, the distinction between *analog* and *digital* would have no meaning in this context. The cathode rays of the picture tube break away from the Western epistemology of the point in favor of an infinitesimally small extension and temporality. At issue here is the temporal mode of the electron as an elementary particle of all electronic media.

> All the physical characteristics of the ultimate elementary particles of matter, particularly of electrons, can be read off from the neighboring field. . . . Then there is no pointlike Now and also no exact earlier and later. . . . The immediate present is not entirely abrupt; there is always a small halo, quickly fading toward the past and toward the future, along with the self-shining light of immediacy.[15]

The television transmission of film images represents a special case. "Telecinematography" was initially based on a discrete series of static images (the mechanical principle of film projection); however, the focus of the scanning cathode rays successfully divided every image successively into infinitesimal elements. The English term for this, the "flying spot scanner," conveys the essence and mode of this process: a point out of time. Manfred von Ardenne's Patent Nr. 174823 describes a television setup in which a photocell receives "momentary" light current depending on the transparency of the *dispositif*—an oxymoron, as a continuous stream can never be momentary. In the context of alternating current, the moment necessarily proves to be an imprecise interval. The fleetingness of the moment—the temporal vanishing point—thus becomes a signature of the electronic image itself. The appropriate mathematics for this is interval arithmetic, in which the limits of an interval are calculated and the value sought is reliable.[16] This is the counterpart to the "forbidden" realm of the binary-digital, where two conditions are distinguished from one another, and it thus represents at the same time a form of analog computer.

In contrast to the diffuse cultural semantics of the image, imaging technology has a concept of the image that is crystal clear. "An image . . . is conceptually a surface on which different brightness and darkness values simultaneously appear at different points."[17] The signal-technical difference between the television camera and the eye also manifests here. "The retina on which the image falls, just as light falls on the ground glass of the photographic camera, is composed of infinitesimally small light-sensitive nerve cells—several hundred in one square millimeter—and each of these infinitesimal cells is connected to the center of sensation in the brain by a special 'nerve conduction.'"[18] To put it another way (following G. W. Leibniz's mathematical aesthetic of *petites perceptions*), this suggests that the human brain perceives the physiological difference between television and retinal images, even when the technical deception remains below the perceptual threshold. According to this argument, the dissonance takes place in the unconscious and leads to cognitive dissonances—a time-critical difference.

The title of Kurt Lipfert's chapter on a characteristic feature of the technological object is striking: "Adding Image to Sound."[19] From the nonlinear perspective of media archeology, it is logical to understand the television image in terms of radio, as it was technically a result of radio broadcasting (just as video recording was a result of tape recording). This formulation also recalls the central time-critical problem of early sound film, which was unsuccessful as long as it was performed through bi-medial coupling—either externally (such as playing a gramophone record on a turntable coupled to a film projector) or on the same storage medium (the needle-sound system). It is difficult to synchronize the delay times of diverse media so precisely that the ears and eyes are not irritated by the asymmetries between the images and sounds. Only electronics—in this case photocells and electron tubes—could confront this (electro)mechanically-conditioned *différance*, as they operated virtually without any inertia. Sound-on-film was developed by a co-inventor of the amplifier tube, Lee DeForest,[20] and similarly in Germany by Triergon. Light is here both the subject and the object of the process. The transitive relationship between the recording and the sound event in mechanical phonography was replaced by a genuine media-technical in-between, a translation: the air pressure fluctuations induced by acoustic articulations are converted into current fluctuations, which modulate a light source that inscribes the converted sound as illumination onto the same optical carrier as the film images. During playback, the same photocells used in electromechanical television scan this optical sound-writing and convert it back into current fluctuations, which then become audible—with the aid of an amplifier—without any time delay with respect to the flow of images. In a time-critical sense, however, the concept of sound film as an "audiovisual" medium breaks down. What is at work here is not metaphysics, but rather chronotechnical *différance*: the "non-conformity of the intermittent movement of the image track and the continuous movement of the sound track."[21] This asymmetry is sublated through the roughly one-second delay between the sound and the position of the image on the film strip. Synchronous sound transformed cinematography into a comprehensive temporal artform precisely through the possibility of a rupture between the continuous audio track and the discontinuous image sequence. Avant-garde directors like Vsevolod Pudovkin consciously raised this technical asynchronism to an aesthetic principle—as technical violence on an apparently natural temporal order.

In electronic signal transmission, synchronization between the sender and the receiver is a constant challenge; the entire communicative meaning of acoustic and optical transmission depends on it. This situation is intensified in color television, as the individual color values must also be orchestrated time-critically. Baird's electromechanical television system was already confronted with the inertia of selenium cells, which are only able to turn light into variations of applied voltage with a delay and thus yield an imprecise image. To compensate for this delay, Baird chose to employ the alternating current in transformers (an entirely electromagnetic time figure), which do not induce gradually increasing current but rather abrupt pulse sequences—a binary signal aesthetic of "on" and "off" *avant la lettre*.

$\Delta T \to 0$: TELEVISION, TIME-BASED AND TIME-CRITICAL

The concept of recording as a form of storage means, first of all, the setting aside of matter, energy, or information for reuse later. Data recording and instantaneous data processing differ not categorically, but only in terms of the time scale of their reuse. In the Von Neumann architecture of computers, data processing always requires the short-term intermediate storage of bits. Long-term storage thus proves to be merely an extreme extension of the "movement" known as the time window of the present. Archival storage is itself dynamic, albeit in extreme slow motion; the time span of the reactualization of recordings—in other words, their frequency—is so low that the intermediate storage appears immobile. Consider the difference between the photographic fixing of a moment in time and the instantaneous transmission of an electronic image as television or video: in the first case, the chemical processing of the light event is expected to have a permanent effect, whereas in the second case, the photonic event disappears from the phosphor-coated screen in a fraction of a second. Yet, this fraction still represents a time interval, a Δt—even when it approaches zero.

In his 1945 essay "As We May Think"—a visionary anticipation of hypermedially connectable data processing and retrieval—Vannevar Bush compares the telegraphic facsimile transmission of images or texts to electronic television. "The record is made by a moving beam of electrons" combined with "a screen which glows momentarily when the electrons hit, rather than a chemically treated paper or film which is permanently altered."[22] Beyond the Gutenberg era of printed letters and records, the materiality of information here becomes fluid and ephemeral; it is a temporary moment—a trace in time rather than a point in space, radically dynamic rather than discrete.

"Live" transmission is the essential feature of electronic (mass) media like radio and television; at first glance, this represents the sheer opposite of the archive. Yet, the deceleration of signal transmission in a channel can already be used as short-term storage, such as the delay line in the PAL color television system or the buffering and literal "recycling" of data in the so-called "delay memories" of early digital computers. There are two forms of storage: dynamic and static. On the one hand, storage refers to the event as the signified (the level of time events); on the other hand, it also refers to momentary time intervals (the media-technical level of signifiers). These temporal modes are not categorically different; rather, both of these extremes represent the same connection. Cinematography lies between with its 24 frames per second.

In a technical sense, the cinematographic motion picture (or "movie") is the function of an "archive in motion": cadres of discrete images on celluloid, which are actualized from a reel through projection. However, the electronic television image is permanently produced (or "refreshed")—an anarchival condition. If the phonographic-cinematographic principle is based on the identical retrievability of stored contents, and it is thus invariant with respect to the passage of historical time, the event of the television and computer image is based on the additional necessity of constant regeneration. A different epistemology opens up here, which Teilhard de Chardin's concept of the "noosphere" grounded in technology: "Farewell to the mechanical worldview in favor of the idea of a force field"[23] in the sense of electrodynamics. The temporary character of the television signal made McLuhan, who only revealed limited technical competence in this context, doubt the concept of the image itself. "The TV image is visually low in data. The TV image is not a *still* shot. It is not photo in any sense, but a ceaselessly forming contour of things limned by the scanning-finger.... The TV image offers some three million dots per second to the receiver. From these he accepts only a few dozen each instant, from which to make an image."[24] The television image cannot be reduced to a two-dimensional grid, as it always unfolds in time—thus, the added depth coordinate t. Television is a time-based and time-specifying medium in terms of both the technology and its programs. Infinitesimal temporal moments at the screen level replace the spatial points of classical panel painting, and a veritable time-image thus emerges. The luminous effect on the screen is proportional to the electrons emitted from the tube, which strike the inner surface. But synchronization is crucial above all else, as television represents the "temporally concurrent occurrence of two analog events."[25] More specifically, it is "a controlling measure, which ensures that the scanning of the image in the sending apparatus . . . is precisely synchronized to the flow of cathode rays in the receiving tube."[26] Fully electronic television and video are time-based media in the microtechnical sense; at the same time, they also function at the mesotemporal level of everyday time perception in that they condition user behavior and its time sovereignty. At its core, however, the electronic image is a function of time. "Televisual perception involves optical-electronic (image) *processes* rather than (image) *conditions*, pixel rasters or mosaics rather than homogeneous imaging units, and accelerations of image transmission and composition to the limit of the speed of light rather than the slow tempos of mechanical methods."[27] As information carriers, signals are physical events, and thus they always already exist in time. Electronic images are spatio-temporal segments of a continuous signal stream that is discretized by line breaks. The time-critical *a priori* of television is grounded in the power of the electronic components and their synchronizing circuit, but it is largely determined by the specific coupling of humans and machines in the temporal domain—in other words, the cybernetic Δt. "In the development of televisions . . . the properties of the human eye prescribe the conditions that must be fulfilled in order to produce usable devices."[28] In cinematography, the mechanics of image transfer were augmented with a disk shutter to reduce flicker; in the electronic image, this becomes a challenge of temporal existence. During the transmission phase, the image signal only exists in latency as voltages that correspond to brightness values; in this regard, it is related to xerographic reproduction technology. The time-critical element here lies in the superimposition of the voltage image and the control pulses. Time control is necessary for the image signals to be reassembled in the same order at the receiving end and thus recognized as images.

A special time-critical effect occurs when the signal information (the latent "image") is modulated onto a carrier frequency in the short-wave range, which is still reflected by the ionosphere (up to ca. 30–40 MHz). "When crossing vast distances, the directly incoming waves would have a shorter delay time than the waves reflected by the ionized layers. If both waves were picked up by the receiving antenna, then the same image would appear on the receiving screen twice with a certain time difference"—the so-called "echo" or "ghost images," literally time ghosts. Carrier waves that behave like optical waves are thus used—the VHF range. Television, in a time-critical sense, "is the art of instantaneously producing at a distance a transient visible image of an actual or recorded scene by means of an electrical system of telecommunication."[29] Although it is falsified by principle of electromagnetic induction, Newton's concept of an immediate *actio in distans* returns here at the level of temporal perception. Above all, however, this transience identifies the character of the electronic image in contrast to the photocinematographic or even painted image. Of prime importance here is the act of transmission itself rather than the ontology of the "live," as films can also be scanned for television transmission. The momentum of the "live" primarily lies in the electrotechnical act of telecommunication—the supremacy of transmission as opposed to the nature of the source material as real-time or stored event. Baird's 30-line Televisor, factory-made since 1930,

> was used in conjunction with a wireless set which received the television signals on the long wave and the output was fed directly to a single neon placed behind the disk. In operation the spinning disk had to be synchronised with the transmitter and this was done by continuously adjusting the knob in the centre while viewing the picture through the lens on the right-hand side.[30]

A differential process that was familiar from the practice of short-wave radio reception, where the constant readjustment of the variable capacitors was necessary to prevent "fading." The difficulty of synchronization on the technical side corresponded to the essence of television transmission as "live": instantaneous visibility (temporally almost equiprimordial) over simultaneous spatial distance (spatio-temporal entanglement).

The fleetingness of time is inherent as a gap in the essence of the electronic image. Momentary conditions can at least be recorded photographically through corresponding time-critical exposure times. "In order for a complete image to be photographed from a recording—without black beam—the shutter speed of the camera must be oriented toward the image frequency of the television system. The television image frequency amounts to 25 images per second. That means 1/25 second for each image. The exposure time for photography thus cannot be shorter than this time."[31] For a long time, however, television images could only be documented as *moving images* in other media, which for their part were time-based and thus represented a memory buffer, such as the Baird Phonovision System, which was modeled on gramophone sound recording, or documentary films. The remembrance of time-critical processes thus required time-based media, and one media memory enabled another—a genuine media memento. When an electronic image functions simultaneously as both subject and object of a broadcast (which was intensified in the "closed circuit" installations of early video art), this results in distortions. The essence of electronics is always already temporalized—a differential identity. The earliest transmitted recording of a program-based television broadcast was the revue "Looking In," featuring the Paramount Astoria Girls, which was aired by the BBC over the 30-line Baird Television System on April 21, 1933 on the frequencies "Vision 262.6 m; Sound 398 m." An amateur recorded this broadcast on an aluminum record using his Baird Phonovision System. Processed and restored through digital filters, "the key to clarity seems to be movement itself," for the reproduction of each of these 30-line television images as lithographic "stills" (literally) in a print medium (the book) necessarily gives a false impression of what was "actually" seen. A new variant of the time-critical comes into play here.

> Though we have only 30 lines and only so much detail along a line, we are missing a crucial element: time. . . . As the image is a still built up from just 30 lines, it is difficult to distinguish between what is image and what is not. A single frame of the Paramount Astoria Girls may be crudely recognisable, but when seen as a moving dynamic television image, the girls come to life before our eyes. There is something strange at work and it has much more to do with what we perceive than what is there in pixels, lines and frames. What we are experiencing is not the detail that the eye sees, but the recognition of movement that the brain sees. . . . our brain somehow builds up a model of what we are looking at. We interpret the model into a real-world scene.[32]

So much for the human level of perception. As a technology, however, electronic television reaches the time-critical limits of quantum physics as it lights upon the crucial point:[33] the electron. The elements of an electronic cathode ray image, which are composed of electrons, are never localizable in the spatial domain as discrete particles; rather, they represent brightness or power fluctuations—in other words, they are pulses in Fourier space. An understanding of these wave images not only in terms of their frequency but also as microtemporal events is provided by their temporal "windowing," resulting in the fast Fourier transformation process—a practical form of quantization.[34] Niels Bohr's model of the atom already applied to the emissions generated by glow lamps in electromechanical television. "When the electron jumps from an orbit that corresponds to a higher energy state to one that corresponds to its normal energy level, this energy difference is emitted, meaning the atom glows. The length of time that an electron can stay at a higher energy level before it emits depends on the inertia of these emission phenomena."[35] In this respect, the process that occurs in electron tubes is only approximately free of inertia. "Inertia phenomena in the emissions process . . . in glow lamps cannot be examined, as stimulation and emission are not spatially separable"[36]—a spatio-temporal imprecision. This is further complicated by the fact that television signals—despite their apparently "live" temporality—suffer from delay distortions during transmission—the echo of electromagnetic worlds.

The human eye immediately recognizes image distortions, whereas the ear is especially sensitive to microtemporal distortions in the frequency range. What is miraculous about the successful video and television image is not the momentary but rather the continuous addressing of the observer's sense of time—a massage of the time-critical sense. The television's actual *joie de vivre* (*Daseinsfreude*) lies in the temporality of its events on a subliminal level rather than the straightforward iconological content of the programs that it transmits. The question of synchronization is all the more delicate here. A constant problem in Nipkow's television system was the difficulty of ensuring and maintaining a synchronous rotation speed for the punched disks on both the sending and receiving sides; the "Phonic Wheel" was employed for precisely this purpose—synesthesia on the technical level. "This wheel is in principle a simple synchronometer whose actuating current is divided into a certain rhythm using a tuning fork";[37] two tuning forks of the same length at the sending and receiving stations thus ensure synchronization. The virtually acoustic essence of the electronic image is thus implemented, although the sonic here is not perceived as music but rather implicitly at work as chronotechnical *mousiké*. The characteristic feature of vibrating signals as the momentary amplitude of a complex oscillation is that they only depend on a single individual variable: time.[38]

In order for the eye to perceive televisually transmitted image pulses as continuous, the apparatus must be able to register thousands of changes every second in all of its components. "Unlike a phototelegraph transmission, which can take any amount of time, . . . television is bound to a definite duration of no more than 1/10 second."[39] This is only possible under completely electronic conditions—that is, with the use of inertia-free components in contrast to the electromechanics of the Nipkow apparatus. The electron tube thus comes into play. If an electron beam accelerates in empty space, it ideally approaches the speed of light—what Paul Virilio calls "intensive time" on the way to zero, the infinitesimal punctualization of the present moment. The chronocracy of the twentieth century actually culminated in the near domination of the Dirac pulse, whose temporal extension ideally equals zero and whose amplitude is virtually infinite—the most compressed form of time signal.[40]

The effectiveness of electronic television with respect to human perception is only superficially revealed on the iconological or narrative level, but it also unfolds on the neuronal, sensory level. This media-archeological insight with regard to the time-critical can be specified more precisely. The actual *media* message of the television image is not the dramaturgy of television shows but rather its temporal essence, as its technical-medial singularity models temporal perception. "Time is thus the challenge of media studies."[41] The time-critical delicacy of the electronic image later migrates into the computer itself, which operates with the smallest time intervals. When an addition in fractions of a second occurs, "the *special time relationship of this machine* appears: it works in the microstructures and microprocesses of time, which cannot be utilized through human actions or thoughts."[42] Electronic television was a model for Norbert Wiener's cybernetic concept of time; in the process of ultra-fast line-by-line image scanning he did not see image contents but rather the electronic implementation of partial differential equations. This mathematical analysis was able to take the temporality of the electronic image into account for the first time[43]—an increase in complexity with epistemological consequences.

Filmic sampling of a moving reality produces time-discrete photographic image sequences. In contrast, video recording involves linear, line-by-line fragmented, point-relocating sampling, which results in the time-shifted reproduction of the same momentariness. In extreme cases, sea waves as an object of television transmission form a moiré pattern with the actual lines of deflected cathode rays—a differential entanglement. This occurs through the superimposition of two presents: the now-time of the object transmitted "live" (the object of the medium) and the now-time of the technical reproduction (the subject of the medium). As a partially mechanical system, the Nipkow disk technologically circulates *between* these two presents. The individual image lines are still perceptible here, but the situation is different for the intertwined half-frames of electronic television. Fully electronic processing speeds lag behind neuronal stimuli.

"Television gave the epistemological biases of the telegraph and the photograph their most potent expression, raising the interplay of image and instancy to an exquisite and dangerous perfection"[44]—the periodic triggering of photographic exposures, which is only possible in the media-technical shift to the electronic. However, the moving image effect depends on the eye, which has a "wonderful defect . . . , this small but crucial inertia."[45] This inertia is "crucial" in a *time-critical* sense. "It *automatically* retains every light impression in memory for a tenth of a second."[46] Eduard Rhein is here (unintentionally) describing Marcel Proust's concept of memory: the subliminal *mémoire involontaire*. When Nipkow disks rotate so quickly "that they make an entire rotation in this critical tenth of a second and thus show all the image points, then none of these points of light has time to slip the 'memory' of the human eye. 'Memory' and new perception merge, shaping luminous points into a luminous image"[47]—a formulation that could have also come from Bergson. With respect to acoustic data, on the other hand, the ear is not sluggish but rather highly sensitive to frequencies. From a media-archeological and time-critical perspective, the conceptually implied unity of "audiovisual" perception breaks apart (physiologically and media-technologically): two auditory signals can already be differentiated after 2 to 5 milliseconds, while an interval of 20 to 30 milliseconds is needed for visual stimuli.[48] An added factor is the genuine media-temporal phenomenon of so-called "time errors" (jitters), measured in microseconds, which are a reminder that the electronic video image is still anchored in the mechanical and thus inertially loaded materiality of the video apparatus.

> The time errors of a video signal emerge through the flutters of the video recorder drives while feeding the magnetic tape, which cannot be entirely avoided. The time error . . . is a measure of the temporal deviation of the video signal coming from the tape with respect to an electronically-generated pulse signal, which comes for example from the central clock of a studio.[49]

The time base corrector (TBC) serves to compensate for such minimal time delays. Whenever one technical time responds to another, the concrete chronopoiesis of technical media is complete.

THE BOUNCING POINT AND THE DISCRETE IMAGE ELEMENT

On the one hand, a media-archeological analysis of the temporal modes of technical images requires a *knowledge*able description of what technologically occurs; on the other hand, it also requires an *epistemological* description of what *actually* occurs in the emphatic sense (to take into account a dictum of the historian Leopold von Ranke). This involves looking beyond the quotidian naturalness of the technological event and instead analyzing its conceptual framework. This shift of perspective can be seen, for example, in the introduction to the brochure accompanying the Televisor Kit—the model replica of John Logie Baird's early electromechanical television apparatus. "It is intriguing to see the picture and then look round the side of the *televisor* to see nothing between the LED and the disc." The image is here a function of invisibility—an intermediate time established by technical media themselves. In this virtually pre-Socratic archaism (a methodological principle of media archeology), the functioning of the television is very simple to read. The introduction then goes one step further in the sense of a Socratic complication. "In practice, things were slightly more complicated," as physically being-in-the-world (*In-der-Welt-Sein*) also means being-in-time (*In-der-Zeit-Sein*). Intermediate space is here also intermediate time. What *actually occurs* in the Televisor is the microtemporal media event. The essential feature of analog sound and image storage media is the creation of presence through the (re)implementation of signal recording, which makes a latency perceptible to the senses (every pick-up is a reader of an archive when in mechanical or electromagnetic use). Transmission media, on the other hand, require and produce synchronization in technical communication. Morse telegraphy lies somewhere in between storage and transmission media, as it was initially decoded by specialists, like a sonic melody, but it ultimately moved into the temporal domain of real time through electronic image transmission. Synchronization between the sender and the receiver is not natural, but rather a technologically induced cultural artifact—a negentropic time figure. "At any point in time, the holes in the spinning receiver disc had to be in the same position as those at the transmitter—i.e., the two had to be synchronized" in order to prevent the "wobbling" of the image. "Any point of time?" This is the moment to show an example of how a "close reading" of the apparatus, which is fundamental for media knowledge, becomes a media-theoretical epistemology. What actually occurs in this formulation is a metonymy of the hole in the Nipkow disk to the subordination of a point in time, which makes a media-epistemic relation transparent: uncertainty. Following Zeno's paradox of the flying arrow, which appears to be standing still at every moment of precise observation, the hole is never at a precisely definable place at a particular point in time, but rather always in phase. A theorem is automatically evoked here, which was formulated in quantum physics at the same time that Baird was conducting his television experiments: Werner Heisenberg's uncertainty principle with regard to the speed and location of electrons at the moment of their (microtelevisual) observation. "Baird's solution to the problem was to send a synchronisation signal that changed the speed of the receiver's disc motor—speeding it up or down by tiny amounts"—a time-discrete rhythmization of image transmission.

> Most of the electronics of the new *televisor* provides synchronization with the CD signal. A circle of black and white stripes on the back of the disc pass in front of a tiny sensor which feeds back the position of the holes to control the speed of the motor and keep the disc in the correct position. As each track of the CD is played, the first thing you will see is a number countdown. This gives the nikpow disc time to synchronize each time. As it does so, the picture wobbles wildly—just like it did on the original when the synchronization was slightly out.

Such negative feedback (an inheritance of "operational research" in World War II) is the central time figure of cybernetic media systems. In this context, however, the chronotechnical temporal mode of television is "live" transmission. The chronotechnical is here understood as the smallest temporal units of rhythm (in the sense of the *chronoi* that music theorist Aristoxenus defined for prosody and dance): long pulses, short pulses, pauses. A challenge for the formation of media-theoretical concepts is to reactualize this approach for the operative algorithms of the digital.

The retina in the human eye processes a complete image impression in parallel; that is, it chemically converts the light impressions concentrated by the lens and conveys them electrically through neural pathways to the brain, where the signals are synthesized as image impressions. In the physical materiality of photography as "the pencil of nature" (Henry Fox Talbot), there is virtually a point-by-point similarity with respect to the photographed object; in contrast, electronic image scanning transforms its object into a signal sequence (that consequently corresponds more to the linearity of alphabetic writing). Instead of the coexistence of an ideally infinite number of image elements in space, image points are generated by the cathode ray as points in time. This is revealed by the quotation marks in descriptions of technological image disassembly and analysis for the purpose of transmission. "In reality, the position affected 'point-like' by the sensor can be infinitely extended. Its surface area determines the size of the raster element. This can be round or square-shaped."[50] With the "scanning finger" (McLuhan) of the cathode ray, the discrete symbolic (the logic of the alphabet) is replaced by a moving point, which is ideally extensionless but actually infinitesimally focused: a meteoric blurring—the transition from Cartesian space to the dynamic time field.

With phototelegraphy the allocation of the image to scanning times is determined by the feed rate per rotation of the cylinder, and it is precisely countable in numerical frequencies; however, "a certain degree of difficulty and arbitrariness exists in the differentiation of lines in 'image points.'"[51] The idea that "the image area can *theoretically* be broken up into a specific number of 'points' (raster elements), whose size, measured in the scanning direction, is consistently the same" is nothing but a heuristic fiction.[52] The technical economy of the time-critical suddenly appears evident: the wider the available frequency range, the shorter the time in which all of the imagined image points can be transmitted. This is exacerbated in the extreme case of a sudden transition from black to white, which corresponds to a jump in the photoelectrically generated "telephoto current" (Schröter) from zero to a maximum value.

> We shall thereby refrain from flattening the increase caused by the finite extension of the scanning spot of light. Such a surge of current can be represented mathematically through a Fourier integral. ... This integral can be interpreted as meaning that the function ... additively consists of an infinite number of partial oscillations. ... An infinitely wide frequency channel would thus be required to replay such a surge.[53]

In this moment, the process reaches the threshold of digital sampling. Digitization initially meant the time-discrete (or rather actively discretizing) scanning of an analog signal; the signal voltage is measured in periodic time intervals—in other words, it is clocked and thus temporalized. Quantization then occurs as the allocation of a binary value for the amplitude of the scanned signal moment in the "sample and hold" mode—a semiotic operation, as it represents the production of an isomorphy. The signal to be scanned is in itself not originally numerical, but rather another form of (implicit) mathematics.

During the digital conversion of analog television images, the individual elements of the video signal (in terms of color and brightness) are not immediately assigned numerical values; the electronic image signal is first broken down to its basic elements in actual "image points." Aristotle's hypothesis concerning the correlation between movement and number ("time" as the measure of movement) thus comes straight to the point. Time is not measured continuously here, but rather counted in pulses; the smaller the timing units, the more accurate the measurement. An analog time signal, on the other hand, actually moves *with the times* as time function $f(t)$, and it thus has a share in the essence of the physical world.

KAIROTIC MOMENTS: LINE BREAKS AND SCREEN CHANGES

Time-critical moments distinguish television and video electronics from the world of the acoustic radio broadcasting. Even though the writing of image lines in the cathode ray tube is continuous and amplitude-modulated, there are nevertheless two discontinuous moments when the electronic moving image approximates clocked computer technology: line skips and screen changes. The moment of a screen change is delicate when video editing cuts from one image source to another. This process involves calculating the absence of the image, and an interval switch ensures that an image changes at the precise moment of the blanking interval between two television images. Changeover delays must therefore always be less than 1/25 second—a dynamic *aisthesis* of absence.

However, the diverse norms of video frame rates require a separate time machine, the time base corrector (TBC), to facilitate technical communication between international visual cultures. Time is already critical in an etymological sense, as it means separating (*krisis*), cutting, clipping. While the borders of film frames on celluloid are still optically recognizable as intersections between frames, the electronically recorded image (signal) is no longer visible when the machine stops, as it remains in a latent state (albeit one that is entirely different in nature than the undeveloped photographic image). This temporal mode is also entirely different from the technomathematically controllable time of the videotape. The cue track of the magnetic tape carries a time code, through which the linearity of the reel is transformed into the mathematical logic of memory addressability. The individual images are literally numbered with the "time stamp" (SMPT code)—an entanglement of time and number.

The digital is already partially at work in electronic television through the line skip pulse, as the pulse-generating sawtooth signal introduces discreteness into image transmission. The Nipkow disk already put the electromechanical time scanning of images to work as sampling—not in the sense of the complete quantization of the image, but rather simply in the sense of the discrete line skip. As an electrotechnical event, television subverts the linear concept of time and breaks it quasi-typographically through the temporal ruptures of lines and pages (half-frames).

In order for the television event to succeed, however, the neurophysiological apparatus must act on people. McCulloch and Pitts conducted a series of tests on the optical scan process involving character reading (shape recognition). The apparatus used to conduct these selective reading tests resembled the schema of a layer in the visual center of the brain. On the basis of this analogy, a theory was discussed at the Macy Conference in 1947 that connected the anatomy and physiology of the visual center, and in this theory the operation of scanning over a set of transformations played an important part. "This scanning process involves a certain periodic time, which corresponds to what we call the 'time of sweep' in ordinary television. There are various anatomic clues to this time in the length of the chain of consecutive synapses necessary to run around one cycle of performance."[54] In this context, Wiener already formulated the idea of computational "cycling units."

> These yield a time of the order of a tenth of a second for a complete performance of the cycle of operations, and this is the approximate period of the so-called "alpha rhythm" of the brain. Finally, the alpha rhythm, on quite other evidence, has already been conjectured to be of visual origin and to be important in the process of form perception.[55]

Marshall McLuhan holds the introduction of the vocal alphabet responsible for this privileging of the sense of sight. The test series discussed by Wiener was actually inspired by the development of an apparatus that was supposed to enable the blind to read the printed page by ear. And this is what distinguishes optical character recognition in the sense of "scanning."

> The production of variable tones by type through the agency of a photocell . . . can be effected by any number of methods; the difficult point is to make the

pattern of the sound substantially the same when the pattern of the letters is given, whatever the size. This is a definite analogue of the problem of the perception of form, of *Gestalt*, which allows us to recognize a square as a square through a large number of changes of size and of orientation.[56]

The time-critical essence of the video and television image is of a philosophical as well as a concretely electrotechnical nature. Therein lies the function of the additional synchronizing pulse broadcast from every television station. Short electrical surges were turned into a clock pulse. "The 'line pulses' influence the line deflection unit so that the electron beam is constantly led to the beginning of the next line at the right moment. The 'image pulses' regulate the activity of the image deflection unit and initiate the punctual start of the next image."[57] If the electronic line break (which is generated in television just as it is in the oscilloscope through sawtooth pulses) is suspended, it is only a single line that pulses, which thus reveals for the first time the temporal essence of the electrotechnical image. Swiss video artist Jean Otth suspended the vertical deflection of a television object for this purpose in his 1974 work "Exercise IV from the Television Primer," and he thus aestheticized an image synchronization error. This media-surgical intervention is radically time-critical. On a media-archeological level—that is, in the oscillographic analysis of the manipulated circuit board—what appears on the screen as a pulsing phenomenon proves to be not simply a random permanent defect, but rather a conscious media-artistic act. Snapshots of the process of restaging this work hold this situation in a technological "double bind"—namely, (a) as a condition of hardware and (b) as a dynamic process—which manifests as photography and diagram. "(A) Reversing the manipulation with clamp and soldering iron to prove that it alone is the cause of the effect and not an additional defect. . . . (B) Oscillogram to document the manipulation. The signal is tapped in the circuit close to the manipulated position."[58] In teletext, the time-critical gap that occurs during a television screen change actively becomes an "intermezzo"[59] of media history. Teletext involves the transmission of data in the form of text and graphics, which are "piggybacked" on a television signal. The data transmission thus takes place during the return flow of the cathode ray from the lower right to the upper left corners of the screen ("blanking interval"). In contrast to the German Bildschirmtext system, however, there is no dialogical feedback channel to the sender of teletext.[60]

The texture of the electronic image recalls media-archeological case histories—and not only in comparison with the line-by-line reading of texts.[61] The electronic image also evokes the technology of mechanized weaving, in which the time-critical and thus technokairotic form of the line break is already familiar. Kairos and time are essentially related etymologically: the ancient Greek word *keiro* can be traced back to "cut" and the root **dai* to "separate."[62] In connection with weaving technology, this refers to "the delimited, designated section, the specific area, and thus also the correct position in space or in time."[63] Linear writing, typographic printing, and other writing mechanisms operate with the line break and the mechanical carriage return, including the enter or return key as the trigger of logical commands in the computer. Such mechanisms no longer lie primarily in the hands of human beings, but rather the human is (in the Heideggerian sense) "framed" (*gestellt*). Gilbert Simondon describes the consequence for human-machine relations: "Man is forced to learn a new function and to find for himself a position in the technical ensemble. . . . It was man who provisionally took the place of the machine before real technical individuals could be made."[64] A previously hand-crafted process was thus media-dynamically automated. Charles Babbage's analytical engine design foresaw programmability on the basis of punched cards, and it was explicitly inspired by Joseph Marie Jacquard's pattern control mechanism for textiles.[65] A characteristic feature of the becoming-media (*Medienwerden*) of machines—namely, the gradual emancipation of their operation from individual human participation—here shifts into the temporal domain of logical processes as the automation of the kairotic moment. Once the data is fed into the machine, it is processed by virtue of the variable or even recursive proper temporality of the mechanism. This corresponds to the so-called "Murray code" in wireless telegraphy (International Telegraph Alphabet No. 2), in which—in contrast to the continuous recording of symbols on paper strips—the line break of the writing machine also became part of the electronic signal transmission as a control character.[66] If the kairotic moment is wrested from the mythographic imaginary and identified as a *terminus technicus*, it constitutes the actual operativity of weaving: the entanglement of linearity and inscription (warp) with intervals and breaks (woof). A media-specific process was prefigured in this proper logic since antiquity, and it was later recursively "revived" by the typewriter.[67] The associated media-temporal chronologic replaces the historiographic concept of technological development with concepts like reconfiguration and reentry.[68] The terms borrowed from the domain of textile technology to describe contemporary methods of image and data processing thus do not simply represent anachronistic metaphors,[69] but rather a media-archeological anamnesis (as opposed to technical-historical forgetting)—the insistence of a genuine media-temporal signifier.[70] When *kairos* knits the woof threads to the linear threads in order to form the fabric, this effectively replaces historiographic narration, whose linear concept of time depends on continuous writing, with time-discrete word processing. The ancient Greeks invented Kairos in order to solve "the dilemma resulting from chronology as the dominant time mode."[71] The temporal-logical linking of elements weaves another, ahistorical model of knowledge. From the very beginning, the weaving loom included the operative textuality that is always expressed in new and at the same time equiprimordial processes—from early modern machines to digital computers. It is also expressed in the electronic image sequences of classical video technology. "The divisions into lines and frames are . . . the opening and closing of temporal windows that demarcate periods of activity within the flowing stream of electrons."[72] The video signal succeeds as an image only because it circumvents the human ability to distinguish such critical time intervals.

TELEVISION IMAGES AND THE FLAT TEMPORALITY OF THE SCREEN

In Claude Shannon's model of communication, the transmission interval also counts as an actual *medium*—that is, a media-technical channel. "Live" television appears to collapse the image and the event, as the electrotechnical bridging of spatial intervals is imperceptible to the human sense of time. The consecutive time signals are disassembled on the surface of the screen to form an image, yet the gaps between the image lines are the actual real in the television image, and they are even more radical than those between the frames in film. As the space is broken up, temporality breaks in. The television image is only apparently iconic, as it essentially consists of time; the crucial point here is the focused cathode ray. The television does not know any real point in time; the electronically advancing focus displays every moment in its difference. The individual image points only became discretely addressable by the computer in the form of the fully digitized liquid crystal display. This rupture in the microtemporal event field also brought about a reconsideration of previous macrotemporal categories. "If the traditional construct of the continuous course of history appears in discontinuous jumps from present field to present field, from time window to time window, then this temporal segmentation is impelled technologically by the dot matrix image of television to the extreme of point jumps in the nanosecond range."[73] Insofar as media-technological images only occur in implementation—in contrast to works of fine art—time is actually their *a priori* (following Kant): the condition of possibility of being-media (*Mediendasein*). However, this involves a flat temporality. Goethe's drama *Faust* describes "past times, in which the present is reflected"; this mirror metaphor is still based on the emphatic discourse of history, as the present is fleeting. The screen represents a reversal of this time aesthetic: the electronic interface makes permanent signals *present*, even if they were recorded in the past, so the *process* at the level of signifiers dominates the signified "past."

The fact that electronic images are time-based is already announced in the concept of the technical *retrieval* of the recorded and transmitted image as a synchronous likeness on the receiver side—in contrast to genuinely active digital imaging. Electronic light pulses are thus only "images" for human cognition; from the perspective of the apparatus, they are pure time objects. An apparently focused yet microscopically diffuse point of light emerges at the place where the electron bundle emitted by the cathode ray tube strikes the inner surface of the screen, which is coated with a fluorescent substance. An image is successfully retrieved when the strength and movement of the electron bundle corresponds precisely to that of the electron bundle "in the analyzing tube."[74]

> There would be no television image to see if the eyes could still distinguish individual image lines and pixels in 1/25 second. There would be no music to hear on the CD if the ears could perceive purely discrete amplitude values at a sampling frequency of 43 kHz. . . . Every interface circumvents perceptual thresholds and only bears the name interface out of pure mockery.[75]

Chronophotography first put mechanically clocked movement analysis to work, before it was standardized as cinematography at 24 frames per second. In electronics, however, these clock pulses become a time field. The technical movement of cathode ray oscillography still seems to correspond to the reading direction of texts.[76] From a technical (internal) perspective, the cultural gap between text and image is thus surpassed.

> Just as the measurement of time and technical movement through the train, punctual departures and arrivals, and the dissolution of the landscape in between constituted the central acceleration experiences in the nineteenth century, the technical (time) management of image points and their reassembly (which tends to be random) constitute our central disruptions in the present—even though they are imperceptible.[77]

The fact that electronic images are time-based is revealed in the chronotechnics of their transmission.[78] Unsuccessful synchronization between the sending and receiving units leads to temporal anamorphoses and image distortions.

> A televisual image has to be established and sustained onscreen moment by moment. With transmission, images and sets of images pass the time and fill out the current: in this sense television is always "live". . . . Scanning cannot deliver an image all at once—its composition is always in process, and a "stable" frame can be instantaneously switched midway through. Although pixels can retain luminosity long enough to await the next scanning cycle and

thereby approximate the succession of discrete filmic images, the fact that no image is ever constituted entirely in a single instant grants television a range of technical options for framing and editing, including incision and torque of the image's surface.[79]

Information theory is able to grasp such media-temporal moments mathematically and concretely: the intensity of a dot-like moment of light (x, y) on a cathode ray screen at time t.[80] The analysis of image content from this close-up perspective passes over to media archeology. Television analysis can examine broadcasts not only discursively but also in terms of the concrete media event itself. In 2001, Juliane Zelwies, a student at the Berlin University of the Arts, designed an oversized screen for the street facade of a Berlin television studio (ARD). The extremely decelerated movement of the cathode ray was supposed to disassemble current news images. What the transmitted signal writes lightning-fast as image was thus supposed to become decipherable to the eye, which reveals that "breaking news" is actually—at the technical level of the line break—the resolution of the output screen in its time-critical condition. In order to make it perceptible as the historiography of the time event, the decelerated cathode rays must necessarily be simulated, as a process from analog image electronics only becomes representable through its digital successor. However, this algorithmized world is no longer based on the linear "historiographic" unfolding of time events, but rather it occurs as logical time.

According to Hugo Münsterberg, the nonlinear jumps that already constituted the time modes of film after the introduction of montage also correspond to the neuronal memory mechanisms in human themselves; instead of objective real time, techno-imaginary time is privileged here. Pudovkin's plea for filmic montage in the early 1940s shows how the culture of discontinuity was liberated from the primacy of narrative. "Separations, omissions of various kinds, sometimes measured according to minutes and meters, sometimes according to thousands of kilometers and dozens of years. Separations and omissions penetrate deeply. What is apparently the simplest movement or action of an actor can prove to be divided up into separate parts."[81] In electronic mass media, such as television, such technical timing culminates no longer in the mechanical manipulation of montage on celluloid, but rather in the immediacy of electronic signals. Media-technical timers have long generated social time. In contrast to film, broadcast media (radio and television) also enable real-time participation in a distant event—the aesthetic of the "live." The literally "spark-like" quality of electronic media lies in the suddenness of electronic time,[82] and the media-archeological basis of this temporal *dispositif* lies in the dynamic essence of the electromagnetic world itself.

The dramaturgical equivalent of line breaks and screen changes can be found in the familiar form of the "cliffhanger"—the abrupt and suspenseful interruption of the narrative arc at the end of each installment of a television series. What Lessing in his *Laocoon* treatise defined for works of fine art as the "pregnant moment," which motivates the observer to complete and finalize a perceived motif independently in his own temporal imagination, is here produced in the domain of moving images.[83]

TELE-VISUAL RELATIONS: TELEVISION AS TEMPORAL SUBLATION

In the updating and resetting of the cathode ray on the television screen, a temporal mode emerges that has nothing to do with phenomenological protention and retention. In electronic television, event, transmission, message, and observation are virtually identical in terms of their temporality, as the traditional intervals between writing and reading implode. "The sender and the receiver occupy the *same* point in time as the event of primary reality and its staging as medial message or media reality."[84] However, this interval compression, which is virtually programmatic for "live" television transmission, is only apparent (to humans) and not technological. Intermediate storage also operates on the most material of all image levels—that is, the image storage tube, which was based on the iconoscope patented by Zworykin in 1923. During an image period (1/25 second), the image stands still and triggers electrons, which are emitted by a line-shaped cathode ray. The temporary time-basis of medial images thus clearly illustrates the dimensional difference between technical and nontechnical images.

The event character of television lies not only on the semantic level but also on the surface of the image itself. In the electronic television image, time contracts to the dimension of a zero point. The television image is permanent reactualization, like the "refresh cycle" of a computer monitor, and the media-aesthetic event lies in its spatio-temporal momentariness—the disassembly and reassembly of the image. The electronic point is also regarded as a temporal contraction of an expansion to a *punctum* (almost) without past or future. The macronarrative historical perception of time is here circumvented on the subliminal level.

> The accelerated medial process thus disassembles all time events, which were conceptualized in the historical consciousness as continuous, into point elements, which are then reassembled like mosaics on the surfaces of monitors. This results no longer in linearly unfolding *"texts"* but rather frayed *"image surfaces,"* on which, to quote Benjamin, "the Then (*das Gewesene*) and the Now (*das Jetzt*) come into a constellation like a flash of lightning"—and then once again disintegrate.[85]

Broadcasting is to be understood here in the media-technical as well as dramaturgical or even emphatic sense, insofar as "the success of a television broadcast always still depends on the moment of the broadcast itself in spite of all preparations."[86] This electronically induced moment corresponds to the anarchival overspending of television on the present. This transitive moment became concretely apparent with the testing of the light pen as input medium on the computer screen; its further development was interrupted in favor of the mouse (thus peripheral cursor control), as it resulted in short circuits with the phosphorous on the ground glass.

An operative form of temporal sublation in television is the recovery of the earliest television images. Donald McLean describes a primal media-archeological scene that took place in early 1981.

> I had borrowed a documentary on audio LP disc from Harrow library. It was a light-hearted history of television narrated by the comedian, John Bird, and called "We seem to have lost the Picture". . . . It included something utterly fascinating. At one point, Bird introduces a strange sound, describing it as "Baird's brain-damaging buzz-saw." Sounding more like a swarm of angry bees, this was supposedly a recording of the vision signal from Baird's original 30-line television system. Finding this fragment was timely: I had just completed building the software to capture audio into my home-built computer and here was something to work on.[87]

High-tech experimental culture is contingent. What is most important is that the signals on Baird's Phonovision records, which were converted by the pickup into electromagnetic waves, were relayed not to a loudspeaker but rather an oscilloscope. The media-archeological epiphany then occurred in late 1981.

> The green flicker of the oscilloscope trace was difficult to decipher. I was looking at what was supposed to be a video signal. I could see that the waveform repeated in a slowly changing pattern every 80 milliseconds, and another pattern repeated within it. This was undoubtedly a signal from out of history: a 30-line television signal with a picture rate of 12 1/2 per second.[88]

What this illustrates is the media-technical rather than iconological essence of the image. In his 1950 lecture on "The Thing," Heidegger explicitly refers to electronic television when he notes that "the frantic abolition of all distances brings no nearness."[89] This could only have been written by someone who thinks through technology in a fundamentally philosophical way without taking into account the actual technical conditions of the apparatus.[90]

The errors that occurred during the reconstruction of Baird's Phonovision records were due above all to the phase and frequency of complex wave forms. These wave forms became analyzable with the help of Fourier analysis, which was able to break them down into their individual sinusoidal components; when recorded in this manner, they became technomathematically controllable. McLean describes the wondrous metamorphosis of constant image signals into information as tabulation and consequently as the diagrammatization of time. "The stream of numbers is created into a list of values that are stored in the computer as a data file holding the raw, unprocessed data. The signal is now . . . the starting point for digital signal and image processing."[91] The image can thus be analytically manipulated. "Line by line, the correction values plot out the profile of errors in the signal's timing."[92] A special form of media studies: when analog signals are recognized digitally, then the computer itself becomes an archeologist of events, which would otherwise remain—like Baird's gramophone video records—ruined monuments of television in the curiosity cabinet of technological artifacts. "Their latent images would remain unseen and the information imbedded in them would still be completely unknown"[93]—a technical use of unconcealment (*Unverborgenheit*) enriched by informatics. This is truly media-archeologically formulated, and (unlike Sigmund Freud in his psychoanalytical investigation of the unconscious) it resists the metaphor of archeology as excavation studies. "Unlike traditional archaeology, the artefacts are not embedded in layers of history but have existed in both private and public collections, largely ignored as curiosities."[94] Media time is latency time. Baird's Phonovision is thus not a "dead medium," in Bruce Sterling's sense, but rather an aggregate that awaits its media-technical implementation; in other words, it awaits its becoming-medium (*Mediumwerdung*)—a temporal form of existence that coincides with the electrotechnical act of induction.

RECORDED TIME AND THE ELECTRONIC TIME FIGURE OF THE "LIVE" BROADCAST

Unlike alphabetic writing and the snapshot, signal storage media like the phonograph are able to reproduce the flow of time itself. A recording of a landscape with flowing water on magnetic tape for television reproduces the entire physical ensemble of the flow of time. Since at least the introduction of magnetic recording, however, the spectator has not been able to determine whether television images are recordings or "live" broadcasts. "There is no recognizable difference between direct transmission and transmission on Ampex with an interval of two minutes."[95] Instead, the authorizing information must be given outside the image—*parergonally*, like the frame of the image—as an archival reference to the temporal distortion. The aesthetics of "live" television are based on a pact of trust between the sender and the receiver in the temporal domain—temporal "high fidelity." The public places its trust in the immediate signal transmission, even though it possibly involves a time-shifted video recording ("live on tape"). This represents a *temporalized* variant of the concept of the original.

> One can no longer distinguish, visually or aurally, between that which is reproduced and its reproduction. Indeed, one cannot even discern *that* or *when* reproduction or repetition, in the manifest sense of recording or replaying, is taking place. We must be informed whether or not what we are seeing is "live". . . . We cannot distinguish through our senses alone between what we take to be simply "alive" and what as reproduction, separated from its origin, is structurally posthumous . . . what Derrida has called the irreducible "iterability" of the mark.[96]

Electronic media thus circumvents the cultural competence to be able to distinguish unambiguously between historical time and the present. "Time is television's basis, its principle of structuration, as well as its persistent reference. The insistence of the temporal attribute may indeed be a characteristic of all systems of imaging enabled by mechanical or electronic reproduction."[97] Television connotes a form of image transmission whereby the observer does not notice that the images only consist of sections—as in cinematography—that are sequentially assembled,[98] and therefore they can no longer reasonably be called images. Although they still appear to be images, they are actually *time structures*.

> While in film each frame is actually a static image, the television image is continually moving, very much in the manner of the Bergsonian *durée*. The scanning beam is constantly trying to complete an always incomplete image. Even if the image on the screen seems at rest, it is structurally in motion. . . . While the film frame is a concrete record of the past, the television frame (when live) is a reflection of the living, constantly changing present. . . . The filmic event is largely *medium dependent*, while television in its essence (live) is largely *event dependent*.[99]

In other words, it is closer to the time-critical essence of the electrophysical signal than that of the conventional image. There is no substantial "in itself" (*an sich*) of the television image—not even at the level of magnetic storage, where it is fixed in the lines of the video recording.

In the temporal disassembly and transmission of image lines, media time operates at the level of the signifier itself. In contrast, television programs represent a symbolic temporal order, and television time here becomes a signified. Even though it consists of abrupt sequences, the classic television program is based on the maintenance of a "flow" (as diagnosed by

Raymond Williams) that completely absorbs the attention of the spectator—a television-syntactic equivalent to the flow of electric current and thus to the physics of the medium. "Flow" undermines the discrete, but "breaking news" interrupts the program schedule; "live" television itself becomes an event. Siegfried Kracauer has already described the affinity between motion pictures and the "river" of life or the "stream" of time,[100] but in television this metaphor is actually electrotechnical. On the other hand, the regular interruption of broadcast formats through commercial advertising (and self-promotion) forms a dialectical relationship with the actual program. Television thus generates a genuinely immanent temporal technology as a cultural form. Literary narrativity is replaced by program flow as the overriding time pattern of broadcasts and their intervals, which was already deconstructed by the recipients themselves as "channel switching" through the invention of the remote control—unlike theater and cinema programs, which are still perceived as self-contained. The discretization of programs was internalized in computer programming—in the algorithm as literal "sequential unities." Beginning with John von Neumann, the aesthetics of programming actually yielded the "flow chart," which symbolically represents logical operations. "The replacement of a programme series of timed sequential units by a flow series of differently related units in which the timing, though real, is undeclared, and in which the real internal organisation is something other than the declared organization."[101]

FROM WRITTEN TO CALCULATED TIME IMAGES: ANALOG AND DIGITAL TELEVISION

In electronic image media, time is prior to space. Electrotechnical images are not only generally time-based but also concretely woven out of points in time and written through sequential scanning (circularly in radar, horizontally in television). Each part of the image corresponds to a different point in time; the temporal form of the radar screen is closer to a gramophone record than a classical panel painting, and it thus pushes the boundaries of the concept of the image itself.[102] There is still talk of quasi-stationary "images" only due to the inertia of our cultural semantics and optical physiology.[103] The dynamic radar screen, which operates in real time, was employed in the networked anti-rocket defense system of the United States as the interface of the Whirlwind computer. It did not produce photographic or filmic recordings of the military situation like those supplied by reconnaissance aircraft, which first needed to be developed and were always already outdated, but rather electronic signals, which needed to be interpreted without delay. The actual event thus became the image itself as the line-by-line or concentric overwriting of the static, unchanging object in order to focus instead on enemy movement. The military paradigm requires not only rapid information processing but also rapid reactions (computer games *avant la lettre*), which led to the parallel development of direct input media like the "light pen," whose electronic marks on the screen represented a new form of time writing.

In addition to intermittent screen changes as a condition of the cinematographic effect, every electronic television image (field) is produced from line skips, interlacing, and refresh cycles. Unlike the diaphanous projection medium of celluloid, the luminescence of the phosphorous on the inner side of the screen is employed. However, only completely computerized images can be chronopoetically designed by human programming. In Jim Campbell's 2001 LED-based media art installation "Running, Falling, Cut," the screen is divided into two time-delayed halves. "Because the cycle time of the clock variation . . . is different from the loop time of the image . . . the relationship between the two parts takes somewhere between a few months and a few years to repeat itself."[104] In connection with Campbell's work, Richard Shiff concludes: "I am evoking a time beyond time."[105] In 1948, Clement Greenberg already identified "the dissolution of the picture into sheer texture" in contemporary painting—the aesthetic consequence of the nuclear era. Hierarchical levels were replaced with "the more and the less immediate"; the medium itself became a transitive statement.[106] This transitional zone between matter and the present was already evident in Claude Monet's oil painting "Stormy Sea" (1884), in which water and air (clouds) break up into zigzag brushstrokes. "His wavy stroke relates more to the position of the canvas edge than to any logic of representation. . . . Monet's zigzag just happens."[107] It is no different from the television or video screen. "The 'snow' effect actively represents the medium as it represents itself in its sensate condition of being 'on,' responding to the ambient electronic environment"[108]—a molecular movement whose ergodicity eludes all historical time.

The operating point of the cathode ray is photographically captured on television through ultra-short exposures or through the superimposition of the monitor with a screened perforated foil that is photographed to produce image sequences composed of discrete lines—Zeno's paradox, the object for infinitesimal calculus (integrals). The delicacy of time-critical processes finds its complete expression in the electronic image. In contrast to the electronic image point, however, the pixel is not connected to the line or the frame, but rather a basic unit of information in and of itself. Proximities are here of a topological rather than physical nature.[109] The classical variant of the digital computer as a single-processor system essentially has a single data channel at its command. Taking this into account, the stored image values from the graphic processor must be read consecutively; the digital image is already time at the operative level of its occurrence. Computed images are thus also images of the computer itself[110]—with all its consequences for the media-cultural world picture (*Weltbild*) and the chronologically calculated time image (*Zeitbild*).

THE VIDEO (TIME) IMAGE: TIME ERROR CORRECTION, TIME CODE, AND ELECTRONIC EDITING

The focus on the specific temporal modes of technical media results in increased attention on new time arts, which are associated with a time-critically intensified concept of mimesis: the image worlds of electronic media (television and video) do not imitate the world, but rather time. The concept of identity in time-critical difference (to borrow loosely from Hegel) applies to the electronic television and video image. Something becomes identifiable that nevertheless always remains elusive in the smallest time-critical moment of the cathode ray. The concept of existence as absolute movement, which means that the river of the world never flows through the same situation twice, was first formulated in a fragment of the pre-Socratic philosopher Heraclitus and later continued in the process philosophy of Alfred North Whitehead. This river has since become electrotechnically concrete. The video image is actually a standing wave pattern of electrical energy—an oscillation system consisting of cascades of frequencies. What people see on the cathode ray tube is the trace of a single, moving, focused point of light from an electron bundle. "In video, a still image does not exist,"[111] which is evident from the flickering of every freeze frame when the PAUSE button is pressed. That is precisely why such images are called *time-based*. There is actually no such thing as a video still; the source of this stationary movement as "steady state" is "the activated constantly sweeping electron beam—the steady stream of electrical impulses coming from the camera or video recorder"[112] (no matter whether the signal is "live" or recorded). At this media-archeological level, the "streaming" of images is closer in essence to electrical current than the iconic concept of the image.

The time base corrector (TBC) is the technical form of videographically active time critique, which is employed to keep signals in sync. In real electrophysics, time comes into play as a worldly veto; mechanical imprecision sometimes allows the signal to arrive temporarily faster or slower. Flawless synchronization between video signals is only possible through an external reference signal. When using a TBC, the associated audio signal must also be delayed by the same amount of time in order to avoid asynchronicity between image and sound. In the time-critical domain, the so-called "audiovisuality" of electronic media falls apart. Before there was digital buffering through

TBCs, analog delay lines were employed that were essentially the same as dynamic delay time memory in early computers. With the help of electronic buffering, the time base of video signals could be manipulated. The option to stretch or compress time differs from cinematographic "fast motion" or "slow motion" in that it is not mechanical but rather differential; it thus represents a practice of memory-augmented presence.[113]

The "temporal rate," which defines the representation of movement, provides an appropriate concept to explain the time axis of video. Image flicker is a function of the physiological time resolution in the human eye (afterimage effect) as well as a specific imaging technology. In contrast to cinematographic projection or cathode ray tubes, the image impression on pixel-based flat screens is kept stable through "sample-and-hold" technology. The translation of different frame rates between different formats (film, video, and digital image) requires a high degree of skill.

When the videographic sampling rate is not sufficient, it gives rise to visual artifacts as "temporal aliasing." Microtechnology here formulates an epistemogeneous concept of time. Mathematical intelligence could only be employed to correct time errors once every image point was made discretely addressable through digitization—such as through time-critical interpolation. In video technology, time errors initially included "temporal fluctuations of the stored signal, which altered the reproduction of the line structure on the screen in the horizontal direction."[114] The phenomenon of time is only revealed in deviations from the natural, and the essence of technical time only manifests in its withdrawal—as deviation. Video technology responds to this with the velocity compensator, which operates with controllable pulse-forming networks. However, analog timing compensators are no longer sufficient for magnetic helical scanning, as a storage capacity of 20 lines is needed. "Such requirements can only be fulfilled with the help of digital technology."[115] Through the inclusion of a clock generator, the now quantized video signal becomes computable and thus chronopoetically correctable—a discrete temporalization of the previous oscillation event known as the image. The entanglement of signal transmission and storage in the technical concept of *stored transmission*[116] technically sublates what until now was an oxymoron in the cultural semantics of time.

DIGITAL IMAGES AND STREAMING VIDEO—TIMELESS?

There is "high fidelity" not only for binaural acoustics, but also for time-critical transmission processes. In television, analog "live" transmission takes place in real time, while digital transmission involves a delay, as the images must first be calculated in (real) time. It makes a difference whether the tempor(e)ality of electronic images is defined as time-based or as a function of time-critical computing processes. In the current media culture, "time-based media" essentially refers to film and video. The classical ontology of the concept of the image in painting was static, but "the form of being (*Seinsform*) of media art is the temporal form of a dynamic system. Instead of a static concept of being, tech-art constructs a dynamic (interactive) concept of state (*Zustandsbegriff*)."[117] For genuinely electronic images, however, this is operative and not simply phenomenological. "*Represented time* and the *time of representation* can no longer be separated."[118]

As mathematical functions in calculating space, digital images finally lose the representational quality that still belonged to photography in favor of a fundamentally different indexicality that is based on software and codes and that calls for a discussion of the arbitrariness of the digital world picture (or time image) as opposed to the analog-physical concept of the world. Sampled, and thus technically interpreted, they become functions of an interactive process and their concrete processualization transforms the classical pictorial narrative into the discrete act of counting. What the digital image loses in terms of indexicality, as compared with the photographic image, returns at the temporal level as a hyperlink in time. The classical homogeneity of the visual, which was provided by its two-dimensionality, implodes in this time; instead, the visual obtains coherence through the time axis. Mike Figgis's film *Timecode* (2000), which splits the screen itself into parallel time scenes, represents a decidedly time-based form of visualization. The temporal unfolding of the sound alone compensates the observer for the loss in optical coherence.

There is a world of difference between the analog recording of electronic images and their technomathematical coding or between the electronic video image and the alphanumerical image matrix. Digital image time is no longer perceptible in linear temporal processes; instead, digital images represent images of *logical time*, which are only images in a topological sense (mapping). Digital images are basically measurements of light, which result from the conversion of light-analog voltages into discrete data. Ludger Hennig and Felix Sattler's 2001 media installation "tidalCURRENT," which was exhibited in connection with the "interactive/narrative" conference at the Bauhaus University in 2001, subsumes its self-description under the title "Points in Time: Frame Goes Pixel." Digital images are created from the signals of a continuously filmed situation, and these images are transphotographic in that the individual image points represent not only a complete image but also all of the different points in time of this situation—the accumulation of time in an image without its own temporal extension. A new intermediate state between photographic, filmic, and computer images thus emerges, and it is grounded in the time channel itself.

When image sequence processing takes place in digital space and shift operators and vectors come into play, the smallest time-economic differences between the past and the present are expressed as Δt. Acoustic microtemporality and optical microspatiality differentiate themselves here. While an image pixel is surrounded by neighbors in all spatial axes, in digital sampling every sound sequence only needs to be analyzed on the single time axis—namely, as moments in the millisecond range between immediate predecessors and successors. An observed value t thus knows what its value was at time $t-1$ and what its value will be at time $t+1$; the isolated moment here transforms (entirely unhistorically) into a microtemporal horizon.

The digital coding of "streaming media," in which images and sounds are transmitted as numerical values, follows a time-critical logic, as it has no other purpose "than to be sent quicker and packaged smaller."[119] Such transmissions in the Internet seek to remain close to the time window that human perception understands as the real time of movement through the interposition of an algorithmic dynamic—namely, effective compression. The goal of video compression technologies is "to reduce the spatial redundancy among the picture elements and to reduce the temporal redundancy between successive frames, i.e. interframe coding";[120] their dominant method is "predictive coding." A microtemporal world thus presents itself in the time-critical horizon. Media archeology also seeks to produce media-epistemological and time-theoretical sparks from high-tech relations and their engineering description.

So-called "codecs" are employed here, such as MPEG for video. They do not represent electrotechnical modules, but rather algorithmic software within the framework of a computer-technical method that enters into the temporicity of the data to be processed from the physical world: digital signal processing (DSP). "MPEG-2 defines a bitstream that tries to reconcile the complicated psychophysical, technocultural, and political-economic processes of seeing."[121] It also reveals how the computer manages time. "Motion video can be manipulated as a form of computer data."[122] This time operation can be formulated using the concepts of informatics. "Algorithmically, MPEG-2 combines several distinct compression techniques," including "converting signals from time-domain to frequency domain using discrete cosine transforms, quantization."[123] Researchers in the field of software studies discuss these algorithms with the necessary technomathematical knowledge, but they seldom address their epistemological implications, such as the question of their temporal essence and the temporal modes that they produce. The media-archeological gaze (at images) takes place here, as it moves away from cultural iconology and turns instead to the techno*aisthesis* of the computer medium. This requires, first of all, a description of what algorithmic processing does at the level of pixel blocks, which is the lowest level of the digital image and thus the media-archeological level. What appears to be uniform to humans from an iconological perspective is calculated by the information aesthetics of the digital computer as a dynamic form of time. "Nearly all video codecs transform spatially extended images into sets of simple frequencies."[124] This temporalization is a form of implicit sonification, although this may seem counterintuitive at first glance. "In what way can a videoframe be seen as a waveform?"[125] A concept of the sonic comes into play here that is completely detached from acoustics and instead refers to the quasi-musical temporal mode of sound.

What ultimately takes place in "streaming media" is a time figure that is already familiar from Husserl's phenomenology of the "internal consciousness" and technomathematically-calculated air defense in World War II. With MPEG-2, internal data compression is joined by "motion prediction." "Interpicture motion prediction compression relies on forward and backward correlations, . . . in particular on the calculation of motion vectors for blocks."[126] Such vectors pluralize the traditional concept of cinematography. Mathematical and thus calculable time arrows shatter the movement image (to freely borrow Gilles Deleuze's terminology) into time crystals. The underlying algorithms thus represent not only a pure mathematical problem but also a challenge to the traditional cultural concept of the image, as they transform images into time functions in calculating space.

VIDEO ART: IN SEARCH OF ANOTHER TEMPORALITY

Unlike celluloid film, whose images are fixed and transparent in photochemical emulsion, the video image lies on the tape as in latency—a concrete virtuality that only becomes an image in the present through the implementation of electromagnetic induction. In *Creative Evolution* (1907), Bergson discusses the strengths and limitations of calculating time with a mathematics of pure symbols: differential calculus is able to analyze movement, but it misses the essence of time. "In the smallest discernible fraction of a second, in the almost instantaneous perception of a sensible quality, there may be trillions of oscillations which repeat themselves. The permanence of a sensible quality consists in this repetition of movements."[127] The perception of the present thus proves to be the operative processing of high-frequency oscillations, which human optics condenses into a perceptual impression in fractions of a second. Consciousness *integrates* these oscillations; in this respect, perception *is* radio reception.

> While film still affects human perception with the comparably harmless frequency of 24 frames per second, video and television images occur in the high-frequency range where the traditional difference between light and matter is

cancelled out. If perception is conceived as a function of timing (an insight of nineteenth-century physiology), then the world is also experienced by different beings—and different apparatuses—as temporally accelerated or decelerated.[128]

Consciousness is never solely in the present; rather, the accumulation of past image impressions and remembered sounds modulates the perception of the present. On this basis, it is possible to extrapolate the now-past in view of what will be; this neuronal form of autocorrelation has long since become a digital method in the algorithms of search engines. Walter Benjamin also articulated these thoughts in his *Arcades Project*. "It's not that what is past casts its light on what is present, or what is present its light on what is past; rather, image is that wherein what has been comes together in a flash with the now to form a constellation."[129] The concept of perception as recording and storage (the photographic and phonographic *dispositif*) transforms into true dynamics. The analogy with electronic image media now comes into play—not as a media-historical but as rather a technical-epistemological consequence. "Video technology and digital recording operate first and foremost like the Bergsonian brain, which means like a switch that introduces a gap and a change in the flow of currents"[130]—just as the signal-carrying control grid of an electron tube modulates an asignifying current.

Video technology is able to address the human perception of time in a way. States of consciousness are not only dramaturgically represented but also genuinely reproduced, as this temporality lies in the essence of the electronic image itself. The inner time consciousness of humans is, so to speak, videographically collected in its own medium. In the electronic medium, media-induced time actually means time fields, which are no longer reduced to a linear time dimension but instead enable a relativistic entanglement of spatiality (in Minkowski's terms). Peter Weibel, a theoretician and practitioner of video art, claims that Lessing's distinction between arts organized primarily on surfaces and those organized primarily in time is explicitly "sublated" in video environments.[131] When Gerry Schum established the Television Gallery (1968–1969) and exhibited the broadcast *Land Art* (38 min.), the emphasis was on the medium in its implementation. A time-based artform was here technically at work: "The Television Gallery is more or less only a mental institution, which only becomes real in the moment of the television broadcast."[132] Video enables the coupling of the televisual present and past time. Richard Kriesche staged this in his installation *Twins* (1977), which placed the observer in two identical rooms and made him unsure of the temporal reality he was confronting: "the present, the past of the twins, or the simultaneity of the video image."[133]

Audiovisual media time requires a dynamic method of reading. In contrast to symbolic fixing (the sublation of current events through writing), electronic recording (magnetic tape and video) performs fixing in the physical real. Even in its digitized and computerized form, electronic signal sublation always occurs as a physical event. The operative coupling of the electron flow and mathematics is only possible in electronic space, which allows analysis and synthesis to take place in the same medium. Although a phonographic sound trace could also be analyzed mathematically using mechanical measuring devices (like the harmonic analyzer), the difference between mechanical and electronic or electromathematical analysis is the time-critical dimension (and real-time option).

Cinematography on celluloid must first chemically crystallize and permanently fix light radiation before it is able to cast light-suffused shadow images on the screen. With electromagnetic images, on the other hand, there is no essential difference between storage and "live" transmission. If the signal generator switches from "live" camera recording to play back from videotape, the character of the monitor image (television) does not change. On the one hand, a camera image transmitted in the present indexically corresponds to the event taking place, and it thus refers to an immediate reality and massages human perception on the neuronal level of the senses. On the other hand, this presence still remains in latency when it is stored on tape. In other words, latency time has two extremes: "live" transmission and remanent storage. A spatio-temporal entanglement here comes into play ("live" transmission bridges distance in space, storage bridges distance in time). It is useful to conceive of both extremes as equiprimordial in terms of a differential function.

The storage metaphor does not apply to the neuronal image that occurs in the imagination of the observer. The brain stores visually perceived reality impressions not as discrete sequences, like a cinematographic roll of film, but rather as a constantly new bringing-forth (*Hervorbringen*), re-production. Proto-visual information is distributed in the neurons, and it can be reconfigured back into image impressions through similarity-based associations—a (quasi-algorithmic) schematism that autopoietically brings forth images that are seen again after external stimulation (or dream-like association). This is an essentially genuine media-dynamic process: the concept of a "historical" past is foreign to technical apparatuses, like magnetic tape, as they persistently bring forth the signal event anew. This does not emerge out of nothing, but rather it requires actual reassembly. The concept of electromagnetic induction (shaped by Faraday on the basis of phenomenology and then mathematically calculated by Maxwell as the temporal asymmetry of forces in the electromagnetic field) thus appertains to the description of the relationship between the actual present and the apparent past (which is simultaneously existential and differential).

Electrotechnical and human dynamics are entangled in the so-called "video-dance."[134] In this art-technical form, dilatory time and time axis manipulation are employed as genuine options of electronic space in order to create movements whose temporal form could only occur in the medium of video due to the physical limitations of real stages: compression and acceleration, "fading in" and "fading out." In principle, these temporal games already began in early photography with time exposures, in which moving people were only visible as faint shadows or streaks compared to static objects (such as architecture).[135]

MICROTEMPORAL AND MACROTEMPORAL SHIFTS ON VIDEOTAPE

From a media-archeological perspective, the video medium is more than simply a technical object system.[136] "In the close interplay between work and leisure, the video recorder can be interpreted as a culture technique."[137] The implementation of Ampex's quadruplex system as the "time-shift machine" for North American television networks made it possible to organize the flow of audiovisual messages as the coordination of different time zones, so that it could be synchronized with the rigidly structured daily processes of viewers across the country. In this context, the definition of media in systems theory is plausible: "We would like to call media the evolutionary achievements that enter at those possible breaks in communication and that serve in a functionally adequate way to transform what is improbable into what is probable."[138] Video transmission between time zones compensates for the time delay, and it is thus not only time-based but also genuinely time-setting. This delay shrinks to zero as soon as the physical time in networked communication is replaced by the logical time of the computer (what Heidegger identified as the explicitly "vulgar" time of the clock).[139] The image components are transmitted between networked computers in the uncompressed binary format and then reassembled in the target computer.[140] The speed limits of classical video coding, which were limited to pure video signals and created signal delays, are exceeded with algorithmic intelligence. Digital, technomathematical time outsmarts historical-physical time. This electrotechnical outsmarting of linear time was already looming when Ampex decided to inscribe the magnetic tape used in the Mark IV (the prototypical video recording machine) no longer lengthwise but rather crosswise (the lengthwise direction was reserved for the sound and sync signals); the multiple pickups no longer behaved passively, as with traditional tape, but rather they were set in motion through rotation contrary to the transport direction of the tape. Tape speed was here replaced by a differential, and the electronic image was *derived* in the act of recording and playback in the sense of mathematical infinitesimal calculation.[141]

While the spectator experiences video as ordinarily consumed media time, its condition of possibility actually lies on the time-critical microlevel of electronic media and its form of time axis manipulation is almost infinitely scalable, such as through the potentiometer in the analog computer called the video synthesizer. Nam June Paik described the time regime of video art using the cybernetic concepts of "input time" and "output time," which are connected by a black box of time transformation. Paik also recalled that Norbert Wiener developed his notorious distinction between "Newtonian" (reversible) time and "Bergsonian" (irreversible) time not as a metaphysical speculation, but rather as "a micro two-way enveloping-time analysis" in connection with the most tangible events ("in his design of the Radar system").[142] Paik carries this process over to video art, which does not passively replay an unchanging form of time delay, but rather actively edits time itself as the actual message of the medium. "There is not really space in the electronic picture and everything is time."[143] Movement is thus not simply an object of technical recording, but rather it is simultaneously brought forth. The observer always also experiences the reflex of his own immanent temporal modes. Paik adopts the black box model of a process from cybernetics and thereby focuses on its time events: "Certain input-time can be extended or compressed in output-time at will . . . and this metamorphosis . . . is the very function of our brain, which is, in computer terms, the *central processing unit* itself. The painstaking process of editing is nothing but the simulation of this brain function."[144]

FREE FLOATING? TIME *IN* AND *AS* MEDIA ART

The Fluxus art movement defined aesthetic objects primarily through their "temporal values." The minimalistic sculpture of the 1960s also emphasized the temporal moment, in which they revealed their processuality to the observer.[145] Paik sought to make the representation of passing time visible; the affine medium for this was—after his experiments with electronic music—the temporal mode of video. Dan Graham's video installation "Present—Continuous—Past(s)" (1974) was an experimental arrangement designed to demonstrate temporal invariance by means of video in its double role as both a transmission and recording medium. Graham staged the time-shifted playback of the mirror image of an observer, whose sense of presence was thus consciously disrupted as he moved through the rooms of the gallery. This *mise en abîme* of time-shifted presence was based on a technically-produced time delay of 8 seconds, which outlasted the 3 seconds that humans generally perceive as the time window of the present moment. Time, which was neglected by the fine arts for so long, entered the spatial arts not just symbolically but in real terms as the fourth dimension.[146] This difference was made possible not by a phantasmatic need, but rather by the technical medium itself; time only becomes truly cybernetic in electronics (in this case, through the "closed circuit" feedback loop of camera and monitor). Video art is a genuinely temporal artform precisely because it does not iconically imitate nature, like painting or sculpture (*mimesis*), but rather it equiprimordially brings forth its own inner time structure (*poiesis*). Video is also a medium of Bergsonian time by the grace of electronics, which allows it to manipulate the parameters of time images such that immediacy and delay are relativistically entangled with one another.[147] The possibilities of video editing come close to the processes of human optical perception, which is not simply a duplication of reality but always also the invocation of stored perceptual impressions. In the timelines of video editing, memory potentially appears behind every image or sound—a process that was expressed in Angela Melitopoulos's video *Passing Drama* (1999).[148]

Real time and delayed time—the two separate time forms of previous vision machines—are entangled in the electromagnetic event of the video image.[149] The video recorder sublates intervals of real time in magnetic latency, and they are once again brought forth equiprimordially at the moment of playback. It is irrelevant to the video image whether it takes place as a "live" transmission from the electronic camera or from magnetic tape. "Once on videotape, you are not allowed to die."[150] In the video *Merce by Merce by Paik* (1975), a video previously produced by choreographer Merce Cunningam is superimposed over Paik's own variations, like different oscillations of sound and time. A text overlay points to the theoretical connection: "Dance of time. Time reversible and irreversible."[151] This sentence has acquired an additional meaning since the actual death of Cunningham. Since the invention of the phonograph, the medial storage of real signal flows has made the unique event repeatable and manipulable, as time can be compressed and dilated. But which media time adheres to the materiality of the video itself? Media knowledge here requires techno(archeo)logical analysis. From a media-archeological perspective, the system time of the video recording is concretely connected to the materiality of the magnetic tape and the technical ensemble of the apparatus with its heterochrony of individual electronic elements. Storage media always adhere to the concrete proper time of their hardware, which explains their temporal oscillation between document and monument. Electromagnetic signals and data—no matter how fleeting and immaterial they may appear—are functions of the physics of their storage. When reconstructing early video art installations in the museum, defective device components cannot simply be replaced with new parts without clearly changing the message of the artwork and its effect. Is the relationship of the technology to the artwork like that of a frame to a painting? Is it also true of ancient video art that the individual technical components "should not contribute to the ideal substance of the artwork, but rather only have a functional meaning"?[152] What the aesthetic gaze of cultural semiotics overlooks is the readability of an artwork's technical conditions. In the world of electronics, signs are not only appearances with meanings but also physical characters, which are perceptible to the senses because they adhere to their materiality. Only material aesthetics addresses this with regard to technomathematical space, the acoustic sound of electronic music, but also the poetic word.[153]

"Media are constant, forms alternate," writes Niklas Luhmann.[154] Technical artworks also ensure the repeatability of observing operations and thus "the actualizability of what is momentarily nonactual"[155] through their material substrates. Audiovisual media art stands and falls with entropy—in the information-theoretical sense of communication as well as the physical-thermodynamic sense of their technical materiality. Video tapes are thus particularly transmission-critical (*überlieferungskritisch*), as they are especially sensitive information carriers; recording formats and playback systems are just as ephemeral, such as Betamax, VHS, and digital Betacam. Homer's *Odyssey*, on the other hand, can still be read equiprimordially after two and a half thousand years (albeit from printed works and not from ancient papyrus rolls), for the actual information units of this work are the coded letters, not their storage and transmission hardware. Alphabetically coded and thus symbol-technical migrations remain invariant, unlike the physical reality of storage media.

NOTES

1. "Geschlechtsorgan der Maschinen: *Playboy*-Interview mit Eric Norden," in *Absolute Marshall McLuhan*, ed. Martin Baltes and Rainer Höltschl (Freiburg: Orange Press, 2002), 47.

2. Zielinski, *Deep Time of the Media*, 31.

3. Ibid.

4. Roland Barthes, *Camera Lucida: Reflections on Photography*, trans. Richard Howard (New York: Hill & Wang, 1981), 80.

5. John Logie Baird, *Television and Me: The Memoirs of John Logie Baird*, ed. Malcolm Baird, (Edinburgh: mercatpress, 2004), 64; Baird is referring here to the earliest television image of a human face.

6. Ibid., 64f.

7. See Martin Gfeller, "Der Referenzgerätepool von *AktiveArchive* an der Hochschule der Künste Bern. Eine Basis für die historisch informierte Wiederaufführung von Medienkunst," in *Schweizer Videokunst der 1970er und 1980er Jahre. Eine Rekonstruktion*, ed. Irene Schubiger (Zürich: JRP Ringier, 2009), 212–21.

8. See R. G. Collingwood, *The Idea of History* (Oxford: Oxford University Press, 1946).

9. See Adolf Lumpe, "Der Terminus 'Prinzip' (*arché*) von den Vorsokratikern bis auf Aristoteles," *Archiv für Begriffsgeschichte* 1 (1955): 104–16.

10. Denis Guedj, *Numbers: The Universal Language* (London: Thames & Hudson, 1998), 97f.

11. See Marshall McLuhan and Quentin Fiore, *The Medium is the Massage* (New York: Bantam Books, 1967).

12. See Jean-François Lyotard, "Das Erhabene und die Avantgarde," *Merkur* 38 (1984): 151–64.

13. Kurt Lipfert, *Das Fernsehen. Eine allgemeinverständliche Darstellung des neuesten Standes der Fernsehtechnik* (Munich: Lehmann, 1938), 17.

14. Ibid., 8.

15. Hermann Weyl, "Zeitverhältnisse im Kosmos. Eigenzeit, gelebte Zeit und metaphysische Zeit," in *Proceedings of the Sixth International Congress of Philosophy*, ed. Edgar Sheffield Brightman (New York: Longmans, Green and Company, 1927), 58.

16. Jörg Pflüger, "Wo die Quantität in Qualität umschlägt. Notizen zum Verhältnis von Analogem und Digitalem," in *Hyperkult II. Zur Ortsbestimmung analoger und digitaler Medien*, ed. Martin Warnke, Wolfgang Coy, and Georg Christoph Tholen (Bielefeld: transcript, 2005), 48.

17. Lipfert, *Das Fernsehen*, 10.

18. Ibid.

19. Ibid., 7.

20. Lee DeForest, "The Phonofilm," *Transactions of the Society of Motion Picture Engineers* 16 (1923): 61–75.

21. Jan Philip Müller, "Synchronisation als Ton-Bild-Verhältnis," beta.see-thissound.at/kompendium/texte. See also Michel Chion, *Audio-Vision: Sound on Screen* (New York: Columbia University Press, 1990) and Dieter Daniels and Sandra Naumann, eds., *See this Sound: Audiovisuology* (Cologne: Walther König, 2011).

22. Vannevar Bush, "As We May Think," in *The Origins of Information Science*, ed. A. J. Meadows (London: Taylor Graham, 1987), 256.

23. Frank Hartmann, *Globale Medienkultur. Technik, Geschichte, Theorien* (Vienna: WUV, 2006), 153. See also Pierre Teilhard de Chardin, *Die Entstehung des Menschen* (Munich: Beck, 1961).

24. McLuhan, *Understanding Media*, 313.

25. László von Szalay, *Moderne Technik* (Berlin: Safari-Verlag, 1954), 568.

26. Ibid.

27. Götz Großklaus, "Das technische Bild der Wirklichkeit. Von der Nachahmung (Mimesis) zur Erzeugung (Simulation)," in *Zeichen(theorie) und Praxis*, ed. Michael Titzmann (Passau: Wissenschaftsverlag, 1993), 101.

28. Heinz Mann, *Fernsehtechnik, Band 1. Die physikalischen und technischen Grundlagen des Fernsehens* (Berlin: VEB Verlag Technik, 1962), 12.

29. S. W. Amos and D. C. Brikinshaw, *Television Engineering: Principles and Practices* (London: Iliffe & Sons, 1953), 1: 13.

30. Michael Bennett-Levy, *TV is King* (Musselburgh: MBL, 1994), 6.

31. Gerhard Lechenauer, *Videomachen. Technische Grundlagen, Geräte, Arbeitspraxis, Erfahrungsberichte* (Reinbek bei Hamburg: Rowohlt, *1979*), 132.

32. Donald McLean, *Restoring Baird's Image* (London: Institution of Electrical Engineers, 2000), 211f.

33. Translator's note: there is a play on words here in the original text, as the German words for "crucial point" (*springender Punkt*) literally mean "bouncing point." The electron is thus the most crucial element of electronic television as well as a literally "bouncing point."

34. See Denis Gabor, "Acoustical Quanta and the Theory of Hearing," *Nature* 4044 (May 1947): 591–94.

35. Heinrich Peters, "Über die Zeitdauer der Emissionserscheinungen bei Glimmlampen," *Fernsehen. Zeitschrift für Technik und Kultur des gesamten elektrischen Fernsehwesens* 1.7 (July 1930): 307.

36. Ibid.

37. Wilhelm Schrage, *Fernsehen* (Munich: Franz, 1930), 11.

38. Amazingly enough, this is also the principle by means of which electronic analog computers are able to calculate: derivatives according to time.

39. Dionys von Mihály, *Das elektrische Fernsehen und das Telehor* (Berlin: Krayn, 1923), 17.

40. See Axel Volmar, "Zeitkritische Medien im Kontext von Wahrnehmung, Kommunikation und Ästhetik," in *Zeitkritische Medien*, ed. Axel Volmar (Berlin: Kulturverlag Kadmos, 2009), 9–26.

41. Stefan Rieger, *Kybernetische Anthropologie. Eine Geschichte der Virtualität* (Frankfurt am Main: Suhrkamp, 2003), 143.

42. Max Bense, "Kybernetik oder die Metatechnik einer Maschine," in *Ausgewählte Schriften*, ed. Elisabeth Walter (Stuttgart: Metzler, 1998), 2: 440.

43. Translator's note: there is another play on words here in the original text, as the German word for "account" (*Rechnung*) also means "computation." By taking the temporality of the image into account, Wiener is thus also literally bringing it into the realm of computation.

44. Neil Postman, *Amusing Ourselves to Death: Public Discourse in the Age of Show Business* (London: Penguin, 2006), 78.

45. Eduard Rhein, *Wunder der Wellen. Rundfunk und Fernsehen dargestellt für jedermann* (Berlin: Deutscher Verlag, 1939), 228.

46. Ibid.

47. Ibid.

48. Pöppel, *Grenzen des Bewußtseins*, 43. See also Mirjam Schlemmer, "Audiovisuelle Wahrnehmung. Die Konkurrenz und Ergänzungssituation von Auge und Ohr bei zeitlicher und räumlicher Wahrnehmung," in *Musikpsychologie*, ed. Helga de la Motte-Haber and Günther Rötter (Laaber: Laaber-Verlag, 2005), 173–84.

49. Lechenauer, *Videomachen*, 123.

50. Fritz Schröter, "Die Zerlegungsmethode der Fernbildschrift," in *Handbuch der Bildtelegraphie und des Fernsehens. Grundlagen, Entwicklungsziele und Grenzen der elektrischen Bildfernübertragung*, ed. Fritz Schröter (Berlin: Julius Springer, 1932), 4.

51. Franz Tuczek, "Leitungsbildübertragung," in *Handbuch der Bildtelegraphie und des Fernsehens. Grundlagen, Entwicklungsziele und Grenzen der elektrischen Bildfernübertragung*, ed. Fritz Schröter (Berlin: Julius Springer, 1932), 388.

52. Ibid.

53. Ibid., 388f.
54. Wiener, *Cybernetics*, 23.
55. Ibid.
56. Ibid., 22.
57. Walter Conrad, *Fernsehen* (Leipzig: Urania, 1960), 61.
58. Johannes Gfeller, "Anmerkungen zum restauratorischen Hintergrund der Ausstellung," in *Schweizer Videokunst der 70er und 80er Jahre. Eine Rekonstruktion*, ed. Irene Schubiger (Zürich: JRP Ringier, 2009), 125.
59. This concept was borrowed from Siegfried Zielinski, *Audiovisionen. Kino und Fernsehen als Zwischenspiele der Geschichte* (Reinbek bei Hamburg: Rowohlt, 1989).
60. See Gerd Klawitter, *Funk-Lexikon. Begriffe aus der Funktechnik leichtverständlich erklärt* (Meckenheim: Siebel, 2001).
61. "Just as our eye reads a text word-by-word and line-by-line, so too does the 'eye' of the television transmitter—a fine beam of light—scan the image point-by-point and line-by-line." Hanns Günther, foreword to *Das grosse Fernsehbuch. Die Entwicklung des Fernsehens von den Grundlagen bis zum heutigen Stand, mit zahlreichen Versuchs- und Bauanleitungen*, ed. Hanns Günther (Stuttgart: Franckh, 1938), 4.
62. See Ellen Harlizius-Klück and Annette Hülsenbeck, eds., *Selfactor. Zeitformen des Textilen—Schnittformen der Zeit*, (Berlin: Edition Ebersbach, 2002), 16.
63. Friedrich Pfister, "Kairos und Symmetrie," in *Würzburger Festgabe. Heinrich Bulle dargebracht zum siebzigsten Geburtstag am 11. Dezember 1937*, ed. Reinhard Herbig (Stuttgart: Kohlhammer, 1938), 138.
64. Gilbert Simondon, *On the Mode of Existence of Technical Objects*, trans. Ninian Mellamphy (London: University of Western Ontario, 1980), 70.
65. See Charles Babbage, *Babbages Rechen-Automate. Ausgewählte Schriften*, ed. Bernhard Dotzler (Vienna: Springer, 1996).
66. See Patrice A. Carré, "From the Telegraph to the Telex: A History of the Telegraph, Early Networks and Issues in France in the 19th and 20th Centuries," *Flux* 9 (1993): 17–31.
67. Benjamin Nübel, "Kairos und die Frage: Wie der Gott in die Maschine kam" (examination paper, Humboldt University, 2010), http://www.medientheorien.hu-berlin.de.
68. See Christian Kassung, *Das Pendel. Eine Wissensgeschichte* (Munich: Fink, 2007), 7–24. This chapter discusses alternatives to technological and media history in the name of science with reference to George Spencer-Brown, Gilles Deleuze, Bruno Latour, Michel Serres, and Michel Foucault.
69. Ellen Harlizius-Klück, *Weberei als* episteme *und die Genese der deduktiven Mathematik* (Berlin: Edition Ebersbach, 2004), 14.
70. For a discussion of "networking logic" (*Vernetzungslogik*), see Bernard Dotzler, *Diskurs und Medium. Zur Archäologie der Computerkultur* (Munich: Fink, 2006), 7.
71. Zielinski, *Deep Time of the Media*, 30.
72. Viola, "The Sound of One Line Scanning," 158.

73. Götz Großklaus, "Medien-Zeit," in *Zeit—Medien—Wahrnehmung*, ed. Mike Sandbothe and Walther Ch. Zimmerli (Darmstadt: Wissenschaftliche Buchgesellschaft, 1994), 45.

74. Max Egly, *Eintritt frei Fernsehen* (Lausanne: Rencontre, 1963), 73.

75. Friedrich Kittler, "Gleichschaltungen. Über Normen und Standards der elektronischen Kommunikation," in *Geschichte der Medien*, ed. Manfred Faßler and Wulf Halbach (Munich: Fink, 1998), 255f.

76. See Sigrid Weigel, "Die Richtung des Bildes. Zum Links-Rechts von Bildererzählungen und Bildbeschreibungen in kultur- und mediengeschichtlicher Perspektive," *Zeitschrift für Kunstgeschichte* 64.4 (2001): 449–74.

77. Peter Gendolla, "Zwischenzeiten. Zur Kultur und Technik der Zeit in der Moderne," in *Zwischen-Bilanz. Eine Festschrift zum 60. Geburtstag von Joachim Paech*, ed. Frank Furtwängler, Kay Kirchmann, Andreas Schreitmüller, and Jan Siebert (Konstanz: University of Konstanz, 2002), http://www.uni-konstanz.de/paech2002.

78. See P. Lertes, "Fernbildtechnik und Elektrisches Fernsehen," in *Kanalarbeit. Medienstrategien im Kulturwandel*, ed. Hans Ulrich Reck (Basel: Stroemfeld/Roter Stern, 1988), 237.

79. Richard Dienst, *Still Life in Real Time: Theory after Television* (Durham, NC: Duke University Press, 1994), 20f.

80. Claude E. Shannon and Warren Weaver, *The Mathematical Theory of Communication* (Urbana: University of Illinois Press, 1963), 4.

81. Qtd. in Hans Beller, "Aspekte der Filmmontage. Eine Art Einführung," in *Handbuch der Filmmontage. Praxis und Prinzipien des Filmschnitts*, ed. Hans Beller (Munich: TR-Verlagsunion, 1995), 25.

82. Knut Hickethier, "Dispositiv Fernsehen. Skizze eines Modells," *Montage/AV* 4.1 (1995): 67.

83. The musical equivalent is the delayed resolution of melody, such as the diminished seventh chord of Richard Wagner's Tristan motif.

84. Götz Großklaus, "Medium und Zeit. Zum Verschwinden des zeitlichen Intervalls," *Kunstforum International* 151 (July–September 2000): 213.

85. Götz Großklaus, *Medien-Zeit. Medien-Raum. Zum Wandel der raumzeitlichen Wahrnehmungen in der Moderne* (Frankfurt am Main: Suhrkamp, 1995), 52. Großklaus here refers to Vilém Flusser, *Für eine Philosophie der Fotografie* (Göttingen: European Photography, 1991), 60f.

86. Egly, *Eintritt frei Fernsehen*, 31.

87. McLean, *Restoring Baird's Image*, xvi.

88. Ibid., xvii.

89. Martin Heidegger, "The Thing," in *Poetry, Language, Thought*, trans. Albert Hofstadter (New York: Harper & Row, 1971), 163.

90. For an analysis of Heidegger's ignorance of radio as a technology, see Wolfgang Ernst, "Das Radiotier," in *Zoologicon. Ein kulturhistorisches Wörterbuch der Tiere*, ed. Christian Kassung, Jasmin Mersmann, and Olaf B. Rader (Munich: Fink, 2012), 327–31.

91. McLean, *Restoring Baird's Image*, 119.

92. Ibid., 93.

93. Ibid.
94. Ibid., xvi.
95. Egly, *Eintritt frei Fernsehen*, 143. Egly continues: "And it should not be forgotten that a great advantage of the video tape recorder was that it reproduced 'live' broadcasts much better than film."
96. Weber, *Mass Mediauras*, 121.
97. Mary Ann Doane, "Information, Crisis, Catastrophe," in *Logics of Television: Essays in Cultural Criticism*, ed. Patricia Mellencamp (Bloomington: Indiana University Press, 1990), 222.
98. Hans H. Hiebel, *Kleine Medienchronik. Von den ersten Schriftzeichen zum Microchip* (Munich: Beck, 1997), 24.
99. Herbert Zettl qtd. in Jane Feuer, "The Concept of Live Television: Ontology as Ideology," in *Regarding Television: Critical Approaches*, ed. E. Ann Kaplan (Frederick: University Publications of America/American Film Institute, 1983), 13.
100. Knut Hickethier, "Überlegungen zur Konstruktion einer Fernsehtheorie," in *Fernsehtheorien. Dokumentation der GFF-Tagung 1990*, ed. Knut Hickethier and Irmela Schneider (Berlin: Sigma, 1992), 26.
101. Raymond Williams, *Television: Technology and Cultural Form* (London: Routledge, 1990), 93.
102. See Albert Abramson, *Electronic Motion Pictures* (Berkeley: University of California Press, 1955), 41f.
103. Lev Manovich, "Eine Archäologie des Computerbildschirms," *Kunstforum International* 132 (November 1995-January 1996): 124–35.
104. Jim Campbell qtd. in Richard Shiff, "Something is Happening," *Art History* 28.5 (2005): 782.
105. Ibid.
106. Clement Greenberg, "The Crisis of the Easel Picture," in *Collected Essays and Criticism* (Chicago: University of Chicago Press, 1988), 2: 224f.
107. Shiff, "Something is Happening," 766. For a discussion of the dispute between d'Alembert und Euler concerning the (non-)intentionality of painterly gestures, see Axel Roch and Bernhard Siegert, "Maschinen, die Maschinen verfolgen. Über Claude E. Shannons und Norbert Wieners Flugabwehrsysteme," in *Konfigurationen. Zwischen Kunst und Medien*, ed. Sigrid Schade and Georg Christoph Tholen (Munich: Fink, 2003), 219–30.
108. Shiff, "Something is Happening," 772.
109. Yvonne Spielmann, *Intermedialität. Das System Peter Greenaway* (Munich: Fink, 1994), 124.
110. See Michael Cebulla, "Rechnender Raum. Bilder rechnen Bilder," in *High Density. Ein Ausstellungsprojekt über die Verdichtung von Räumen*, ed. Fritz Balthaus (Berlin: Neue Gesellschaft für bildende Kunst, 1999), 56ff.
111. Viola, "The Sound of One Line Scanning," 158.
112. Ibid.
113. Johannes Webers, *Handbuch der Film- und Videotechnik. Die Aufnahme, Speicherung, Bearbeitung und Wiedergabe audio-visueller Programme* (Munich: Franzis, 1991), 484.

114. Ibid., 485.
115. Ibid., 488.
116. See Fritz Schröter, "Gesetze, Probleme und Grenzen des Fernbildrasters," in *Handbuch der Bildtelegraphie und des Fernsehens. Grundlagen, Entwicklungsziele und Grenzen der elektrischen Bildfernübertragung*, ed. Fritz Schröter (Berlin: Julius Springer, 1932), 95.
117. Peter Weibel, "Transformationen der Techno-Ästhetik," in *Digitaler Schein. Ästhetik der elektronischen Medien*, ed. Florian Rötzer (Frankfurt am Main: Suhrkamp, 1991), 205.
118. Gottfried Boehm, "Bild und Zeit," in *Das Phänomen Zeit in Kunst und Wissenschaft*, ed. Hannelore Paflik (Weinheim: VCH Acta Humaniora, 1987), 23. Boehm locates the temporality of the image in the moment of seeing. "We see constantly new ways in which an image becomes 'integrated' into a whole and then 'differentiated' into succession." Ibid., 21.
119. Christina Vagt, "Zeitkritische Bilder. Bergsons Bildphilosophie zwischen Topologie und Fernsehen," in *Zeitkritische Medien*, ed. Axel Volmar (Berlin: Kulturverlag Kadmos, 2009), 125.
120. Tomas Fryza, "A Complete Video Coding Chain Based on Multi-Dimensional Discrete Cosine Transform," *Radio Engineering* 19.3 (September 2010): 421.
121. Adrian Mackenzie, "Codecs," in *Software Studies: A Lexicon*, ed. Matthew Fuller (Cambridge, MA: MIT Press, 2008), 50.
122. Ibid.
123. Ibid.
124. Ibid., 51.
125. Ibid.
126. Ibid., 53.
127. Bergson, *Creative Evolution*, 317. See also Maurizio Lazzarato, *Videophilosophie. Zeitwahrnehmung im Postfordismus* (Berlin: b-books, 2002), 25f.
128. See Karl Ernst von Baer, "Welche Auffassung der lebenden Natur ist die richtige? und wie ist diese Auffassung auf die Entomologie anzuwenden?" in *Zeitkritische Medien*, ed. Axel Volmar (Berlin: Kulturverlag Kadmos, 2009), 44–60. See also Siegert, "Das Leben zählt nicht," 161–82.
129. Walter Benjamin, *The Arcades Project*, trans. Howard Eiland and Kevin McLaughlin (Cambridge, MA: Belknap Press, 1999), 462.
130. Lazzarato, *Videophilosophie*, 47.
131. Peter Weibel, "Video als Raumkunst. Arbeitsnotizen," in *Video—Apparat/Medium, Kunst, Kultur. Ein internationaler Reader*, ed. Siegfried Zielinski (Frankfurt am Main: Peter Lang, 1992), 148.
132. Gerry Schum qtd. in Wulf Herzogenrath, *Videokunst in Deutschland 1963–1982. Videobänder, Videoinstallationen, Videoobjekte, Videoperformances, Fotografien* (Stuttgart: Hatje, 1982), 56.
133. Richard Kriesche qtd. in "Zwillinge," *Media Kunst Netz*, http://www.medienkunstnetz.de/werke/zwillinge.
134. See, for example, *Topic I & II*, directed by Pascal Baes (1990).
135. See Hartmut Rosa, ed., *Fast Forward—Essays zu Zeit und Beschleunigung. Standpunkte junger Forschung* (Hamburg: Körber-Stiftung, 2004).

136. Siegfried Zielinski, "Audiovisuelle Zeitmaschine. Thesen zur Kulturtechnik des Videorecorders," in *Video—Apparat/Medium, Kunst, Kultur. Ein internationaler Reader*, ed. Siegfried Zielinski (Frankfurt am Main: Peter Lang, 1992), 91–114.

137. Siegfried Zielinski, *Zur Geschichte des Videorekorders* (Berlin: Wissenschaftsverlag Spiess, 1986), 330.

138. Niklas Luhmann, *Social Systems*, trans. John Bednarz, Jr. (Stanford: Stanford University Press, 1995), 160.

139. See Katharina von Falkenhayn, *Augenblick und Kairos. Zeitlichkeit im Frühwerk Martin Heideggers* (Berlin: Duncker & Humblot, 2003), 127.

140. Georg Trogemann, "Einrichten im Dazwischen," in Medienhochschulen und Wissenschaft. Strukturen—Profile—Positionen, ed. Karl Friedrich Reimers and Gabriele Mehling (Konstanz: UVK, 2001), 107.

141. See Friedrich Engel, Gerhard Kuper, and Frank Bell, *Zeitschichten. Magnetbandtechnik als Kulturträger* (Potsdam: Polzer Media Group, 2008), 465.

142. Nam June Paik, "Input-Time and Output-Time," in *Video Art: An Anthology*, ed. Ira Schneider and Beryl Korot (New York: Harcourt Brace Jovanovich, 1976), 98.

143. "Nam June Paik talkin with Angela Melitopulos and Maurizio Lazzarato," *Mixed Pixels: Nam June Paik*, Zentrum für offene Kommunikation, http://www.cultd.eu/paik/interv/_interv2.htm.

144. Paik, "Input-Time and Output-Time," 98.

145. Rosalind Krauss, "Video: The Aesthetics of Narcissism," in *New Artists Video: A Critical Anthology*, ed. Gregory Battcock (New York: E. P. Dutton, 1978), 13–64. This was also the topic of the "Temporal Values" exhibition at the Städtischen Galerie in Karlsruhe (December 2003–April 2004).

146. See Michael Baudson, *Zeit. Die vierte Dimension in der Kunst* (Weinheim: VCH/Acta Humaniora, 1985).

147. See Heike Helfert, "Raum Zeit Technikkonstruktionen. Aspekte der Wahrnehmung," *Media Kunst Netz*, http://www.medienkunstnetz.de/themen/medienkunst_im_ueberblick/wahrnehmung.

148. See Angela Melitopoulos, "Timescapes," in *Lab. Jahrbuch 1996/97 für Künste und Apparate*, ed. Kunsthochschule für Medien (Cologne: Walther König, 1997), 172–82.

149. Virilio, *The Vision Machine*, 66.

150. Paik, "Input-Time and Output-Time," 98.

151. See Edith Decker, *Paik. Video* (Cologne: DuMont, 1988), 170.

152. Julia Meuser, "Urheberrecht und Werkintegrität in der Video-Kunst," in *Wie haltbar ist Videokunst? Beiträge zur Konservierung und Restaurierung audiovisueller Kunstwerke*, ed. Bärbel Otterbeck (Wolfsburg: Kunstmuseum, 1997), 79.

153. M. M. Bachtin, "Zur Ästhetik des Wortes," in *Kontext. Sowjetische Beiträge zur Methodendiskussion in der Literaturwissenschaft*, ed. Rosemarie Lenzer and Pjotr Palijewski (Berlin: Akademie, 1977), 139.

154. Niklas Luhmann, *Art as a Social System*, trans. Eva M. Knodt (Stanford: Stanford University Press, 2000), 129.

155. Ibid.

Chapter 7

The Media Timing of NonLinear Communication

A WORLD OF SIGNALS AND THE ALPHABETIZATION OF TIME

In 1906, Charles Sanders Peirce questioned the conceptual world of his dynamized semiotics. "All my notions are too narrow. Instead of 'Signs,' ought I not to say *Medium*?"[1] The technological world breaks down into signs (logic) and signals (physics). Time-critical media processes are embodied not in symbolic signs but in indexical signals. Media theory thus shifts from semiotic aesthetics to the media-technical time event. Unlike discrete symbols, useful as well as interfering signals are physically grounded in matter and energy, so they are not only indexically related to time, but they also already embody continuous or discrete tempor(e)alities.

This characteristic feature is connected to the form of the signal itself. In the form of electrical telegraphy, for example, the succession of symbol sequences is a necessary condition for the signal to be able to be sent over a single line;[2] the alphabet thus becomes signal-technical. "With telecommunications signs take on the character of signals as the transmission of energetic pulses, be they discrete, as in the case of the 'dots' and 'dashes' in telegraphy, or analog, as in the case of the telephone, which converts the boundary between sign and non-sign into the ratio between signal and noise."[3] As an information-carrying variable, an analog signal can assume an infinite number of values; its mathematical model is thus the real function, which assigns a real value (y) to every point on the time axis (x). However, mathematizable communication media require signals whose variables can only assume a finite number of values, and in the extreme case of binary systems these values are limited to only two. In order to become a veritable digital signal, it must nevertheless be not only value-discrete, but also time-discrete.

This rhythm is familiar since the times of phototelegraphy. This occurs in electronic image transmission when the scanning of a line does not happen as a continuous voltage curve, but rather as a series of "on" and "off" switches that indicate each brightness and darkness value; the image point here becomes a point in time.

With the translation of symbolic letters into combinations of short and long electrical impulses (which were visible as dots and dashes on a paper strip), the alphabetical code became part of communications engineering. With the necessary interval sign, time itself—critically intensified—became an operator of symbolically coded communication. In Gutenberg's printing press, the typographic distinction between symbol and space, as opposed to the *scriptio continua*, was already included as the space between words, the *spatium*. In letter cases and on typewriter keyboards, it acquired its own sign. It was first employed time-critically in telegraphy as the interval between electrical impulses.

In terms of Claude Shannon's mathematical theory of communication, telegraphy is based on a discrete source, as the message is conveyed through individual, consecutive signs—a strict order in (linear) time. Within this sequence, signs are regarded as probabilities—depending on language and context—resulting in a time-critical economy: since the standardization of Morse code in 1865, the most frequent letter in German and English, the letter "e," is given the shortest possible code, a single dot, which thus literally represents a point in time. A time figure that is unique to linear symbol sequences here comes into play: transition probabilities or rather frequencies, as A. A. Markov Sr. analyzed in 1913 on the basis of consonant and vowel sequences in Pushkin's *Eugen Onegin*. First order Markov processes have since been defined as stochastic processes, in which—starting from a fixed point in time—the future development of an observed system depends only on its current state rather than its history prior to this point in time: antihistoricism in the time-critical domain.

Time-discrete telecommunications actually began with telegraphy. "For the first time, sheer duration in a standardized time defined the symbolic meaning of dots and dashes. The pure, immaterial message invented a code for itself."[4] With Morse code, alphabetical symbols were situated in time and articulated as electromagnetically induced time events, which could then revert to phonetic intonations. In contrast to the purely statistical examination of frequency distributions of linguistic elements, a consideration of their duration—the becoming-event (*Ereigniswerden*) of linguistic articulation—leads to its quasi-telegraphic analysis on the time axis. "We think of the flow of speech as a function of time, as it is constructed out of discrete elements (such as syllables). A certain time interval belongs to every element."[5] What is one-dimensionally ordered in time t and is distinguishable as the time interval

Δt can be quantified in terms of communications engineering. David Albert Huffman optimized this process in 1952 through the development of what has since been named "coding,"[6] which follows the model of Morse code by assigning frequent signs to short signals (and vice versa). In media-archeological terms, the momentary point in time thus becomes the extreme articulation of entropy-coding *avant la lettre*, which is electrotechnically real rather than purely symbolic. Time-serial currents become tangible above all in the form of one-dimensional signal recording, whether on gramophone records or magnetized tapes. "The law of what can be said" (Foucault's definition of an "archive") here becomes media-technically concrete. Even before such practices become high-tech media *strictu sensu*, and thus objects of media studies, their epistemological conceivableness is frequently practiced in the form of cultural techniques[7]—not as linear progress, but rather as nonlinear paradigm shifts. Time-discrete communication media are prefigured in the symbolic practices of culture, which is why the topic of the vocal alphabet will be discussed in this context. An alphabetical script is usually not time-critical during the act of writing (unless performed by a writing machine), although the situation is different with respect to the intermittent, interval-determined reading process of human eyes. In the song-like performance of ancient Greek poetry, however, letter sequences are time-critical and a true Morse code *avant la lettre*. According to Aristotle, time can first be perceived as discretely (and thus numerically) measured movement; Étienne-Jules Marey experimented very concretely with such time series using chronometry.[8] With ancient Greek prosody, poetic performance was measured according to the length and shortness of syllables (*chronoi*). It was possible for this chronopoetic rhythm to be handed down over time precisely because it was based on a time alphabet. "All verse metre was based on this binary opposition, whether it was spoken verse or sung."[9] In contrast to absolute mathematical timing, this rhythm is nevertheless articulated in verse and song as musically relative time; the poetic meter avoids exact determination through the concrete sensomotoric variance of the performance. However, the prosodic time alphabet is completely time-critical when its phonetic articulation is technically analyzed as a frequency mixture, whose signal fidelity is sufficiently calculable and transmissible with a sampling rate of 44,100 Hz. Media archeology goes beyond discourse analysis in that it examines communication at the level of not only cultural practices, but also technically concrete microphysics.

TIME-CRITICAL TELECOMMUNICATION

The act of transmission becomes time-critical when its temporal form is just as crucial as its bridging of space. Electrophysical oscillations can function as dynamic carriers of discrete information through coded modulation. In their simplest form, signals are one-dimensional functions that are dependent on time. This relation becomes the medium of the message in the form of so-called "time signals," for example, which were transmitted from the German long-wave radio station near Frankfurt.

When technically implemented, Morse code represents a special case of time-critical signals, "whose meaning solely arises from its time index";[10] in 1890, Ladislaus Fiedler called this "time telegraphy." On the other hand, telegraphy was also employed to allow time to communicate itself—that is, to send time signals, which electrically synchronized clocks.[11] For phototelegraphy in particular, "the most precise simultaneity is the prerequisite for clarity."[12] The "sync" signal in the Internet is a continuation of the synchronization signal in television, which was a necessary condition for successful electronic image transmission.

The multiple use of communication lines—the media economy of telecommunications at the time-critical level—involves frequency-division multiplexing (analog) and time-division-multiplexing (digital), which are based on the temporal interconnection and interleaving of signals. Temporally alternating multiple transmission means "that periodically alternating short-term signal functions are connected to the line."[13] Multiplexing enables more telephone conversations to take place simultaneously on a single line through the time-critical shifting of signal chains (a counterpart to the "time sharing" of a central computer by a multiplicity of users). The frequency hopping process developed by the actress Hedy Lamarr and the composer George Antheil became a cryptographic method in wireless message transmission. Time hopping represents a digital escalation of this time-critical process, as the individual coded segments of speech are not only transmitted in short segments of time, but they are also constantly varied. These slices of time are defined by a random pattern, which must be known in order for the receiver to be able to make sense of pure time signals.

MATHEMATICAL TIME VERSUS TRANSMISSION TIME: TELEGRAPHY AND INTERNET

Fritz Heider distinguished between tightly coupled elements and batches—as "thing" and "medium"—on the basis of their frequency. The temporality of a "thing" is defined by its natural oscillation; "in the medium, on the other hand, connections are grouped around false units, forced oscillations."[14] The natural oscillation of the materially continuous thing is indivisible, unlike discrete symbol chains, which are first expressed as combinatorial time objects. "The principle underlying flag signaling and Morse code is the combination of independent elements. The arrangement of units occurs not only in space, as with letters, but also in time."[15]

In order for them to be broadcast over a signal channel, two-dimensional texts and images must be broken up into sequential series, which inevitably involves temporalization. The linear one-dimensionality of telecommunications is already familiar from the phonographic sound event; however, the alphabetical order shifts into the fourth dimension when time itself functions as a medium for communicating a finite character set. Digital signals are defined as information because their discrete values stem from a fixed alphabet, and they are thus coded. In clocked systems, like digital computers and coded telecommunications, information is derived from the *temporal* position of pulses in a particular time period; these pulses thus represent a virtual "time alphabet," which in the case of binary signals is reduced to only two radically discontinuous values. Temporal behavior literally "counts" here, and it is grounded in well-defined points in time.

However, human speech consists not of a well-defined alphabet, but rather a mixture of sounds and noises, whose frequencies essentially represent a continuum. The acceleration of speech with the help of technomathematical intelligence requires discrete coding as well as a statistical measurement of the information source. The concept of "ergodicity" refers to this time-critical world induced by communications technology.[16] This coding is actually considerably more time-consuming for the encoder, but data compression and advance calculation allow the message to be sent faster than the amplitude- or frequency-modulated signals of analog radio.

The question of the time relations of the present leads to the heart of digital communications technology. Since the technical implementation of the mathematical theory of communication in the twentieth century, the conceptual understanding of communication is no longer dominated by the postal principle of space-bridging transmission; for the purpose of channel-friendly transmission, coding is what literally *counts*. The capacity of the part of the transmission chain that Shannon's design actually refers to as "the medium" is here measured according to the rate of transmissible "bits" per second—a temporal signal economy that is already familiar from the Baudot code in telegraphy. In this bottleneck, time and medium are entangled in both a communication-theoretical and communication-technical sense, which defines the temporal modes of the current media economy.

Jean-François Lyotard recognized in the telegraphic principle the temporal *non*-essence (*Unwesen*) of information-processing media. Through technical coding, physical signals are largely detached from the world and thus from time. "After they have been put into digital form, these items of data can be synthesized anywhere and anytime. . . . They are thereby rendered independent of the place and time of their 'initial' reception, realizable at a spatial and temporal distance: let's say telegraphable."[17] The most radical escalation of the formerly telegraphic time relation of technical communication is the proper time of cybernetic networks in the Internet—a time-critical medium on the microtemporal (and largely hidden) level of implementation. In neurobiology, the logical connection of signal pathways for a long time shaped the interpretation of the virtually binary operations of nerve cells; this theory proved to be "sufficient for all purposes in which the exact time for impulses to pass through the whole net is not crucial."[18] On the technical level of the smallest operative moments, however, every logical network becomes time-critical. With the permanent availability of vast quantities of data, the address structure in the Internet is not only hypertextual, but also hypertemporal. On the microtemporal level, it involves logical circuit connections rather than hermeneutic understanding and time-critical decisions rather than the continuous "flow" that defined both the signals and programs of earlier mass media. The "flow of time" is thus discretely canceled.[19]

In digital "online" networks, the classic function of telecommunications—spatio-temporal transmission—is replaced by the virtually immediate replication of information, which legally culminates in the oxymoron "original copy." Use and consumption are here already acts of copying. "Every time you use a creative work in a digital context, the technology is making a copy";[20] however, this copy can no longer be distinguished from an original through the passage of time. Uniqueness with regard to place and time, which Walter Benjamin described as the essence of the work of art as an original and the characteristic feature of the historical index and historical witnessing,[21] is thus sublated. In the case of digital copies, it is no longer possible to read the context in which the original emerged, and a new type of genuine media-historical context assumes its place. In order to be able to transmit information embodied in numbers over a media channel, the sender and the receiver must agree on two things: a finite alphabet (the symbol set) and a form of coding. These agreements take place in time and are constantly changing due to the high speed of development in digital space. In a hermeneutically familiar way, certain *historically* based conclusions regarding the context of coding can still be drawn from the process that is used. "It must be stated, however, that . . . this is far removed from the historical witnessing described by Benjamin. Witnessing does not exist in this form."[22]

IN THE STREAM OF TIME: STREAMING MEDIA

In the era of the World Wide Web, media use differs from previous broadcast media in terms of distribution—unicasting rather than simulcasting—but it imitates this earlier form of distribution in the time window of reception. "Streaming" denotes a form of transmission (primarily of audio and video signals) via the Internet, which is received by the end user without any noticeable time delay and thus under the time-critical threshold of the receiver's perception of the present. The technological condition necessary to convey the impression of an audiovisual data *flow* instead of discrete cascades and blocks is a broadband Internet connection; from a time-critical perspective, fiber optic cable is here literally more transparent than the copper cores of the classical telephone network. Insofar as it conceals the act of digital copying and thus the archival principle, streaming media appears to share the fleeting temporality of analog radio and television, whose signals immediately dissipate with each broadcast. In terms of this fleeting transience, it makes no difference whether the transmission involves "live streaming" or "streaming on demand"—a form of media-technical time sovereignty that is familiar since phonography—that is, the availability of signal streams at any given point in time.

The categorical distinction between retrieval from storage and direct transmission, which has been validated by the Western philosophy of time until now, implodes in the technodynamic modes of digital buffering;[23] this represents the assault of storage logistics on the present. With "live streaming" there is actually no previously stored file to be retrieved from the distributing server, yet the data is nevertheless organized as files in order to be processed through software. An art postcard by Joseph Beuys once announced that "name equals address." Addressing serves to identify a data stream, which can then be reached on a server through an encoder in real time (thus minimally buffered).[24] It is thus based on digital recording and minimally time-delayed playback. While the document description HTML has the rather geometric function of formatting texts and embedding graphics for communication in the Internet, a time-critical variant was developed for the technology of streaming, the Synchronized Multimedia Integration Language (SMIL), whose objects are time-based media formats (audio and video). The integration of these formats requires constant synchronization. This is associated with "advanced video coding"—that is, intelligent algorithms that make compressed blocks of data appear as a coherent sound and image stream. Unlike on-demand films from video recorders, streaming is based not on the storage of data in advance by users, but rather on a dynamization of the storage process itself; in an accelerated form, it becomes batch bulk buffering. Such buffering enables the quasi live-archival management of the continuously arriving data stream (and, at the same time, the skirting of copyright). The shift from a stored to a dynamic format no longer leads back to an economy of storage media, which utilized sound carriers and videotapes as products.[25] In the LZ77 algorithm for data compression, the time figure of the future past is articulated almost chronopoetically. "We employ the concept of encoding future segments of the source-output via maximum-length copying from a buffer containing the recent past output."[26] That accelerated time horizon, which once led to the development of historical awareness in the era of industrial and political revolution, is here intensified at the time-critical level of immediate protection and retention; however, this is no longer social time or the inner time consciousness of human subjects, but rather technomathematical operativity in its own right.

Streaming media—the online transfer of audiovisual data formats—is subject to a strict time economy. Network communication is characterized by time-critical moments, as it necessarily involves time analysis through the disassembly of information into individual data packets and their asynchronous transmission.[27] The so-called "cyberspace" is composed not of "space" in the sense of Kant's *a priori*, but rather of constant, complexly branching calculating processes;[28] in contrast to the spatial metaphor, the Internet can only be considered a spatial structure in the sense of mathematical topology. However, topology only becomes a communication medium through its implementation as a function of radically time-critical processes. Data constantly change their locations as discrete packets and are temporarily stored (cache) in a series of nodes (relays). Their trajectories can be read no longer in the sense of classical cartographies, but rather as the tracking of rapidly fading traces, like the tails of comets. So-called "traceroutes" function here as advance scouts for the broadcast; an ultra-short-term historicity called "Time To Live" (TTL) is inscribed in the data packets themselves—as a premonition of the end in the times of high-tech communication.[29]

OTHER SPACES: NETWORKED TIME

Marshall McLuhan interpreted electric communication as a contraction of the world to pure telepresence. Techno- and neurocybernetic systems are essentially time relations; this does not mean that humans are alienated, but rather that they come to themselves as a whole "in the electric age, when our central nervous system is technologically extended . . . to incorporate the whole of mankind in us."[30] The colorful world of local times was first replaced by Greenwich Mean Time in the nineteenth century due to the synchronization requirements of networked traffic and communication systems, such as railways and telegraphy. Acceleration in the time scale thus took place simultaneously at the level of physical transport and media-based communication. When time units are extracted from the power grid itself, then the technical world acquires a radical proper time that is already familiar from the synchronization of electromechanical television. The rotating Nipkow disks were equiprimordially synchronized on the sender and receiver sides through the application of alternating current from the grid; this is also how the frame rate of fully electronic television was derived (50 half-frames per second in Europe, 60 in the United States).

Communication relations are always also time relations. While pure interpersonal dialogue requires the presence of both partners, storage and transmission media generate new forms of temporally and spatially shifted communication that disrupt the classical time structure of conversation. In the Internet, these temporal modes are reflected at all levels: from the software used to verify time-critical connections in the sending of data packets between servers to the supply of real-time messages (chat rooms and instant messenger) to time-shifted discussion in long-term forums, in which simultaneity is no longer a condition for communication. This change in emphasis from places and transmissions to immediate addresses based on mathematized telecommunications also introduced new models of a completely autonomous Internet time that was detached from seasonal and daily rhythms of nature. The idea of an "Internet time" was endorsed in the media art project "XTime" (1998). Shortly afterward, it was also promoted by the watch manufacturer Swatch,[31] which sought in vain to establish its own global time base. However, the most vivid reminder of the power of technodigital proper time (for the power of clocks has migrated into the computer itself) was the "millennium bug"—the second change from December 31, 1999 to January 1, 2000.

Does media time massage the human perception of time in a way that liberates it from previous cultural experiences of time? "Submerged in the mediascape we can easily forget time—watching a film, emerged in the audioscapes of a CD, hooked on television, lost in cyberspace."[32] The audiences who listened to Homer's songs were already presumably situated in another temporal space (that of the Trojan War or the *Odyssey*) during the interval of narrative time. However, the experience of being lost in another time is more intense in the case of electronic media, as they affect the human sense of time more profoundly than all previous symbolic systems and thereby induce a new epistemological concept of time. Humans are trained to live autonomously in different temporal states and to alternate between them by "time hopping." In the time-critical intensification of multiplayer online computer games, latency times in data processing (measured in "ping" times) literally become a question of (symbolic) life and death.[33] In addition to individual storage-intensive memories, a dynamic form of regeneration increasingly emerges between networked computers in the temporal mode of real-time Internet, which processes data objects no longer in their entirety but rather in algorithmic compressions.

"The great obsession of the nineteenth century was, as we know, history."[34] Does time now appear as only "one of the various distributive operations that are possible for the elements that are spread out in space"?[35] In a rare moment when he writes about computers, Foucault underestimates their time-critical character and reduces it to a mere principle (archeology, but not *media* archeology). The mathematical *dispositif* of networked electronic computers is in fact graph theory. Foucault thinks diagrammatically about the topological archive, but this is only possible in the case of the Internet by considering the time figures of technical signal processing. It is characteristic of real heterochrony that it is "linked ... to time in its most flowing, transitory, precarious aspect. ... These heterotopias are ... absolutely temporal."[36] Translated into technical practice, the Internet thus proves to be a form of ecstasy. Topological operations are here not pure spatial orders, but rather a superimposition of time-critical algorithms. Communication and knowledge networks are thus increasingly dominated by time planning. The temporal economy of the Internet privileges momentary stimulation over duration.[37] The lived experience of "net time" replaces the homogeneous present with intensive now-times (*Jetzt-Zeiten*). "It's not clock time, and it's not Swatch time, it's a time of lags and latencies, of waiting and clicking through, of fast and slow. It is the experience of differing speeds and asynchronicity."[38]

AM, FM, PCM, PPM: MODULATION *IN* TIME AND *WITH* TIME

In the era of technomathematical media, communication (understood as the transmission of information signals rather than a vehicle of physical transportation) has become a chronotechnical drama. Time itself is swallowed up.[39] Following the launch of a French railway line in the mid-nineteenth century, Heinrich Heine predicted that time would "kill" space;[40] Paul Virilio's "dromology" updated this diagnosis for the era of high-tech telecommunications.[41] The temporal interval of a train journey is still experienced as linear, which corresponds to the historical sense of time, whereas telegraphy is experienced as simultaneous, and immediacy becomes the nonlinear message of the medium. Communication and transmission are thus no longer entirely different categories, but rather different time-critical emphases with regard to one and the same interval.[42] In contrast to the storage economy of accumulation, a temporal vector—namely, the importance of punctual transmission—is inscribed into the logic of capital: the exchange of goods seeks not only to eliminate every spatial barrier to movement but also "to minimize the amount of time that movement from one place to another costs."[43] This led to the electrotechnical escalation of the gap between space and time in terms of traffic. A contemporary witness from the Telefunken Company diagnosed this electrotechnical escalation in the following words: "Thanks to the mysterious and, for us corporeal entities, transcendental powers of electricity and magnetism, we are well on the way to attaining 'terrestrial omnipresence'. . . . It was indeed an eventful evening . . . when the first official telephone conversation was held between Berlin and Buenos Aires, which was 12000 km away."[44] This is successful communication in the technically embodied noosphere.[45] Time delays, which were a limiting factor for earlier forms of communication, virtually disappear in the era of "live" media due to the compression of emphatic time intervals. In wireless communication, electronic speed no longer requires a mechanical vehicle, as the media world is capable of transporting (and transmitting) itself. "Wireless waves are electromagnetic processes, which are not bound to matter or energy. Like light, they penetrate . . . the empty space where there are no electrons."[46] In the technical tempor(e)ality of analog communication media, the interval Δt *ideally* tends toward zero, but in actuality, the delay still comes into play as delay time—a function of the materiality of electrical lines. The difference between analog and digital can thus be understood as the *time difference* between delay-line "live" transmission and digitally-computed real-time transmission.

With classical amplitude modulation (AM), radio signals were completely continuous; frequency modulation (FM), on the other hand, introduced the low-frequency signal (speech or music) *as a time event* into the formerly equidistant carrier frequency—chronotechnical immediacy. However, an entire media-epistemic world lies between continuously modulated alternating currents and a signal form that consists of periodically consecutive pulses.

A French patent application submitted by A. H. Reeves in 1938 describes the fact that it is not necessary to transmit all of the relevant oscillations in order to communicate over the telephone, as extracts of these oscillations are adequate to reproduce the signal clearly at the receiver's end. Samples were then taken from the signal at the sender's end "in sufficiently short temporal intervals"[47] (thus in the strictest sense time-critically) and transmitted. The necessary condition was that the sampling rate had to be twice as high as the highest frequency component to be transmitted; this sampling theory is the *time scale* of successful communication in the era of digital telecommunications media.

As the pulse amplitude modulation (PAM) of a high-frequency carrier oscillation was still highly vulnerable to distortions in the transmission of signals, Reeves developed a symbolic operation: pulse code modulation (PCM). In the context of his definition of the electronic image as a technological sound, video artist Bill Viola described PCM as a "shift from analog's sequential waves to digital's recombinant codes."[48] This combination of temporal clock pulses and binary melodies results in the tonality of digital message transmission.

Signal transmission in pulse phase modulation (PPM) is temporally delicate. The phase of the carrier pulse is altered such that the time of the pulse—in relation to the time pattern—displays the momentary value of the controlling signal oscillation.[49] The moment counts here. When coded information is reduced to points in time in this way, different messages can be independently transmitted on the same line through nesting and time-shifting. Such time-multiplex transmission requires time selection. Instead of providing access to constant signals, digital transmission represents a fundamentally existential conversion—namely, the coding of a physical signal as a randomly defined sequence of symbols, which is similar to the relationship between spoken language and a typewritten text. Compared to analog systems, the time relations of digital transmissions are "entirely different."[50] Media archeology focuses on such time-sensitive breaking points.

ECSTASIES OF TRANSMISSION: THE TIME-CRITICAL ESSENCE OF THE INTERNET

With the rise of "packet switching," which was developed by Paul Baran and Donald Watts in 1963, and the transmission control protocol (TCP), which was developed by Bob Kahn and Vinton Cerf and later accompanied by the Internet protocol (IP), the transmission-oriented media of the postal era transformed into virtually immediate, address-oriented media. IPs serve to reassemble fragments of data packets, which requires, above all, synchronization. This involves the transformation of not only space-bridging transmission technologies but also traditional time-critical channels. PCM made transmission independent of the physical data line (whether it be telephone lines, fiber-optic cables, or satellites) through the mathematization of communication, which first turned signals into information. "The bit canceled the spatial constant in the calculation of the expenditure of redundancy for the channel. Distance, that is, no longer played any role in the relationship between capacity and rate of transmission. This in turn meant that communications theory no longer had anything to do with (letter) mail."[51] This is embodied in the fate of the electron tube. As an analog signal amplifier ("repeater amplifier"), it enabled the establishment of the first transcontinental telephone line in the United States in 1915, but continuous signal transmission was later replaced by the pulse—a discrete and thus numerically manipulable physical unit. The amplifier element was then employed discretely as a "regenerative repeater." The quantization of space-bridging telephony was accompanied by a new media-technical quality.

The network became a machine. No longer was the network a passive device, for repeater amplifiers actively added energy along the route. This change decoupled the wave that represented the conversation from its physical embodiment in the cable.... Electricity in the wires was now merely a carrier, separate from the message or signals it carried.... Now voices became signals.... The message was no longer the medium; now it was a signal that could be understood and manipulated on its own terms, detached from its physical embodiment.[52]

This electrotechnical configuration escalated as it was combined with Shannon's radical mathematical concept of communication, which was based on binary-coded information rather than constant signals and thus made communication networks calculable. However, this required another technological modification. When amplifying signals, the "regenerative repeater" no longer includes noise, but only clearly separates binary "digits." The concept of the signal as energy is here replaced by the concept of the signal as information. This gives reason to refer back to Michel Serres' distinction between techniques and technologies. He opposed the "hard" techniques of the industrial revolution, which work at the entropic level—in short, the mechanics and thermodynamics of machines—to the "soft" technologies of data carriers, which work at the negentropic level. Serres reserved the term "technology" for "those artifacts that deal with signs and thus with the logos"—in contrast to "techniques," whose energy range was 10^{16} times greater.[53]

Is it simply a quirk that the current concept of the "smart grid" makes the power grid reactive in the time-critical domain through the reading of individual power consumption at precise moments as bits in digital counters? The instantaneous calculation of power consumption makes its logistical distribution in space and time controllable. The readjustment principle, which previously maintained the stability of the 50 Hz frequency of alternating current, is also replaced by microchronological feedback in real time. The basis of these autodynamic time responses is no longer statistics concerning relative power consumption (the archive), but rather immediate knowledge of the network itself. Telecommunications provided a preview of this shift around 1960 with the introduction of digital networks, which were able to recognize the capacity of the network and enable temporally effective responses.[54]

The technomathematical theory of communication has a dual origin; on the one hand, it was developed to reduce the noise of telephone lines for the purpose of improving communication (Nyquist and Bode); on the other hand, it was also developed for military purposes due to the predicted escalation of conflicts between real and potential opponents in the immediate future. The human voice was thus no longer the actual media-determining event, but rather it became a peripheral phenomenon in the modulation of diverse carrier frequencies. Like radio, classical telephony also became the transmission of abstract signals. "The Bell System became not merely a set of voice channels but a generalized system capable of carrying any signal as a new currency: information."[55] What remains of the human in transmission? A special case of communication is a section of the frequency spectrum. In the twentieth century, transmission is only a metaphor of what it once meant until the late nineteenth century: the transportation of messages from A to B. The concept of media as a space and time-spanning vehicle—in other words, the bridging of distance—is purely nostalgic, as technomathematics makes the traditional medial concept of the channel itself obsolete. With the rise of binary-coded communication, transmission implodes into the algorithmized operations of machines.

The classical transmission metaphor is also undermined from the other side. In the "sonic" phenomenon of (electro)physical resonance, the oscillations of a vibratory object are induced in another object with the same natural frequency virtually without the transfer of energy. An alternative to the historical concept of knowledge transmission can be derived from the technical model of reciprocal coupling. However, the material veto in the act of transmission remains as yet irreducible. Transmission channels, such as the lines between electronic components, always already imply delay and dead times, even when they involve the shortest routes. What is erased by mathematical intelligence is the transmission and reaction time. This also changes the concept of tradition from the historiographic to the archival *dispositif*, as the elementary schema of communication is no longer transmission, but rather the modification of a configuration.[56] "*Archiving* is one of the main applications for a lossless code."[57] In the uncompressed transmission of complex audio and video data in PCM mode, every sampled value is stored consecutively and then condensed into groups or "frames." Codecs analyze such intervals and develop real-time and channel-efficient functions. The enhanced approximation of the course of time also requires more intensive calculation; time approximation replaces the physical time axis in the digital. With regard to time, the classical linear transmission channel also splinters into dissipative structures. The synchronous transmission of classical broadcast and "live" media, which was essentially based on the emissions of electromagnetic waves themselves, was replaced by asynchronous communication. Internet communication is thus linked to the pretechnical era of postal communication in the form of e-mail, which similarly involves storage and time-shifting, but it simultaneously undermines postal communication through electronic immediacy.

In terms of media theory, the temporal processuality of the Internet can only be understood through new concepts that make the temporal mode of the medium a crucial criterion of analysis, such as the concept of the "real-time web" in the communicative mode of "instant messaging" instead of periodic "updates." The message of the medium is here a temporal form: the creation of (pseudo)co-presence. The concept of cyberspace as a "docuverse" is replaced by extremely accelerated information processing in cybertime. "In earlier times it was about the creation of spaces. . ., today it is about time itself"[58]—*chronos* and *kairos*. The Internet is thus more than simply a spatial-topological extension of the archive; rather, it is a chronopoetic "compression of time."[59] The time-critical efficiency of search engines like Google is based on the fact that the universe of websites is indexed and calculated in advance by crawlers (the "page repository") in order to allow current search questions to be almost immediately fed back to users according to the "PageRank" algorithm.[60] Insofar as it is not predetermined if and when a communication occurs, the real-time network requires the ability to recalculate answers virtually instantaneously and equiprimordially with regard to the stored data—a mixture of tables (archive) and dynamics (real-time calculation). The concept of the time-critical thus gets to the heart of online communication.

With electronic scalability, time falls to pieces. Digital real time is no longer simply the time of the clock, but rather that of a discrete time field. On the media-archeologically hidden level of the Internet, a zigzagging exchange of packets always already occurs between the client and the server during TCP connection establishment. The user datagram protocol (UDP) developed by Jon Postel in 1980 already coexisted alongside TCP; in time-critical applications like voiceover IP (VoIP) telephony and video transmission, the speed of interaction between two concrete computers in the network has priority over the generally ensured arrival of a data packet.[61]

In 1983, the International Organization for Standardization defined seven protocol levels (the actual "archive" in Foucauldian terms) for successful communication in the Internet. Layer 2 is the "data link layer," which divides the data to be sent into packets. These packets follow the technical philosophy of the classical transmission channel, as every packet is considered individually and transmission takes place independently of previous or subsequent packets. This procedure is radically time-critical, as the "time to live" field indicates the maximum length of time (in seconds) during which a packet is permitted to be in the Internet. A counter continuously counts down during these routes; if the counter reaches zero before the packet reaches its goal, the packet expires, so that it does not remain stored on a server. The files in which the progress of chat conversions or navigations is stored is called "history." However, history is also the name for the short-term memory of the most recent program commands. Delay-time technologies, which make digital data processing a function of the smallest intermediate memory worlds, thus go hand in hand.

Digital media is divisible. The system is not only time-critical in terms of "time sharing"—that is, the simultaneous use of the computing time of a processor by multiple users at terminals; the time-critical moment also becomes actually operative in the "ping" signal of the Internet (metaphorically derived from sonar). The Internet manifests here as a structure characterized by *différance*—in other words, it is literally poststructural. When the wikipedia.org server is addressed not through the search word "ping" but rather through its electronic referential—namely, the "ping" signal itself—then the answer is not a text but rather a time protocol.[62]

The "ping" is a communication practice that entirely ignores humans. A source computer sends small data packets of the "echo request" type to the target computer; no intact letter is sent, but rather a fragmented body of data searches for the target according to temporal effectivity—in other words, in the time-critical domain rather than postal space. When the packets reach their target, it is configured to answer with an "echo reply." It can thus be verified whether an online data connection between the two machines was in principle (and thus on a media-archeological level) established. This is the first step in looking for network interruptions. The so-called "ping of death" is a media-archeological relic: for a while many TCP/IP stacks were sensitive to oversized "ping" packets, and they crashed so radically that they carried the entire machine along with them. This eschatology is inexorably inscribed in every data packet as a signal-technical intensification of Heidegger's anthropocentric diagnosis of Dasein as "being-toward-death" (*Sein-zum-Tode*).

Time-critical processes also take place at the "physical layer" of the Internet, which is the most basic layer of bit transmission. This layer is the interface to the respective material or electromagnetic transmission medium (copper cables, wireless networks, and optical fibers), and it embodies very concretely the interplay of logic and matter that the concept of technology itself already implies. What is specified here is which voltage level corresponds to a logical 1 and a logical 0. The function of the bit transmission layer is to convert the signals in the physical transmission channel into information, which is then transparently relayed to the second layer of the system. The success of this transformation of signals into data depends on the time-critical domain: "particularly signal frequency and duration, synchronous or asynchronous timing,"[63] as well as the choice of serial or parallel data transmission.

Asynchronous data transfer undermines a crucial criterion of mass media community building—simultaneous experience—and is linked instead to the "peer-to-peer" relations of classic telephony. Due to effective compression processes, audio files can be streamed so rapidly in the Internet that the delay required for computation can actually be minimized to 10–20 milliseconds; "real-time" digital transmission is thus experienced as equivalent to "live" analog-electronic transmission, although the extreme usage of channel capacity prohibits the simultaneous addressing of the public on a massive scale. Simulcasting first fulfilled the criterion of "Internet radio" as a format. To take the concept of "net time" literally: "There is one important difference between ordinary logic and the automata which represent it. ... Every network or nervous system has a definite time lag between the input signal and the output response. A definite temporal sequence is always inherent in the operation of such a real system."[64] In digitized communication networks, the "time stamp" counts in every sense of the word. The network time protocol (NTP) is an implementation of the TCP/IP for synchronizing the devices connected to a network. Dave Mills attempted to synchronize the system times of networked computers for the first time in the 1980s; today, this protocol operates in the picosecond range. Special protocols like open sound control (OSC) also operate on this basis; in an OSC packet, a "time tag" determines the temporal relationship between the computer and the server. Formulated in the time logic of a 64-bit operating system, this means that the first 32 bits specify the number of seconds since midnight on January 1, 1900, while the last 32 bits specify fractions of a second. If the "time tag" consists of 63 zeros followed by a one, this triggers the special case of immediate execution.

BREAKING WITH HISTORY: KNOWLEDGE IN THE AGE OF THE INTERNET

Siegfried Zielinski's *Deep Time of the Media* argues that the interface-oriented appearance of the Internet sublates all previous media history. "In the Internet, all earlier media exist side by side."[65] In its media-archeological actuality, however, the Internet is already post-history in the dual sense of "after history" and "after narrative."[66] Hypertextual links and network structures represent a mode of contingency rather than meaning. Historicity returns instead through the technological back door, as Harmut Böhme explains in light of the countless gaps, absences, construction sites, and instabilities in networks: "In addition to the time required for actual processing, they also require considerable time and energy for maintenance, repair, reorganization, self-stability: therein lies the historicity of networks."[67] However, this historicity points to the fleeting temporal mode of the medium rather than history, as it is virtually constitutive of websites that they are never definitively completed. Lessing's *Laocoon* treatise defined such temporal gaps as "pregnant moments," which always require the observer to actively anticipate the next moment of action. Beyond such tendencies to bring about narrative coherence, however, nonlinear trajectories of data in digital space require an explanation of other temporal conditions.[68]

The Internet is endogenously defined by a temporal mode that represents not linear development, but rather permanent dynamics and thus a virtually homeostatic dynamic equilibrium in terms of biological systems theory (Bertalanffy). It can only be historicized through subordination to an exogenous time arrow—that is, by bringing it into a linear temporal order. However, this embodies no longer a causal sequence, but rather a form of dynamic stability. The message is *temporary order* or order in fluctuation, as "the temporal stability of structures is only a result of temporal observation"[69] and thus a dynamic archive rather than history.

In the medium of print newspapers, information is linked to periodic moments in time. With the development of the high-speed press in 1812, and later the rotary press, newspapers and periodicals became synonymous with a form of time that was symbolically clocked by print media, while the symbol carrier itself remained entirely non-temporal. The program structure of radio and television inherited the symbolic structuring of human time planning from print media in the form of series and periodic news broadcasts, but with one crucial difference: in their technical substance, radio and television were themselves temporal modes.

The aesthetics of contemporary computer interfaces are still dominated by the office metaphor of the "desktop"; in contrast, David Gelernter designed the "lifestream" as an alternative interface that allows users "to maneuver around any repository of time-ordered electronic data."[70] Marshall McLuhan defined the efficacy of media as the way in which technologies massage and modify people's modes of perception. In this sense, they amplify not only the actual sensory channels, but also the implicit sense of time (Barthes' *punctum*). The present archive of knowledge is thus also transformed. "A text or image fixed in writing 'does not run away,' as access to the information source remains 'stationary.'"[71] Electronically stored knowledge (analog signals and binary data) appears to be different, as the fleetingness of electrons is an obstacle to magnetic remanence. In the age of Wikipedia, the online encyclopedia, knowledge has become time-critical—a function of access times and constant, always shorter actualizations. The time stamp necessary for all citations from Internet documents, which identifies "access times" down to the second, illustrates the degree to which knowledge has become time-critical *online*. The automated delivery of transmission material from broadcast media also comes not from the program archive, but rather from temporary storage, which constantly retains the predominant portion of musical formats. In the electronic field, the previously inert archive becomes dynamic.

Theodor Holm Nelson originally designed hypertext as a "docuverse" that preserved all of the respective versions of a text, including revisions of individual lines or even words (a project that is still ongoing in his Xanadu concept). Such a mnemotechnical layering results in a palimpsest, which allows not only spatial but also temporal cross-referencing: "hypertime."[72] Informatics responded to this with delta-coding, which was designed above all for the purpose of "online backups." In this process, successive variants of a document are not stored in their entirety; rather, only the differences between them are stored or transmitted. A protagonist in the temporal mode of high-tech media, the mathematical interval symbol Δt, becomes a linguistic concept in such "deltas." The archive (or rather its dataset) is thus understood no longer in terms of identity, but rather quasi-cinematographically in terms of minimal successive difference.

The concept of the encyclopedia still refers to the panoptic regime and the spatiality of typography as *dispositif*, but it has now been replaced by a dynamic order of networked time—an ephemeral chronotopology in dynamic equilibrium.[73] The technical *dispositif* makes a difference here. The printing press represents the permanent, non-rewritable fixing of encyclopedic knowledge in the form of read only memory (ROM). Photo albums already implied an entirely different order of knowledge, such as the photo panels from Aby Warburg's *Mnemosyne Atlas* in their operative form.[74] If the medial message of the electronic image—namely, its essential fleetingness—yielded a true media art in the form of concise video works under the corresponding term "Fluxus," then the temporalization in the World Wide Web now also seizes the spaces of knowledge. Topologically dynamized media are not only the technological form of contemporary communication; they also define the form of their content. In contrast to the closed format of the printed encyclopedia, the Internet orders knowledge in the form of an *open* encyclopedia. It no longer hierarchizes this knowledge, but rather correlates it with so-called "chaotic storage" (a term borrowed from the economy of warehousing). "The more serious, longer-range obstacle is that much of the information on the Internet is quirky, transient and chaotically 'shelved'"[75]—a nightmare of all librarians. On account of its open access use, the Internet represents no longer a social memory (as defined by Maurice Halbwachs), but rather a fleeting temporary storage—in principle, a random access memory (RAM). There have been attempts to counter this, such as the periodic storage of "snapshots" of the Internet by archive.org in the United States, but the name "archive" is only rudimentarily appropriate for this apparently total copy. The Internet is more like an archive at the technomathematical level—that is, as a hidden or secret archive (*archivum secretum*), which distinguishes it from the archives of public libraries. If the archive is understood not in an institutional sense but rather in terms of Foucault's concept of "the law of what can be said," then the archive that operates behind dynamic forms of knowledge like Wikipedia are the protocols of their addressability and availability.[76]

Hegel's philosophical *Encyclopedia* provides a theoretical model for this through the systematic distinction between technical and appropriative memory. With regard to the mechanical act of memory operations, Bergson emphasizes in his work *Matter and Memory* that the actualization of the past takes place in consciousness not as the identical retrieval of fixed and addressable information from storage, but rather as constant variation. Memories of the past and actions in the present are no longer mutually exclusive, as memory latently coexists with action as its storage-addressable shadow. The media-technical justification (thus the media-archeological condition) for the relative entanglement of these two temporal modes is the electronic circuit.

Not only is the infrastructure of the Internet radically poststructural, and thus equipped with time-critical vectors, but also the economy of knowledge on which it is based. In scientific essays, Internet contributions are cited with not only their year of publication (in the tradition of classical print publications) and their logical address (URL), but also their precise retrieval time down to the minute or even second. Constant mobility (the credo of modernity) is increasingly displaced by discrete temporal mobility—a kind of "time hopping." If the task of the archive has until now been to preserve legal claims and knowledge permanently in a fixed form and symbolic order, the field of knowledge called the Internet introduces a highly different dynamic of actualization in permanence. The economy that rules this fundamental temporalization of knowledge is of a time-critical nature, as fleetingness is the price to be paid for continuous actuality—"flow" and "streaming" as opposed to monumentality. This dynamic chronologic amounts to something provisional, as temporal finitude is included from the start. The traditional concept of the encyclopedia refers to the panoptic regime (and typographical geometry as its *dispositif*), but it is now replaced by the dynamic order of networked time—a chronotopology in dynamic equilibrium.

TIME CRITIQUE AND REAL TIME

A *process* is the transportation and transformation of matter, energy, or information. In informatics, more specifically, it refers to algorithmically executed operations involving the surveillance and control of technical processes, in which the computer must be able to react immediately. A process computer is thus equipped with a high-precision timing mechanism, which allows actions to be triggered or interrupted at defined points in time. Special programming languages, such as PEARL (Process and Experiment Automation Realtime Language), have been explicitly formulated for time-critical processes. Process control ultimately seems to disappear in interactive human-machine communication, as the computer conceals its temporal mode in order to enhance its creative effects. "What is made visible to humans on the surface is computed by the program behind the scenes at such a high speed that our senses cannot detect the slight delay while the image is being calculated."[77]

In the practice of digital signal processing (DSP), the concept of "real time" is a computer operating mode in which programs are always ready to process incoming data and the processing results are thus immediately available within a given time frame.[78] In human-machine dialogue, the concept of real time is anthropocentrically oriented—that is, it is determined by the human perceptual window of the present. In a broader sense, however, it generally means that the process defines the amount of time required; as a result, the system can be utterly machinic. The purpose of time critique is thus not only the nanotemporal processuality of neuronal communication. Calculated real time in the digital mode generates a temporal "as if"; as differential calculators, however, electronic analog computers offer insight into dynamic problems that are synchronized and thus chrono-contingent on values that can be altered *during the operation*. Real-time programming only allows the same interactive dynamics today due to the exponential increase of processor performance. As a model of differential calculus embodied in a technological form, the electromechanical analog computer known as the differential analyzer, developed and realized by Vannevar Bush, was able to solve dynamic problems immediately by referring to a library of default mechanical settings and practicing mathematics kinetically and virtually instantaneously up to the vibrational limits of the apparatus.[79] Real time is the equivalent to the dramatic "event" on the microtemporal level. In the neuronal time window, the stage is set for an operative rather than performative microdramaturgy.

> The preplanning of a movement . . . can only be accomplished within a time frame of up to three seconds. . . . When we speak with one another, our brain preprograms sentences of up to three seconds, which are then processed with word chains of ten to twelve syllables. This applies . . . independently of the respective grammar. The rhythmic structure of speech can also be found in the temporal structure of poems. In all languages, a spoken line of verse lasts up to three seconds. The brain thus predetermines the working platform, within which the poet's statements are realized.[80]

This is the chronopoetic verse time of oral epics and all musicality since Homer's time. However, fractals (nonlinear "chaotic" systems) seamlessly connect microscopic and macroscopic worlds with one another. Infinitesimal fluctuations can here cause large changes in effect; the irreversibility of such processes points not to a historical development of long-term causation, but rather to short-term threshold values. History is suspended at the microlevel of conditions of probability and time windows of the present, which correspond to another concept of time. The issue here is no longer a difference that first becomes visible on a linear time scale, but rather the microdramas that play out within an interval of real time. Clock time divides the apparently linear flow of time in a time-discrete way. This was supplemented by a new technology of parallel time-sharing: the simultaneous access of multiple users to the same delay time of a computer. In time-sharing mode, the idle time of the central processing unit is interpolated by the processing of tasks of other users.

The concept of real time refers to the implementation of a process within a punctual time interval. Real-time systems are able to respond to a signal either immediately or within a given timeframe.[81] In addition, real time also refers to a completely computable time beyond chronology, unlike the "live" transmission of analog broadcast media. "In real-time mode, strict time conditions are associated with the processing of a command. In other words, the calculation of results must be completed immediately or at least within a given time period, which can lie in the millisecond range,"[82] such as measurement and control tasks or the changing of traffic light phases in automatic traffic control systems—cyberneticized time.

Time-critical systems are dominated by a specific conceptual terminology of time. The key words here are real-time capability, real-time extension, reaction time, latency time, and interrupt.[83] Such process control differentiates two techniques: in the non-preemptive mode, each active process decides when to return control to the operating system; in the pre-emptive mode, on the other hand, process management can interrupt the active process at any point in time, and urgent processes can thus be given priority. Local time-critical intelligences in the form of digital signal processors enable the recording and online processing of measurement data with the highest sampling rates. The real-time capability of these systems enables rapid reaction times and thus "timeliness" (*Rechtzeitigkeit*).[84] In the context of automated processes with ultra-fast cycles, the concept of the time-critical has now become normal. It can even be found in the operating system of the computer that wrote this text.

A TIME-CRITICAL ECONOMY AT WORK

In the industrial age, the perception and practice of time depends on the clock pulses of machines.[85] In the United States, Gilbreth's time-critical study of work processes—the successor of Muybridge and Marey's chronophotographic analyses of movement—was directed toward the individual elements of the process.[86] In an economy of time constraints, only finite events are defined; the slogan that applies here is the same as the one that applies to the operational mode of the digital computer: treat time as discrete. When mechanical inertia is replaced by the speed of electric current, however, the stretching of time transforms into a kind of negative time, which is "unstable, flowing, inaccessible, uncontrollable."[87]

Money was once the embodiment of something that endured. In the form of its electronic equivalent, however, money is decoupled from its material stability (metal or paper), and it thus becomes extremely time-critical. Because the relative value of currencies is constantly changing from moment to moment, the exact time of day of a money transfer is crucial. The tendency toward fragmentation simultaneously gives the concept of time a more powerful meaning; the time-critical moment thus replaces the linear concept of historical time. Time itself is atomized into cascades of instantaneity; its disappearance in the moment makes it all the more valuable. "Speed, not cost, is the dominant factor."[88] In Alexandre Dumas' novel *The Count of Monte Cristo*, a protagonist transmits a (false) report of political unrest in Spain over an optical telegraph, which causes an opposing banker to promptly sell his stocks. He soon loses millions, as the price doubles following the correction of the report. In telegraphic communication, however, what counts is no longer semantics or even truth content, but rather the frequency of the messages themselves; high-frequency trading turns stock exchanges into radio. In the world of high-tech media, time is a precious economic resource (as are the rare minerals for its electronic hardware). Since 2005, the German stock index (the DAX) has been recalculated every second. The intervals between such time series tend toward zero.

The first "stock ticker" (invented by an engineer named Edward A. Calahan) was introduced in a stock trading office in New York in December 1867. Like an electric telegraph, this apparatus transmitted names and prices on paper strips; the interception of these messages at relay stations, which were necessary to amplify signals sent by wire over long distances, represented a potential gap in knowledge. The stock ticker thus transformed stock trading, which had long been performed verbally, into the new temporal form of "uninterrupted flow."[89] (The telegraph also transformed journalistic communication in the same way.) This transformation was based on the standardization of clock time and the establishment of a central timer. In contrast to statistical time diagrams, which recorded general trends over years or months rather than slight deviations, the ticker made price fluctuations visible minute-by-minute. Such fluctuations thus proved to be surprisingly autonomous with respect to major changes in the political or economic landscape—an autopoiesis of tempor(e)al subsystems. Conversely, television news channels now continuously display current stock exchange data as the *basso continuo* of all political news.

The introduction of the telegraphic stock ticker resulted in the instantaneous feedback of data flows and thus the cyberneticization of time. "These codes, required by the need to react promptly to price information, reinforced the bonds between investors and stockbrokers."[90] In the meantime, the mathematical fantasy of the financial market has developed time figures of nonlinear "prediction" that have long since eclipsed the aesthetic forms of chronopoetics. In a mixture of calculus and contingency, the practice of so-called "short selling" trades in falling market values as future options—a complete virtualization of the economy and its technical surrender to time. The widespread use of algorithmic models for the pricing of stock options is no longer a tool but rather the temporal essence of the stock market itself.[91]

NOTES

1. Qtd. in Richard J. Parmentier, *Signs in Society: Studies in Semiotic Anthropology* (Bloomington: Indiana University Press, 1994), 23. For a discussion of this statement, see Frieder Nake, "Zeigen, Zeichnen und Zeichen. Der verschwundene Lichtgriffel," in *Mensch-Computer-Interface. Zur Geschichte und Zukunft der Computerbedienung*, ed. Hans Dieter Hellige (Bielefeld: transcript, 2008), 124.

2. Wolfgang Riepl, *Das Nachrichtenwesen des Altertums. Mit besonderer Rücksicht auf die Römer* (Hildesheim: Olms, 1972), 100.

3. Michael Franz, Wolfgang Schäffner, Bernhard Siegert, and Robert Stockhammer, eds., introduction to *Electric Laokoon. Zeichen und Medien, von der Lochkarte zur Grammatologie* (Berlin: Akademie, 2007), xiv.

4. Friedrich Kittler, "Am Ende der Schriftkultur," in *Der Ursprung von Literatur. Medien, Rollen, Kommunikationssituationen zwischen 1450 und 1650*, ed. Gisela Schmolka-Koerdt, Peter M. Spangenberg, and Dagmar Tillmann-Bartylla (Munich: Fink, 1988), 296.

5. Wilhelm Fucks, *Mathematische Analyse von Sprachelementen, Sprachstil und Sprachen* (Cologne: Arbeitsgemeinschaft für Forschung des Landes Nordrhein-Westfalen, 1955), 19.

6. David Huffman, "A Method for the Construction of Minimum Redundancy Codes," *Proceedings of the IRE* 40.9 (1952): 1098–101.

7. See Ana Ofak and Friedrich Kittler, eds., *Medien vor den Medien* (Munich: Fink, 2008).

8. See Étienne-Jules Marey, *La méthode graphique dans les sciences expérimentales et principalement en physiologie et en médecine* (Paris: G. Masson, 1894), 156.

9. M. L. West, *Ancient Greek Music* (Oxford: Clarendon Press, 1994), 130.

10. Christian Kassung, *Das Pendel. Eine Wissensgeschichte* (Munich: Fink, 2007), 323.

11. Ibid., 327.

12. August Foerster, "Das Telegraphon," in *Die Pariser Weltausstellung in Wort und Bild*, ed. Georg Malkowsky (Berlin: Kirchhoff, 1900), 400.

13. H. Raabe, "Untersuchungen an der wechselzeitigen Mehrfachübertragung (Multiplexübertragung)," *Elektrische Nachrichtentechnik* 16.8 (1939): 213.

14. Fritz Heider, "Ding und Medium," in *Kursbuch Medienkultur. Die maßgeblichen Theorien von Brecht bis Baudrillard*, ed. Claus Pias, Joseph Vogl, Lorenz Engell, Oliver Fahle, and Britta Neitzel (Stuttgart: DVA, 1999), 329.

15. Ibid., 328.

16. "A stationary stochastic process is ergodic with respect to mathematical expectation when the average over the time is equal to the average amount." Gottfried Ehrenstrasser, *Stochastische Signale und ihre Anwendung* (Heidelberg: Hüthig, 1974), 10.

17. Jean-François Lyotard, "*Logos* and *Techne*, or Telegraphy," in *The Inhuman: Reflections on Time* (Stanford: Stanford University Press, 1991), 50.

18. Warren McCulloch and Walter Pitts, "A Logical Calculus of the Ideas Immanent in Nervous Activity," *Bulletin of Mathematical Biophysics* 5 (1943): 119.

19. Hartmut Winkler, "Zugriff auf bewegte Bilder, Video on Demand," in *Medien und Ästhetik. Festschrift für Burkhardt Lindner*, ed. Harald Hillgärtner and Thomas Küpper (Bielefeld: transcript, 2003), 318–31.

20. Lawrence Lessig, *Remix: Making Art and Commerce Thrive in the Hybrid Economy* (London: Bloomsbury, 2008), 98.

21. "For the aura is bound to his presence in the here and now. There is no facsimile of the aura." Walter Benjamin, "The Work of Art in the Age of its Technological Reproducibility," trans. Harry Zohn and Edmund Jephcott, in *Selected Writings*, ed. Howard Eiland and Michael W. Jennings (Cambridge, MA: Belknap Press, 1999), 4: 260. Bernhard Vief postulates a corresponding "inflation theory." See Bernhard Vief, "Die Inflation der Igel. Versuch über die Medien," in *McLuhan neu lesen. Kritische Analysen zu Medien und Kultur im 21. Jahrhundert*, ed. Derrick de Kerckhove, Martina Leeker, and Kerstin Schmidt (Bielefield: transcript, 2008), 213–32.

22. Thomas Kollbach, "Das Kunstwerk im Zeitalter seiner digitalen Reproduzierbarkeit" (examination paper, Humboldt University, 2010), http://www.medientheorien.hu-berlin.de.

23. The title of a book by social historian Eric Hobsbawm is programmatic in this context. See Eric Hobsbawm, *Zwischenwelten und Übergangszeiten* (Cologne: Papy-Rossa, 2009).

24. See David Austerberry, *The Technology of Video and Audio Streaming* (Oxford: Focal Press, 2002). In a broader sense, see also Stefan Andriopoulos, Gabriele Schabacher, and Eckhard Schumacher, eds., *Die Adresse des Mediums* (Cologne: DuMont, 2001).

25. Stefan Heidenreich, *FlipFlop. Digitale Datenströme und die Kultur des 21. Jahrhunderts* (Munich: Hanser, 2004), 204.

26. Jacob Ziv and Abraham Lempel, "A Universal Algorithm for Sequential Data Compression," *IEEE Transactions on Information Theory* 23.3 (1977): 337.

27. This temporal branching is the essence of logistical "routing," and it was an everyday phenomenon in the times of written correspondence. Aurelia Jurtschitsch

recalls: "Perhaps the delay of posting a letter . . . also secretly has the 'meaning' of better timing." Aurelia Jurtschitsch, e-mail message to author, November 25, 2011.

28. See Zuse, *Rechnender Raum*.

29. Analogous to this, the last words of the replicant Roy Batty in the film *Blade Runner* are "time to die." See *Blade Runner: The Director's Cut*, directed by Ridley Scott (Burbank, CA: Warner Home Video, 1997), DVD. I am grateful to Paul Feigelfeld for this observation.

30. McLuhan, *Understanding Media*, 4.

31. See Geert Lovink, "Net.Times, Not Swatch Time: 21st-Century Global Time Wars," in *Dark Fiber: Tracking Critical Internet Culture*, ed. Geert Lovink (Cambridge, MA: MIT Press, 2002), 142–59.

32. Ibid., 143.

33. This technomathematical time competence is communicated in the Internet forum for the online game *World of Warcraft*. "It is impossible to react as fast as everything is transmitted." http://forums.wow-europe.com/thread.html?topicId=1958327347&sid=3.

34. Michel Foucault, "Of Other Spaces," trans. Jay Miskowiec, *Diacritics* 16.1 (Spring 1986): 22.

35. Ibid., 23.

36. Ibid., 26.

37. Georg Franck, *Ökonomie der Aufmerksamkeit. Ein Entwurf* (Munich: Hanser, 1998).

38. Robert Hassan qtd. in Geert Lovink, "Die Indifferenz der vernetzten Gegenwart. Über Internet-Zeit," in *Zero Comments*, ed. Geert Lovink (Bielefeld: transcript, 2008), 179.

39. On the use of this metaphor in the nineteenth century, see Roland Wenzelhuemer, "'Less Than No Time.' Zum Verhältnis von Telegrafie und Zeit," *Geschichte und Gesellschaft* 37 (2011): 592–613.

40. Qtd. in Wolfgang Schivelbusch, *The Railway Journey: The Industrialization of Time and Space in the Nineteenth Century* (Berkeley: University of California Press, 1986), 37.

41. Paul Virilio, *Negative Horizon: An Essay in Dromoscopy*, trans. Michael Degener (London: Continuum, 2005).

42. See Régis Debray, "Pour une médiologie," in *Manifestes médiologiques* (Paris: Gallimard, 1994), 21–33.

43. Karl Marx and Friedrich Engels, *Ökonomische Manuskripte 1857/58* (Berlin: Akademie, 2006), 438. On the dynamization of geographical spaces through information technologies, see Roland Wenzlhuemer, "Globalization, Communication and the Concept of Space in Global History," *Historical Social Research* 35.1 (2010): 19–47.

44. Dr. Ing. Klimke, "Transozean-Telephonie," in *Handbuch für Funkfreunde*, ed. Telefunken-Vertreter-Gemeinschaft (Berlin: Telefunken, 1927), 13.

45. In the present context, this concept was shaped by Pierre Teilhard de Chardin. See Pierre Teilhard de Chardin, *Die Entstehung des Menschen* (Munich: Beck, 1961), 25–35.

46. W. T. Runge, "Elektronische Geschwindigkeit ist keine Hexerei," *radio-tv-service* 77/78 (1967): 2895.

47. Johannes Webers, *Handbuch der Film- und Videotechnik. Die Aufnahme, Speicherung, Bearbeitung und Wiedergabe audio-visueller Programme* (Munich: Franzis, 1991), 160.

48. Viola, "The Sound of One Line Scanning," 161.

49. Webers, *Handbuch der Film- und Videotechnik*, 154f.

50. Ibid., 159.

51. Siegert, *Relays*, 261.

52. Mindell, *Between Human and Machine*, 112.

53. Michel Serres, "Der Mensch ohne Fähigkeiten. Die neuen Technologien und die Ökonomie des Vergessens," *Transit* 22 (Winter 2001–2002): 194f.

54. Friedrich L. Bauer, *Kurze Geschichte der Informatik* (Munich: Fink, 2009), 111.

55. Mindell, *Between Human and Machine*, 107.

56. Pierre Lévy, "Die Metapher des Hypertextes," in *Kursbuch Medienkultur. Die maßgeblichen Theorien von Brecht bis Baudrillard*, ed. Claus Pias, Joseph Vogl, Lorenz Engell, Oliver Fahle, and Britta Neitzel (Stuttgart: DVA, 1999), 529.

57. This is how Josh Coalson describes the use of archiving for the complex "Free Lossless Audio Codec." See Josh Coalson, "FLAC," http://flac.sourceforge.net/comparison.html.

58. Geert Lovink, "Was uns wirklich krank macht," *Frankfurter Allgemeine Zeitung* (June 21, 2010): 27. Lovink here refers to the work of Italian media theorist Franco Berardi.

59. Ibid.

60. See Amy N. Langville and Carl D. Meyer, *Google's PageRank and Beyond: The Science of Search Engine Ranking* (Princeton: Princeton University Press, 2006).

61. Thomas Kollbach, "Echtzeitinternet. Die Bedeutung interaktiver und zeitkritischer Prozesse im Internet" (examination paper, Humboldt University, 2010), http://www.medientheorien.hu-berlin.de.

62. For an "antique" sample output of pinging en.wikipedia.org under Linux see the entry "PING (networking utility)" from December 21, 2010: https://en.wikipedia.org/w/index.php?title=Ping_%28networking_utility%29&oldid=403557765.

63. Christoph Neubert, "Elektronische Adressenordnung," in *Die Adresse des Mediums*, ed. Stefan Andriopoulos, Gabriele Schabacher, and Eckhard Schumacher (Cologne: DuMont, 2001), 41.

64. John von Neumann, "Probabilistic Logics and the Synthesis of Reliable Organisms from Unreliable Components," in *Automata Studies*, ed. Claude E. Shannon and J. McCarthy (Princeton: Princeton University Press, 1956), 44.

65. Zielinski, *Deep Time of the Media*, 31.

66. Translator's note: the German word for "history" (*Geschichte*) also means "narrative."

67. Hartmut Böhme, "Netzwerke. Zur Theorie und Geschichte einer Konstruktion," in *Netzwerke. Eine Kulturtechnik der Moderne*, ed. Jürgen Barkhoff, Hartmut Böhme, and Jeanne Riou (Cologne: Böhlau, 2004), 24.

68. See Lev Manovich, "Database as Symbolic Form," *Convergence* 5.2 (1999), 80–99.

69. Armin Nassehi, *Die Zeit der Gesellschaft. Auf dem Weg zu einer soziologischen Theorie der Zeit* (Opladen: Westdeutscher Verlag, 1993), 214.

70. David Gelernter, *Machine Beauty* (New York: Basic Books, 1997), 113.

71. Burkhard Stangl, *Ethnologie im Ohr. Die Wirkungsgeschichte des Phonographen* (Vienna: WUV, 2000), 71.

72. See Theodor Holm Nelson, *Literary Machines* (Sausalito, CA: Mindful Press, 1981).

73. See Wolfgang Ernst, "Temporary Items. Die Beschleunigung des Archivs," in *ephemer_temporär_provisorisch*, ed. Immanuel Chi, Susanne Düchting, and Jens Schröter (Essen: Klartext, 2002), 77–88.

74. See Aby Warburg, *Gesammelte Schriften Band II. Der Bilderatlas MNEMOSYNE*, ed. Martin Warnke (Berlin: Akademie, 2000).

75. "The Internet: Bringing Order from Chaos," *Scientific American* 276.3 (1997): 49.

76. Alexander Galloway, *Protocol: How Control Exits after Decentralization* (Cambridge, MA: MIT Press, 2004).

77. Frieder Nake, "Zeigen, Zeichnen und Zeichen. Der verschwundene Lichtgriffel," in *Mensch-Computer-Interface. Zur Geschichte und Zukunft der Computerbedienung*, ed. Hans Dieter Hellige (Bielefeld: transcript, 2008), 134.

78. See Georg Jongmanns, "Gute Zeiten, schlechte Zeiten. Das Echtzeit-Real," in *Authentizität als Darstellung*, ed. Jan Berg, Hans-Otto Hügel, and Hajo Kurzenberger (Hildesheim: University of Hildesheim, 1997), 253.

79. See Claude Shannon, "Mathematical Theory of the Differential Analyzer," *Journal of Mathematics and Physics* 20 (1941): 352f.

80. Ernst Pöppel, "Drei Welten des Wissens," in *Weltwissen/Wissenswelt*, ed. Christa Maar, Hans Ulrich Obrist, and Ernst Pöppel (Cologne: DuMont, 2000), 33.

81. See Paul Virilio, *Fluchtgeschwindigkeit* (Frankfurt am Main: Fischer Verlag, 1999), 27ff.

82. "Realzeitbetrieb (Echtzeitbetrieb)," in *Grundbegriffe der Medientheorie*, ed. Alexander Roesler and Bernd Stiegler (Paderborn: Fink, 2005), 65.

83. Gregor Burmberger, "PC-basierte Systemarchitekturen für zeitkritische technische Prozesse" (PhD diss., Technical University of Munich, 2002), http://tumb1.biblio.tu-muenchen.de/publ/diss/ei/2002/burmberger.html.

84. For an epistemological reflection on this concept, see Julian Rohrhuber, "Das Rechtzeitige. Doppelte Extension und formales Experiment," in *Zeitkritische Medien*, ed. Axel Volmar (Berlin: Kulturverlag Kadmos, 2009), 195–211.

85. For a discussion of Aleksej Kapitanovich Gastev's corresponding "Time League," see Zielinski, *Deep Time of the Media*, 240.

86. Frank B. and Lillian M. Gilbreth, *Motion Study for the Handicapped* (London: Routledge, 1920), 7. See also Siegfried Giedion, *Herrschaft der Mechanisierung. Ein Beitrag zur anonymen Geschichte* (Hamburg: Europäische Verlagsanstalt, 1994).

87. Boris Groys, *Unter Verdacht. Eine Phänomenologie der Medien* (Munich: Hanser, 2000), 17.

88. This phrase is from a US Signal Corps technical report published in 1957. Qtd. in Friedrich Wilhelm Hagemeyer, "Die Entstehung von Informationskonzepten in der Nachrichtentechnik. Eine Fallstudie zur Theoriebildung in der Technik in Industrie- und Kriegsforschung" (PhD diss., Free University of Berlin, 1979), 363.

89. Alex Preda, "The Stock Ticker," in *Making Things Public: Atmospheres of Democracy*, ed. Bruno Latour and Peter Weibel (Cambridge, MA: MIT Press, 2005), 624.

90. Ibid., 625.

91. Fischer Black, Myron Scholes, and Robert C. Merton invented such a mathematical formula in 1972. See Donald MacKenzie, "The Imagined Market," *London Review of Books* 24.21 (2002): 22–24.

Part III

RE-THINKING
"MEDIA HISTORIOGRAPHY"

Chapter 8

The Heterochronic Being-in-Time of Technical Media

THE AHISTORICAL INTERPRETATION OF MEDIA EVENTS

The essence of technical media is only evident in their operative implementation. Operational modes are generally subsumed under the powerful concept of time. However, a media-archeological view of the temporal modes of media almost inevitably leads to a critique of the totalizing collective singular "time" itself. It is thus appropriate to employ concepts that do not always already bind technically signifying time figures to a transcendental signified and burden them with an imaginary called history.

On the one hand, the development of techno-mathematical media is part of the field of man-made culture, which Giambattista Vico defined as the world of history. On the other hand, the inherent logic of high-tech objects— namely, the materiality and mathematical modeling of electrophysics—also constitutes a field of knowledge in new and at the same time equiprimordial ways. For one thing, these objects yield time-critical processes that occur below and above human sensory thresholds and therefore circumvent the conscious perception of time. In other words, the concept of media time is inextricably bound to media history, but it is largely invariant with respect to the discursive relativizations of cultural history, like the self-conception of the natural sciences. What is it about technical media that eludes history? In order to discover this *other heterochronous temporality*, it is necessary to perform the most precise investigation of concrete technical time processes; their solid knowledge and highly developed *termini technici* help to formulate new concepts for technical things that are no longer embedded in the narratives of classical media histories and time philosophies by questioning the concrete temporal modes and specifications of operative media.

According to the law of technology, media structures have their own inherent temporal logic. The examination of specifically technical proper times and media-induced modes of temporality with respect to human cultural activities is decidedly media-archeological. The concrete temporal modes of media, as well as their *inner historicity*, thus become objects. Temporality is the condition of possibility of media during their implementation; technologies are grounded not only in material objects and immaterial information, but also in instantaneous and temporally extended events. The essence of technological media should thus be understood in terms of their temporality, which requires another discourse. When understood as technical escalations, photography, phonography, cinematography, radio, television, tape recorders, video, and the computer (analog and digital) represent autopoietic and autonomous temporal forms. This is compounded by an epistemological criterion that marks a turning point in the technological analysis of time: Since the seventeenth century, the invention and technical implementation of infinitesimal calculus has made it possible to describe physical processes as functions of continuous variables and to analyze them discretely as time signals. Analog computers thus modeled the very dynamics that constitute the temporal modes of technical media themselves.[1]

Technical *chronopoiesis* is not simply time-based, but also time-synchronizing. Just as diagrammatic and mathematical analyses of images and sounds turned into technical syntheses (image- and sound-giving processes), which were conceptualized electronically in the synthesizer, so too was time not simply subordinated to high-tech apparatuses, but also produced by them. The role that minimal time moments play in signal- and data-processing resulted in a firm separation between technical media and previous cultural techniques. Chronotechnologies have emerged in which time is available at the symbolic level as well as in real signal processes. On the one hand, this availability is techno-mathematical, as it is realized in the active and passive components of electrotechnical circuitry; on the other hand, it is a dramaturgy—that is, an order in time. Aristotle's *Poetics* defines dramaturgy as that which brings *logos* to the confusing fullness of being. This *logos* is literally tempering.[2] Plato previously defined "rhythm" as a form of temporal structuring ("the order of motion").[3] It is precisely in the practices that Bergson disqualified as mathematical and spatialized time that *chronopoiesis* appears and explodes the concept of time itself. The present text concerns the dramaturgy, processuality, and proper temporality of such techno-temporal economies, which span the cosmos between *kairos*, *chronos*, and *aion*.

Ilya Prigogine suggests viewing time no longer as simply the parameter *t* of Newtonian physics, but rather as an operational time in the theory of irreversible processes, which is designated by the operator *T*. Hans-Jörg Rheinberger extrapolates the consequences of this. "According to this, every system of material units . . . that can be considered self-reproductive would have its own *inner time*. The inner time is not a dimension of its existence in space and time. It characterizes the sequence of system states, in so far as it can be understood as cycles of non-identical replication."[4] High-tech systems are dominated by a time that does not simply exist *a priori*, but rather only comes into existence during the implementation of media as soon as their feedback loops (with humans and other media) are short-circuited. The distinction between time-based, time-critical, and time-discrete media is essential. The contents stored in a printed book do not change depending on the point in time when they are read, as the symbol strings remain identical in accordance with the invariant alphabetic code; this is the condition of possibility for legal works and other canonical texts. In a similar way, musical automata mechanically preserve concretely programmed sounds. This results in a macrotemporal double nature, as they embody a historical index yet they are also metahistorically equiprimordial.

> Mechanical musical instruments uniquely distinguishes themselves in that they represent an unaltered acoustic document from the time of their production and dissemination, provided that changes caused by aging and storage conditions . . . are factored into the equation. The music is presented to our ears as directly as it was to the ears of people in the distant past, as mechanical musical instruments do not require the storage and conversion of the musical signal.[5]

What distinguishes printed letters and pinned barrels from electronic storage on magnetic tapes is therefore the dynamics of reading, but both of these temporal modes—that of the symbol and that of the signal—converge in the reading and writing mechanism of the digital computer.

The separation of the "internal consciousness of time" (Husserl) from the macrotime defined as history occurred long ago not only through the symbolic ordering of historiography but also in the addressing of the human sense of time through technological media. Time has always been ordered by means of chrono-cultural techniques like the calendar. Humans were entirely responsible for such operations with time symbols in the past, but ever since the automation of geared clocks media machines (defined as systems that include the input, processing, and output of energy and signals) have become capable of generating their own time events.

The anchoring of media analysis in the processuality of the technical object itself represents a qualitative leap, as it does not focus exclusively on the historical ordering of media knowledge, but is instead oriented toward the technical material and its logical *time* relations. The proper time of media thus reveals constellations that were obscured by the supremacy of the historical model. In Faraday's experiments with optical stroboscope effects and electrical induction, for example, "the history of radio and cinema momentarily crossed paths"[6]—and media *history* was thus crossed out. What occurred instead was the "internal dynamics of escalating references between technological media."[7] The more precisely media analysis focuses on the temporal congruousness of physics and mathematics rather than historical discourses and cultural contexts, the more it is done in the name of another epochality that is genuinely induced by media themselves. If history is involved at all, then it is the "anonymous history" to which Sigfried Giedion refers in the subtitle to his book *Mechanization Takes Command* (1948). However, a critique of media-historical time is difficult for people in Western societies, who in the narrative narcissism of the modern era have used history to reassure themselves of their own collective being-in-time (*Dasein in der Zeit*). The actual critics of historical time are technological media, and knowledge of their temporality thus helps to provide critical distanciation from the concept of history. This revelation should be understood in a media-archeological sense as a continual condition of possibility (*arché*) rather than simply a chronological-historical origin or beginning. Such a model of time follows the proper time of technical media in the sense of time compression—an elliptical contraction between genesis and current implementation. Every medium that is operative today also *realizes* at the same time transhistorical iterations of technological infrastructures, symbolic circuits, and electrophysical laws. During implementation, therefore, a medium does not yield a simple historical being.

Historiography is idiographic in that it describes the historically and discursively unique aspects of technical inventions, while media archeology is nomothetic in that it focuses on the techno-mathematical rules that first established the scope of so-called "media history" according to the law of a secondary, culturally acquired nature. The gap between virtually invariant structures and an eventful historical time has been a subject of discussion among twentieth-century historians themselves.[8] For the analysis of genuine media time relations, this time window needs to be shifted—like the "cursor" and "tongue" of a chronological slide rule—in order to separate it partially from the historical world. In his discussion of the temporality of the Mediterranean world, for example, Fernand Braudel initially distinguishes between a long-lasting geographical-physical space-time (*longue durée*) and a meso-time of gradually variant sociocultural environments in slow cyclical rhythms. Then comes the familiar history of events: "Surface disturbances, crests of foam that the tides of history carry on their strong backs. A history of brief, rapid, nervous fluctuations, by definition ultra-sensitive; the least tremor sets all its antennae quivering."[9] Braudel's metrological metaphor links back to the media-induced perception of time. Galvanometers are able to measure the slightest electrical events,[10] yet electromechanical apparatuses respond to microtemporal moments too slowly, too late, or not at all. There are thus times that elude the cultural-technical prosthetics of perception.

The idea that there is a difference between the concept of time and the concept of history, which does not adequately describe temporal processes and particularly the past as such, is central to Heidegger's *Being and Time* (1927).[11] Heidegger's philosophical method of questioning forms of temporality remains valid for the analysis of media-induced temporal processes, although these technologies respond to it in another way. The technological *arché* is not revealed in the absolutizing philosophy of time, but rather in the concrete temporal modes of media and thus in a dynamic field modeled on electromagnetic induction—*techniques du temps*.[12] A media archeology based on this expands the archeological-structural components of knowledge by adding the dimension of media-temporal processuality.

MEDIA TIME IS MEDIA IMPLEMENTATION

The media-archeological concept of "time critique" has a double meaning. On the one hand, it refers to chronotechnical moments that are literally "critical" (in the ancient Greek sense of the word) for the occurrence of a process—whether they are the smallest temporal moments or cycles that overlap with the perceptual window of the present. On the other hand, it is also a critique of historical time, which is difficult to perform for people who use history to reassure themselves of their own meaningful being-in-time. Only epistemological sympathy for the proper temporality of technological media helps to foster critical distance with respect to the emphatic concept of history. "Operative time is a medium," writes Kay Kirchmann;[13] as a provision for medial things, this categorization is sufficient. Media-technological being (*Dasein*) is always already being-in-implementation (*Sein-im-Vollzug*), and it yields its medial existence *qua* time.

Media archeology expands the philosophical perspective through another understanding of the connection between being and time. Pythagoras delighted in the mathematical elegance of the harmonic intervals of the monochord, but he did not consider the decay of oscillations—in other words, the transience of the signal—which was later overcome by electronic oscillators through the creation of undamped oscillations (such as the Meissner circuit of 1913). The *ideally* infinite sine oscillations out of which sound is composed in Fourier analysis were divided through the temporal windowing of the oscillation event in the form of wavelet analysis.[14] Short-wave radio receivers based on electron tubes (perfectly manufactured by Telefunken and Siemens) remained technically intact within the *delta-t* in which the infrastructure of broadcasting provided the *dispositif*; in the moment of reception they constituted an "enclave" of historical time[15]—a suspension in the sense of the ancient Greek concept of *epoché*. Within the time window of that infrastructural interval, the technical capability to process signals remains intact with respect to the objective passage of time until physically conditioned aging and malfunction cause an active electronic component in the receiving system to abruptly break down. Such an intrusion of historiographically datable time into the time-invariant essence of radio technology is techno-traumatic.

Technologies are critics of time insofar as they bring forth what discourse calls time through measuring and partitioning (ancient Greek *krinein*). While media history subordinates its objects of study to the discourse of history—and thus a symbolic model of time—media themselves are in a radical way time setters. When John Barwise and Alexander Bain patented their "Improvements in the Application of Moving Power to Clocks and Timepieces" in 1841 (British Patent No. 8783), the mechanical clock was imbued with media time by means of the transmission medium of electricity itself. "The application of the vibration of a pendulum for making and breaking the electric circuit at proper intervals, so that an electric current . . . may be transmitted as moving power to clocks and timepieces."[16] Time telegraphy (as formulated by Ladislaus Fiedler in 1890) is the prelude to frequency-modulated radio: time-critically coded electromagnetism. Everything that incurs in a media-technical channel (even a computer processor) is subject to a radical temporalization. It was evident in the development of phototelegraphy, for example, that in the process of transmission—in the media-operative moment—the so-called content of a medium is converted into time. In other words, images themselves are temporalized when transmitted in this way. This presents an epistemological challenge—namely, "what is an image when it is being transmitted, or aren't images something entirely different in the mode of their transmission?"[17]

The being (*Dasein*) of media-technical objects is not of a static ontological nature, but rather a purpose, an "in-order-to"—in short, it is a form of implementation. Heidegger once defined history as "the *mode of happening of this Dasein*."[18] It is the essence of operative media that they only occur as media during their implementation and thus they not only exist *in time* but also consist *of time* (which becomes very concrete in the "clocking" of the computer). The art of media archeology lies in moving beyond these verbal concepts and instead diagrammatically describing these modes of implementation. Mathematics has developed powerful tools for this, which operate under the name of analysis (integral and differential calculus); there is also electrical engineering, which is able to implement such analyses synthetically in the physical world and thus in real time. A mathematical algorithm is not able to bring about processes on its own; in an operative computer, however, mathematical formalism (in Hilbert's sense) and time (in Turing's sense) coincide.

Against all media-theoretical nominalism, therefore, the essence of media lies in their implementation. The analysis of such process-oriented articulations requires operative diagrammatics, just as chronophotographic plates enabled the analysis of bird flight by providing "an immediate, reliable diagram of the motion to be studied."[19] The temporal essence of the technical medium is realized in this *immediacy*.

PRIMAL SCENES OF MEDIA TEMPORALITY: HISTORICAL AND/OR MEDIA-ARCHEOLOGICAL ARTIFACTS

The ahistorical formulation of the macrotemporal embedding of technologies is methodologically grounded in media-archeological miniatures. Measuring media, which operate below human perceptual thresholds, revealed for the first time a microcosm of temporality, which was established as an object of knowledge and thus an epistemic thing. However, the engagement with this time-critical level also led to a closer examination of macrotime in the name of history.

Is it possible to imagine the past of technological things in a temporally complex way without using the model of history? The time-ordering model of history is only *one* level in the scaling of media-temporal processes. The materiality of technical structures is always subject to historical time, as electrotechnical components have a maximum usability period based on the laws of chemistry and physics, yet the symbolic order evades this physical entropy, as circuit diagrams remain largely invariant with respect to such half-lives. Symbolically notated circuits differ from their real-world implementation through their logical temporal mode. The relationship between technological media and temporality is divided.

For a long time, narrative was the dominant cultural mode of articulating sequences of events. As archives and writing systems became electronic, however, an entirely new information aesthetic came into play, which described the becoming of history in terms of static transition probabilities.[20] The media-archeological perspective is thus closer to the stochastic concept of the event in mathematics than the concept of hermeneutics in the humanities. Media archeology also assumes that another logic is at work here, which is the logic of media themselves.

The media-archeological perspective not only involves a "closing reading" of media-technical apparatuses, but it also pays attention to the temporal processes that distinguish them from traditional cultural artifacts. What does this mean when dealing with devices that are not yet old enough to be media-archeological fossils but are no longer new enough to be simply used functionally in the present?[21]

Suppose a rusty alternating current generator from 1900 falls into your hands. The required crank handle is missing, as it was borrowed for an old Edison phonograph. The fact that this machine can once again emit sparks, as if a century has not passed, shows that technical media have their own era. The techno-baroque splendor of Leibniz's four-species calculating machine also raises the question of the relationship between its mechanical functions and its baroque form (discourse and ornament), yet regardless of these (historically contingent) external features the machine still obeys a strictly mathematical logic that is instead media-archeologically effective. When faced with this machine, knowledge of its historical contextual is immediately visible to the contemporary viewer; the thinking and working time of the past crystallizes in such an artifact. The sheer application of the mechanism, however, sets it above its containment in the conceptual mode of historical time. As a calculating machine, it does not simply store all possible table values, but rather it brings each of them forth once again through concrete operations. The same also applies to printed circuit boards (electronic conducting paths) as the diagrammatic memory of connections that are once again ahistorically operative when they receive current.[22]

In opposition to cultural-historical time, the same physics of the oscillating string of a monochord appealed to the inquisitive ear equiprimordially for Pythagoras in ancient Greece and for Marin Mersenne in the early modern period. Harmonic analysis—a mathematical method of conceptualizing signal curves—is itself an echo of this technically implicit knowledge. Each discursive horizon of knowledge is actually different; depending on the historical index, things resonate differently in humans and yield different concepts. However, the medium event is always one and the same; it constantly induces scientific or aesthetic curiosities, through which humans respond to technological relations in new variations.

THE PRESENCE OF "HISTORICAL" MEDIA IN IMPLEMENTATION

The concept of the artifact, which once referred to an archeological-material thing, has since shifted to the realm of media technologies, where it now refers to a mistake or glitch that emerges from the digital sampling of analog images or sounds. Electronic media place the concept of the past itself in a new time field. In the "replay" of a sound recording from the twentieth century, the clear perceptual distinction between the archivally organized traces of the past and the documents of the present is replaced by the immediate presence of the signal event. The experience of listening to this recording is no longer an experience of distance, as it appeals to the ears in the present at the level of the signal. When a magnetic charge is converted back into sound by the tape recorder, the signal actually takes place in the present—from the perspective of both human listeners as well as the machine itself. A gap that is hardly reflected in the philosophy of history thus opens up between the senses, which immediately perceive such signals, and cognition, which is constantly aware of their "history." In 2008–2009, the Institute for Media Archeology in the Austrian town of Hainburg organized a museum exhibition of "Magical Sound Machines."[23] This exhibition featured historical audio media that could be brought into operation for a short time, which enabled an entirely different experience of frequencies and dynamics at the level of acoustic archeology than the supporting documentation was able to convey in writing. It is only through their sonic implementation that archaic sound generators prove to be actual media; the presentation thus focused on the temporary reactivation of the apparatus (particularly in the commissioned compositions of contemporary composers), which shifted the museum's emphasis from historicization to realization. For the purpose of conservation, such electrotechnical artifacts cannot be repeatedly reactivated, but once the sound event is reproduced it can be digitally sampled, so that it remains available in the future. The sonic information thus becomes timeless information, which is free of the materiality of a specific storage medium. This makes previous media history subject to ahistorical contingency.

TIME AND IMPLEMENTATION

The past is preserved differently in texts, which rely on pure symbolic coding. "Artifacts constitute the only class of historical events that occurred in the past but survive into the present. They can be re-experienced: they are authentic, primary . . . material available for first-hand study."[24] Nevertheless, the objective presence of a media-archeological object, like the early Tesla 4002 television from the former Czechoslovakia, embodies the withdrawal of the medium insofar as it does not process actual video signals. The chassis, which is equipped with active and passive components, here represents the condition of possibility—the *a priori* in a media-archeological sense—as the apparatus is in principle still operable: "Unlike the formal or historical *a priori*, the technical *a priori* implied technical things—but what were technical things? And were they so different from historical things that they required a new *a priori*?"[25] The technical object persists in a liminal state of latency—a literally medial in-between. "The differential of the technical *a priori* was necessary in order to synchronize the space of media and the time of history."[26] Medial epiphany (the electronic image) and technical "being-in-time" (*In-der-Zeit-Sein*) are here inductively entangled and constitute a time field rather than a historical context. In this context, the concept of the past has a peculiar double meaning. The past belongs irretrievably to an earlier time and is entangled with that event, yet nevertheless it can also be present-at-hand (*vorhanden*) like the remains of an ancient temple in a Mediterranean landscape, in which a "bit of the past" is "present."[27] The difference between a classical-archeological artifact and the mode of existence of media-technical time is as follows: not only is the past still present as a condition in the remnant of a technical antiquity, but the technical antiquity also creates the present in its processuality as a medium *in implementation* (provided that it still works when it has power). A characteristic feature of the specific essence of technical media is that it is only revealed in its operative performance. Even media objects in museums are pure presence when brought back into operation. The ancient statue of a deity in a museum is lost to the world, yet an ancient radio is still in its element when in receive mode. Heidegger described how the presence of a god in an ancient temple was produced through the "delimitation of the precinct as something holy.... It is the temple work that first structures and simultaneously gathers around itself ... paths and relations."[28] The structure of a technical medium differs from that of a temple, however, as the assemblage of high-frequency signals that constitutes an image or sound event is dynamic. The presence of a historical document is monumental, but the presence of an ancient technical artifact is processual—a microworld of time *in implementation*. As an alternative to the discourse of history, technical media themselves *create* presence. In contrast to the reading of archival documents, the grooves of an ancient shellac record do not appear to the eye as sound; they can only be heard when the mechanism of the gramophone implements them. In this moment, technical media are able to engage humans at the level of their temporal sense of existence. However, when the realization of stored signals or electronically coded texts no longer depends exclusively on humans, then this once again raises the question of the relationship between memory, technical media, and historical time. Operative media are virtually incapable of being in the past. The discourse of history breaks down when it comes to the penetration of their time-critical processes (time axis manipulation, Markov chains, and ergodics). Media archeology is thus rooted in a different time base.

THE PROPER TIME OF THE INSTRUMENT-BASED WORLD

Until now, I have avoided the concept of *history*. Is it appropriate to leave it entirely out of physical, chemical, and biological systems and to reserve it for the *temporal structure of cultural systems*? This depends on whether it is possible to agree on what history means.[29]

In the temporal domain of the organic, which is characterized by rhythms, periods, and fields, the concept of "differential reproduction" is recommended instead.[30] But what does this mean for that hermaphrodite of nature and culture—technological media—which are products of past human knowledge but strictly obey mathematical-physical laws? "For epistemological reasons, the proximity of archeological history to the natural sciences is . . . indispensable."[31] As symbolic knowledge, the imaginary proper time of the historical and thus man-made world articulates itself at the cognitive level as historiography; it is out of step with the real proper time of technological apparatuses and the sensory experience of temporality coupled with such technical media. A voice from the phonograph is heard by the ear as present. In 1903, a recording of Emperor Franz Joseph I in Austria made the event of the phonograph itself a subject of discussion; in 1945, however, the sound of the voice of Emperor Hirohito coming from public loudspeakers in Japan signified the shock of the acoustic-real. In a synesthetic shift, audiences bowed because they were not permitted to look at the emperor. A medial third body thus emerges *between* the "two bodies of the king" analyzed by Ernst Kantorowicz. With the rise of mass media, this media-induced presence has become universal and quasi-historistically ubiquitous. "The present created by media themselves operates in a special temporal structure."[32] This leads to a special form of present oblivion: special mass media (whether auditory or visual) reproduce "spatial relations and occurrences, including the exclusive perspective of time."[33] The required operating voltage of signal-critical news media—their "bias of communication" (Harold Innis)—lies not only technically in their temporal essence. Their incessant need to be up-to-date also leads to the constant creation of presence. According to their characteristics, news and stock prices (which are computer-based and distributed with commentary by mass media) are local artifacts of autonomous now-time and no longer primarily functions of an all-encompassing historical time that refers back to long-term storage.[34] Conversely, the discourse of history (as anticipatory) already receives the present as "historic," thus undermining the otherwise clearly separate momentariness of the current event. Heidegger sees the dissemination of mass media (which also applies to the Internet) as the climax of historicism;[35] Götz Großklaus diagnoses the heterochronous simultaneity produced by such media systems as "museum presence."[36]

The relationship between the proper time of the instrument-based world and the macrotime of history is proportional to that between self-referential systems and the so-called environment. During his literary studies at Cambridge, Marshall McLuhan's attention was drawn to the electrophysiological research of Keith Lucas and Edgar Douglas Adrian, who deciphered the code with which information was transmitted through the nervous systems of living organisms.[37] McLuhan recast this aesthetic mechanism as a media theory, according to which the message is the medium and electricity represents an extension of the human nervous system. McLuhan was thus also urging for media effects research, but in a technically and neurobiologically concrete sense. The massage of the sensory organs induced by the apparatus takes place at a microtemporal level, which acts asymmetrically with respect to the sequence of events at the macrotemporal level; this is analogous to the dual nature of a television image as both a technological and an iconological event depending on whether it is seen from a media-archeological or cultural perspective. The proper time of electronic images is closer to the essence of mathematical analysis than the mechanics of cinematography ever were. "Cinema history began with stroboscopic integration over time from image to image, but in the electronic image it reaches the level of pixels."[38] The cathode ray, which scans line-by-line with the help of saw-tooth generators, creates a time-critical image on the monitor that basically represents not a pictorial space (a surface) but rather a dynamized vanishing point—a quintessential signal image as opposed to the interval image in film.[39] Television thus represents the rendering of time as an image, which is far more radical than the time image in the cinematographic sense. If the model of "history" (according to Vilém Flusser) was a direct function of linear writing (literally historiography), then time is written differently today. In their book *Anti-Oedipus*, Gilles Deleuze and Félix Guattari describe how the stability of traditional writing evaporates as a result of its electronization—in other words, its conversion into points of light on monitors. The writing of an event is thereby transformed into a genuinely medial historiography. The time that is revealed in this process remains a challenge for media.

The theory of autopoietic systems formalizes the processes and interactions of control and communication, which is characterized by circularity, recursivity, and organizational unity. This also implies the temporal autonomy of systems, whose operations (which are always also external causes, like the "interrupt" in the computer) are continuously oriented towards their own *self-imposed* temporal structures. Focusing on these autopoietic-temporal structures leads to the concept of proper time. Electronic apparatuses bring into play not only a technical proper time, but also a logical timelessness—namely, the timelessness of mathematics, which is epistemologically understood as a metahistorical *mathesis*. Foucault's discourse analysis was an attempt to determine why similar ways of thinking emerged in entirely different disciplines without any verifiable contact between researchers—a phenomenon of the simultaneous co-emergence of questions, knowledge, and artifacts that was previously represented by the concept of *Zeitgeist* (or Plato's doctrine of forms). In his book *The Pattern on the Stone*, W. Daniel Hillis mentions "one of the more dazzling examples of synchrony in science—Turing, Church, and another British mathematician named Emil Post all independently invented the idea of universal computation at roughly the same time. They had very different ways of describing it, but they all published their results in 1937, setting the stage for the computer revolution soon to follow."[40] As a theoretical machine, mathematics is ahistorical because it deals with logical times that largely elude the entropic time of physics.

MEDIA-ARCHEOLOGICAL ANTIQUITIES IN THE MUSEUM

The establishment of prescribed time did not first appear with high-tech programmable media. A premedial setup—namely, the museum—already constitutes an architectural sequence of rooms, which suggests to the visitor a certain symbolic temporal order—mostly historical and linear—that is passed through in microdimensioned real time. The prescribed direction, which represents the temporal structure of an architectural framework, determines the course and the effect of walking through it. The pace (including moments of pausing, strolling, and acceleration) is left to the individual observer's discretion, just as the video recorder later emancipated viewers from the proverbial program and thus from the time restrictions of television. In a different way, cinema also presented itself as a *musée imaginaire*: the experience of time and space is here subject to the editing and montage, which (once stored in a fixed sequence on celluloid) inexorably and literally unwinds in the linear chronology of the film screening.

Media conditions are temporal relations. Is there a museum condition for early media technology? The media-archeological excavation of devices acquired from antiquarians—especially radios and televisions—makes it apparent: if the chassis of the earliest television from the former German Democratic Republic (the Rembrandt from the Sachsenwerk in Radeberg) is removed from its wooden architecture, it loses its historical index, which is always associated with a particular era. In a museum of technology, however, such a television is usually exhibited like a piece of furniture; the same applies to the art nouveau designs of early tube radios. The styles change, but the technical framework speaks to us in the present and constitutes a common family that operates not in terms of history but rather in terms of functional synchronicity following an independent logic with respect to historical time. If an ancient television is in principle still electrotechnically intact and set for reception, it will receive (via antenna or cable) the current program rather than a historical one, such as images from 1960 (following the stamp on the chassis of the television). A Sony video recorder from 1968 behaves differently, as the signal emanating from the reels is actually a recorded television broadcast from 1968 with an extremely dilated time interval. The latency of the signals stored on magnetic tape is a variant of the act of transmitting through the ether. This recalls fictional narratives about the reception of early television signals from space, such as the Hollywood film *Contact*: as electromagnetic waves, television signals fluctuate between immediate presence, which is close to the speed of light, and "historical" time, which is light years away. It actually involves two extremes of the same electromagnetic process, which represents another operational mode of temporality as it anticipates and joins historical thought.

Museums of the history of technology are presented as materialized architectures of memory; however, they actually reveal not the past but rather always only the presence of things. The so-called historical museum thus involves not the addressing of the past as such but rather another mode of being-present (*Gegenwärtigsein*). To paraphrase Bergson's *Matter and Memory*, "this past does not represent something that has been, but simply something that is and that coexists with itself as present."[41] From this perspective, every perception of the present is conversely already a function of stored perceptual schemata and in this respect nothing but the tip of an iceberg that neuronally underlies the latent memory of sounds and images—in other words, cognitive and symbolic data. The literal *anachronism* of storage media also lies at this temporal level. What exists here is time sublated in hardware architectures and software architectures; it is actually impossible to preserve the past in storage media. The technically mobilized museum depository thus more closely resembles the "random access memory" of the computer. Cultural memory does not constitute a stable canonical "read only memory"; rather, tradition is positioned according to all the rules of telecommunications.

Where does the memory of media lie in the museum depository? It is entrusted to technical signifiers, but what makes them a memory and how is their temporal dimension revealed? Heidegger radically raised this question of the time frame of media-technical artifacts in the museum. The possibilities and limits of media historicity can be seen in antiquarian media, whether in the depository or in museum exhibitions.

> The "antiquities" preserved in museums . . . belong to a "time past," and are yet still objectively present in the "present." How are these useful things historical when they are, after all, *not yet* past? . . . Such useful things can only, after all, be *historiographical objects* because they are somehow in themselves *historical*. . . . With what justification do we call these beings historical when they are not yet past? Or do these "things" "in themselves" yet have "something past" about them although they are still objectively present today? *Are* these objectively present things then still what they were? . . . The tools have become fragile and worm-eaten "in the course of time." But yet *the* specific character of the past that makes them something historical does not lie in this transience that continues even during their objective presence in the museum. But then what is past about the useful thing? What *were* the "things" that they no longer are today?

They are still definite useful things, but out of use. However, if they were still in use, like many heirlooms in the household, would they then not be historical? Whether in use or out of use, they are no longer what they were. What is "past"? Nothing other than the *world* within which they were encountered as things at hand belonging to a context of useful things and used by heedful Da-sein existing-in-the-world. That *world* is no longer. But what was previously *innerworldly* in that world is still objectively present. ... A Da-sein that no longer exists is not past in the ontologically strict sense; it is rather *having-been-there*.[42]

The temporality of media not only lies in the "historical" time embedded in technology, but it also derives time from technological movement.[43] "What . . . applies to every tool that is actually a thing but serves a purpose that lies beyond itself"[44] applies even more acutely to operative media, which only exist in implementation, in the being of media (*Mediendasein*), in being-in-the-world-as-time (*Welt-als-Zeit-Sein*). At a conference on mechanology, Gilbert Simondon urged British curators to restore early steam engines from the era of the industrial revolution to working condition (a proposal that has since been realized at the Science Museum in London). "There is something eternal in a technical schema . . . and it is that which is always present and which can be preserved in a thing."[45]

THE BEING OF MEDIA IN TIME

In connection with Johann Gustav Droysen's theory of history, media historics is alert to different modes of the presence of the past. All of these modes revolve around a temporal abyss—namely, the *per definitionem* absent former event (*res gestae*). The functions of this absence include the material traces left behind from the past (*vestigia*), the active memory that lays the groundwork for traditions (*memoria*), and historiography itself (*historia rerum gestarum*). Droysen limited the concept of "sources" to oral and written records created for the express purpose of establishing tradition, which thus consciously carried within them the historical vector of time.[46] This differs from so-called "remains," which typically fall under the discipline of archeology. However, the material remains of the past "only appear as traces of something when they are specifically evaluated according to their history";[47] they thus call for a Peircean interpretation in the *process* of historicizing semiosis. Is the time of media itself able to be an interpreter instead of the cognitive construct of history?[48] The historical narrative here reaches the limits of its plausibility. If the events of the past are grasped not in the writing of historiography, which suggests linearity, but rather in diagrammatic mathematics, such as graphs, then they almost appear as retro functions of *vestigia*, *memoria*, and *historia*. According to the model of differential equations, the assumed so-called *res gestae* can be derived as a discontinuous curve (thus as a function) composed of fragmentary traces that infinitesimally approach an assumed historical time; the curve is thus asymptotic, like in the frontispiece of Joseph-François Lafitau's *Customs of the American Indians Compared with the Customs of Primitive Times* (1724), where Clio's pen and Chronos' scythe virtually converge.[49]

In the practice of mathematics, the phenomenon of so-called "delay differential equations" involves the appearance of functional values or derivatives in the past (a temporal dimension of the second order) next to a function and its derivatives according to time at point x. The approximate (alpha)numeric approach to the historicity of so-called historical media corresponds to a method of researching history that Droysen defines as an "inquiring understanding"—in other words, a concept of hermeneutics that is entangled with the research methods of the natural sciences. The "endlessness of the task" lies "in the restless exploration of tradition";[50] understanding is always only successful in the dynamic process of an asymptotic approximation as "sampling." The annulments and sublations of historical distance through reactualization—the "mode of temporality of recursion"[51]—is a media-specific phenomenon that induces a fundamental questioning of history. Of course, experimental archeologists are also able to settle down in the largely intact ruins of Pompeii and reproduce the everyday cycles of ancient Romans. However, this is different from the emulation of an ancient computer on a contemporary platform. With the help of microprocessors, computers are able to reproduce time itself as a process—something that is not possible in traditional history museums, which simply place their moribund objects on permanent display. When it comes to time, therefore, the sites of memory and the temporal modes of technological media diverge. Such operative pathways illustrate not only how neurological memory occurs from a constructivist perspective (images are always newly assembled), but also how a temporal event is reproduced identically in the medium itself. The media-archeological basis of these "traces" is not a metaphor, but rather a technical diagram.[52] The media-technically sublated past is obviously more than simply the screen memory of the causes of the existing traces. According to Hegel, "an internal common ground" is "collectively brought forth" by the *historia rerum gestarum* and the *res gestae*;[53] the actual challenge of media theory is to define the mechanics and electronics of this equiprimordiality for techno-mathematical worlds.

INQUIRING UNDERSTANDING? THE REENACTMENT OF MEDIA

For the concept of antiquated media, "inquiring understanding" means in the narrow sense of the term the *reenactment of the medium* (meant here in the sense of the *genitivus objectivus* as well as the *genitivus subjectivus*). Reenactment also introduces the option of responding to the question of the status of technical artifacts with Heidegger's interpretation of *mimesis* as *poiesis*. If the essence of technical media lies in their implementation, then they are radically committed to temporality. The classical archeological-historical method conceives of both intentionally and unintentionally recorded sources as tradition, including written sources, material remains, and abstract remains (institutions, infrastructures, etc.). However, "this approach only applies to the time before the invention of modern techniques of documentation, such as photography, film, and tape recordings."[54] What are *vestigia* in the age of technical reproducibility? This is where media archeology comes in. The scene of the Soviet flag being hoisted above the Reichstag in Berlin by the Red Army in 1945, which is repeatedly shown each year to commemorate the end of the war, was actually a reenactment, as were many other photographic and cinematographic icons of our collective image memory.[55] The so-called "historical" event is here constituted in media time rather than historical time. However, the practice of reenactment first became fully technical in the self-referencing of media themselves. The technical-historical double bind of media temporality was evident, for example, in the revival of the first British television play from July 1930 (Luigi Pirandello's *The Man with the Flower in his Mouth*).

> In 1967, an edited version of the play was remade entirely in 30-lines and recorded onto a stereo tape recorder. One track held the 30-line video signal (with the innovation over 1930 of having sub-black synchronising pulses, as used in analogue broadcast television), the other track held the audio. Bill Elliott, at that time with Granada TV in Manchester, used his own modified Televisors acting as camera and monitor. However, the most exciting feature of this re-make is that it was supposedly re-produced and presented by the original producer, Lance Sieveking, supported by the original art-work . . . and music recording.[56]

A *remake* of past technology with a durable apparatus here appears in the *reproduction* of a drama with the original director. The former is functionally equiprimordial to the 1930 television event, which is due not to historical time but rather to the synchronizing pulses, while the latter depends on variance, as it involves other actors and another interpreted staging. Reenactment is also a historical method of experimental practice, which is in reality an ahistorical restoration of the temporal relations of the apparatus. Reenacting an experimental setup for the purpose of education or research means not simply reproducing a time-invariant logic, but rather revealing the algorithm of the experimentation—the processualized logic itself as a temporal event.[57]

THE VECTORIZATION OF BEING THROUGH TIME

The modern era of technical media resulted in an internal escalation in the form of clockwork, which Heidegger subsumed under the concept of "vulgar time." Since the invention of this mechanism, ontology no longer passes by the concept of chronotechnical operations, which leads to a reevaluation of the relation of being and time.[58] The resulting idea of being as a process thus brings it closer to technical procedures as well as music, which like media always only emanate in implementation. "Music has no 'being' or solidified 'state,' but rather only a state of emergence or becoming, which is shaped by process, movement, and tension. The work exists not in the fixed text of the composition, but rather in the performance, which progresses in time and is bound to the succession of individual moments."[59] Christiaan Huygens' pendulum clock structured the measurement of time down to the level of the second. Through his double-clock experiment of 1655, however, the being of physics itself became subject to a measuring temporality—not the time of emphatic history, but rather one that opens up a world of microtemporal economies of synchronization.[60] In his chapter on clocks in *Being and Time*, Heidegger shifts from a fundamental-ontological description to a virtually media-archeological description, which reveals a dilemma of the history of technology. "A history, which is essentially time, intersects with another history, through which the machines of time-measurement themselves pass. Clocks are ontic devices, thus subordinated to fundamental ontology, which nevertheless bring about historically different ontologies."[61] Operative media constitute relational, dynamic modes of temporality, as they exist in time, they are endowed with their own proper time, and they also yield time; their being is always already integrated into the contexts of events and temporal procedures that can be written down—particularly as algorithms. Unlike Husserl, Heidegger starts not from the phenomenological view, but rather from understanding. Following this, understanding the being of media means reenacting a temporally constituted and thus *a priori* postal structure. This poststructural fate became a media reality in the mathematical-technical communication theory of the twentieth century. Heidegger does not offer any instructions on how to define the era of high-tech media realities, as his questioning of the essence of technology hardly addresses their concrete temporal worlds or their techno-mathematical conditions; however, his questions provide mental tools that help to linguistically grasp the genuine temporal modes of media. According to this, it is impossible to formulate the question of the being other than in its temporal and operative modes of implementation, and this being-in-implementation can be defined by a mathematical symbol that applies to it: the vector drawing as a finite interval. It is particularly true of technical media that their being is revealed processually rather than in the nominalizing description of an objective given: foreordinations whose essence lies in operations and temporal modes. The concept of "essence" (the ontological question) thus shifts to the concept of "modes" and referential connections, which emphasize the temporal moment and move the Heideggerian identification of the "in-order-to" of tools closer to Charles Sanders Peirce's semiotics.[62] In this way, the being of technological media is an operation in the world and thus a function of time. Technical media represent operative diagrams, understood here in the sense of Peirce's "existential graphs," which are indexical rather than purely symbolic forms—traces of the real. "Existential graphs . . . index themselves as diagrammatic machines."[63] Peirce himself linked this to the circuit diagrams of electronic apparatuses in a letter he wrote to Allan Marquand on December 30, 1886. "It is by no means hopeless to expect to make a machine for really very difficult mathematical problems. . . . I think electricity would be the best thing to rely on."[64] In the era of alternating current and binary information processing, diagrammatic machines embody an endlessly alternating time: temporalizations and dynamic interactions by the grace of electromagnetic induction, algorithmic formalization, and cybernetic feedback.

The conscious perception of the radical temporal structure of technical media increasingly disappears under the time-critical threshold; this *dissimulatio artis temporalis* is the condition of its efficacy. In computer games, the interactive effect of minimizing temporal human-machine relations depends precisely on the dissimulation of the temporality of data processing (as well as neuronal signal processing). The challenge facing media studies lies in conceiving of time and being poststructurally and thus differentially. Following this, it also suddenly becomes clear how the design of dynamic networks culminated in the Internet—networks that are not simply present-at-hand as static topological spaces, but rather represent chronotectonic models of operative time. If such a being-in-time (*In-der-Zeit-Sein*) proves to be the basic state of electromathematical being-in-the-world (*In-der-Welt-Sein*), then this media-archeological basis also has logically necessary media-epistemological consequences. The task of media theory is to force *mimesis* on the temporal modes of such processes. The being of technical media as operative implementation is incompatible with a motionless ontology, which is still based on the subordination of non-temporal being. Operative media embody the program of a temporality that produces the ecstasies of time out of its techno-mathematical functions. In Turing's model of a machine for calculating computable numbers, the Hilbertian "yes-no question" became the so-called "halting problem" and the symbol-processing machine was even aware of its own inherent temporality; in 1936, Turing explicitly wrote that in every discrete state it was "conscious."[65]

THE "IN-ORDER-TO" OF MEDIAL TOOLS

Media-technical tools are defined as medial through their continuous process of referencing—or, more concretely, their hardwiring. In the era of electrification in particular, they primarily exist as temporal, electrotechnically implemented structures. Technical tools never appear as such, but rather they are realized and assimilated almost entirely in their dynamic function (apart from the resistance and veto power of their materiality). Medial tools can therefore only be experienced phenomenologically when working with them. Being and object are linked in the syntactic structure of the "in-order-to," which has become a circuit; the temporal structure of every syntax (prolepses and retentions) thus emerges. However, the working context of a tool also implies invisibility; when the hammer is absorbed in hammering, it is ready-to-hand (*zuhanden*) and thus extinguished as a conscious thing. The more a tool is ready-to-hand, the less it emerges from its structure of references and the more it functions and processes commands unconsciously. The more interruptions, disturbances, and discontinuities it evokes, the more insistently it is made conscious. As a pointed form of temporality, malfunctions thus reveal the technicity of the medium. The "in-order-to" refers to the operative part of the being-in-the-world of technological media—in other words, the movement of its being. Being is realized at the level of media reality as actual implementation within the wholeness (*Ganzheit*) of a tool. The meaning or directionality of such being is therefore temporality; if this temporality is halted, then the being of media is itself suspended. An electronic television image cannot stand still.

Existential analysis (*Daseinsanalyse*) focuses not on the external features of technological media, but rather the essence of their design. The identification of the temporality of operative media is concretized in the analysis of their referential connections. Readiness-to-hand refers not to objects that are present-at-hand in their sheer material reality, but rather their vector; the dynamic structure of all media is thus implicitly defined. In the restoration of media art, museums distinguish categorically between the preservation of the substance of the works and the preservation of their function.[66] An antique, hand-wired, and completely tube-equipped television is ideally able to show current television programs via a cable signal feed, just like a brand-new device; in the moment of successful reception, the difference between the historical and the contemporary device is sublated, until the signs of aging in the antique television manifest visually or audibly, like a veto of physics with respect to its technical functionalization. In accordance with the second law of thermodynamics, it is certainly unnatural when ancient capacitors are expected to merge with real electronic components in their constant functionality. The fatigue of crucial components reveals the physical-technical temporality of the medium, which undermines the time-invariant nature of logical circuits. The preservation of "signal integrity,"[67] which is a decisive criterion in the revival of earlier media art, must take this conflict in the media-specific nature of their implementation into account.

ANTIQUARIAN RADIO: AHISTORICAL?

From a media-archeological perspective, a history of radio begins not with its transformation into a mass medium but rather with the scientification of electricity. According to its temporal essence, however, electricity is fleeting, which poses a challenge to the transfer of knowledge of radio. The generation of electricity through the rubbing of amber with textiles in ancient Miletus, which was followed by a discharge of sparks, already enabled the transmission of electromagnetic waves, although it was not until the nineteenth century that it was first modeled mathematically by James Clerk Maxwell and verified experimentally by Heinrich Hertz. Radio's media-archeological condition of possibility was thus discovered in the laboratory, and it was first transformed into a communication medium by engineer-entrepreneurs like Giulielmo Marconi, who combined the Hertzian apparatus with Branly's coherer and Popov's antenna in order to create a functional instrument for the wireless transmission of Morse code. However, the primary experimental system already "knew" (or suspected) it. In a variation of an expression of Douglas Kahn, it is possible to say that radio already invented itself before it was invented.[68] The key scene underlying this invention was Oersted's discovery in 1819 that a current-carrying conductor generates a magnetic field. News of this "momentous discovery"[69] spread as quickly as the astonishing effect of induction itself. In the scientific world at that time, however, something had to resonate in order to produce such an instantaneous reaction, which triggered the subsequent research of Faraday, Maxwell, Hertz, and then Popov. This could not be justified on the basis of discourse alone; rather, discourses are here also grounded in media. As a result, it is also easy to *operatively* replicate and thus equiprimordially revive Ferdinand Braun's nineteenth-century crystal radio receiver in the present using commercially available electronic components. "It is remarkable that we can still follow Ferdinand Braun's design principles today, even though they were established 100 years ago and we are now many developmental stages ahead."[70] In the moment of this radio reception, its history is itself sublated in the implementation of media. When we are involved in the temporality of the medium, this effect loses its "remarkable" character; it is only "remarkable" from the perspective of the historical discourse.

Strictly speaking, we know nothing about radio yesterday or radio fifty years ago. We only know at best fragmentary recordings by means of other media, such as optical sound tracks, wax discs, and audiotapes (actually in that order). Radio is thus lacking a homogeneous, coherent unity of conversion, storage, and transmission. The unity of an epic poem involved the conversion of speech into rhythmic meter, its storage in the head of the singer, and its transmission from Sparta to Athens. Writing is a "unity of the vocal alphabet as converter of phonemes, storage on papyrus, and transmission by messenger. ... Film, records, and tapes are storage media that involve conversion and transmission. Radio, on the other hand, leaves no traces behind in a given monitoring system."[71] The cultural-technical coupling of transmission media and memory (as still embodied in the written letter) diverges in the electronic transmission media of radio and television, and they thus rupture the discourse of history.

> Based on the absence of storage and all the traces that storage brings with it, something resists the medium of radio—something that we usually simply call history. History, as Hegel famously says in his *Lectures on the Philosophy of History*, is based on the "ability . . . to preserve." Time periods without preservation, thus without writing, "whether we suppose them to be centuries or millennia . . . are on that very account destitute of objective history." From this perspective, the radio would also be "destitute of objective history," as it leaves no traces behind. However, this is precisely not the case. History is constituted by the media of history, including epics, writing, papyrus, codices, books, films, and audiotapes.[72]

However, storage media are actually a necessary but not sufficient condition for the discourse of history, as history is in fact only constituted by the performance of narration.

> If history exists, and that is also one of the beautiful and great truths of Hegel's lecture, then it updates itself, so to speak, for everything is history and nothing is without history—even a medium that came into the world 100 years ago and is structurally lacking the technical ability to preserve or store. What I . . . suggest is therefore the reverse. It is actually characteristic of radio not that it is without a history—it is utterly impossible to imagine such a situation—but rather that it compensates, replaces, and conceals this absence of historicity. What is the storable material that I hear? What is the material from which what I hear comes? We will get to know the material today. It was called, and is sometimes still called, the ether. The radio substitutes its storage deficit and thereby generates not only pre-fascistic ether-cosmogonies, but also very constructive things.[73]

From when, or when still, is a crystal detector in a battery-free radio receiver historical? "I just received this crystal holder with a galena crystal as a gift. It lay many decades in a cellar and had nothing to do. But the picture proves it: everything still functions as it did on the first day. The component tester in the oscilloscope shows a diode's characteristic curve."[74] With radio transmission in the technical mode of high fidelity, the medium virtually disappears in the process of conveying electromagnetic waves to the ears, yet the price of this bridging of distance is a complete translation of the physical sound event into another (electromagnetic) space. This is just as dramatic as the process of digitization (a further escalation), which is even more imperceptible because the sampling theorem deceives the senses. A profound, almost ontological difference opens up precisely when the difference between analog and digital signal processing is no longer audible—in the complete dissimulation of technology.

The era of analog electronic broadcast media is currently coming to a close. It can be described in terms of media history, but this deprives it of its essence. Its physical and technical basis will still constitute a necessary condition for transmission communication in the future, but it will be displaced by the transformation of physics into information—in other words, the digital, which *technically* still does not represent an alternative to classical radio and television.

Handling artifacts from the era of analog electronic media produces a certain degree of confusion. On the one hand, radios equipped with electron tubes, like Mende radios from the late 1930s, are in principle still ready to receive signals with a 220-volt connection and a sufficient antenna cable. Unlike an authentic medieval document in an archive that reports news from the Salian era, however, these radios do not broadcast programs from the late 1930s. This era was stabile with regard to its technical infrastructure: a technological interval that on a principle level experienced no history, no transformation, but rather the progress of Hertz's laws of radio. This era now comes to a close through the reallocation of international broadcast frequencies in favor of digital broadband, where radio and television are supposed to be transmitted on a spectrum. Even the stronghold of analog radio and direct contact with the physics of the ionosphere (electronic weather)—namely, amateur and short-wave radio—are gradually being replaced with digital interfaces and digital radio mondiale (DRM).

It is possible that a Start television from the 1960s (the first televisions with printed circuit boards, although they still contained exceptional electron tubes), when activated today, might briefly show an outline (in black-and-white negative reversal) of the image of a news moderator from a current television program supplied by cable, which then immediately disappears again. This is because the unstable electronic equilibrium in the apparatus shifts according to temperature and age (in the case of expired capacitors, it is irretrievable). This is actually the unique phenomenon of a distance, however close it may be (Walter Benjamin's definition of the aura), and at the same time a core definition of television: a ghostly appearance, in which something that is temporally past or spatially distant becomes eerily present.[75] The presence of an absence is symptomatic of the posthistorical time consciousness.

However, unlike an archeological site, which only becomes part of the historical imagination through ancient textual sources, the absence of the past in an antique electron tube radio is only a historical absence in one respect: past radio broadcasts. Is radio resonance historical or in the present? When conceived in terms of the concepts of historicity, it is astonishing that an antique tube receiver is able to receive a current broadcast but not the sounds and images that were broadcast at the time it was manufactured (despite the fantasies of literature and Hollywood films).[76] Even when the signal input of such a radio is supplied with recordings of historical broadcasts, it processes these signals as if they were contemporary—just as visual memories in the neurological sense are not retrieved but actually newly assembled and thus regenerated each time they are recalled. Insofar as a radio is capable of receiving current programs in the present, even though it is a deeply historical technical device, it recalls Walter Benjamin's concept of history, as something actually "comes together in a flash with the now to form a constellation"[77]—a flash that becomes literal in the electric spark. Benjamin's concept of the "now" as a quantum of time is more of a temporal short circuit in the electrotechnical sense than a "historic" leap. The apparatus is implemented in the present as long as it still remains ready to receive. Media theory is concerned with the question (which Heidegger did not really answer) of the instrumental nature of a technical medium that has museum status but that is still operational in the electromagnetic field and thus ahistorically "occurs." A time machine is needed to put it in its media-specific place in time—such as the "German small receiver" (Deutscher Kleinempfänger DKE 38 in various variants), a radio from the National Socialist era that has a swastika engraved not only externally over the tuning thumbwheel but also internally in the Telefunken triode, which is thus concealed and largely invisible to users at the time.[78] When connected to a 220-volt outlet this device is still ready to receive current medium-wave broadcasts, like Deutschlandradio, which is proof of the infrastructural stability of media whose systems are still technically intact after decades. A so-called "people's receiver" (Volksempfänger VE 301 W), which was developed in 1933 by order of the ministry of propaganda and available in 1938 as a variant of the "small receiver," is in current rather than historical condition when it *has power* (as an electrodynamized diagram) and receives signals from the strongest local transmitters. Free of all protective and designer casings, the electrotechnical wiring of components on the metal chassis of the apparatus openly reveals itself in an act of media-archeological unconcealing (*Entbergung*); the gaze then focuses no longer on the historicizing form, but rather on the analysis of the radio as a technical medium. The entire difference between a still resonating and a non-intact radio is first revealed by applying power.

The historical status of electromagnetism as a media-archeological condition of radio broadcasting and receiving is one of the key questions of media studies. Even more central for media knowledge, however, is an understanding of the time of this event. When it occurs, an electromagnetic field behaves in a way that is invariant with respect to historical time; it is thus genuinely ahistorical, and it appertains more to media archeology than history. In principle, technical storage and receiving media exhibit the same behavior at different points in time. The parameters of their electronic components always comply with identical mathematical descriptions. If their differential equations remain constant, then their functions remain equiprimordial. In radio, the electrophysically conditioned temporal delay of a received input signal actually leads to linear distortions of the output signal, yet the temporal form of the signal remains largely intact. In an antique radio, the behavior of each electrical resistor is largely autonomous with respect to macrohistorical time. "Microhistorical" changes occur instead at the time-critical level: the voltage difference increases slightly through the warming of the resistance in the current flow. This electronic *différance*, which is familiar to phenomenological human ears as the "drift" in the reception of radio waves, follows the proper time of the system. Under the same starting conditions, the same voltage difference prevails at all times. If such processes are analyzed using an electronic analog computer, their modeling must strive to minimize the drift of the computer's own active and passive components (its own proper Δt).

Media historiography thus necessarily assumes another form, as the diagram replaces the narrative mode of writing. A graph depicting the progressing historicity of an antiquarian medium diverges from the functional progress of a media technology over time; parts of the curve are mostly linear, but there are also discontinuous leaps (eras). *Superheterodyne* radio receivers, which are based on electron tubes, remained largely similar for decades with respect to the basic components of their circuitry. They thus followed the ahistorical logic of the series rather than the historical logic of evolutionary development; the shift to transistor radios actually marked a qualitative leap with respect to social media use (portability through miniaturization), yet they still remained functionally equivalent to the previous technical architecture.

Media archeology offers accounts of origins, but in an ahistorical form. The literally printed circuit has a *documentary* character when it appears as a circuit diagram in a printed book, but the radio that is subsequently constructed is the actual document of this diagram. Circuit diagrams and other technical drawings cannot be reduced to a function of cultural discourse or even visual rhetoric, as they are not readily available. They only function when put to test, and in this moment the temporal modes of operative diagrams are revealed. The double nature of technical media thus becomes apparent—namely, their simultaneous existence as symbolic notation and actual embodiment.

> Copy the design of the circuit diagram on the last page of the instruction booklet. You should then . . . cut this page out and paste it into the radio. In this way, you will still be able to understand the design years later and eventually carry out repairs. Such diagrams are also kept in old tube radios. Repairs are still possible after many years because the diagram can always be found in the radio.[79]

The circuit diagram *in* the radio is a fortunate expression, as it can refer to the diagram as subject as well as the soldered radio as object—or, even better, as realization, reification. Printed circuits thus prove to have a double essence, as they are both historical and ahistorical. Two key components are needed in order to understand an ancient radio today: the documentation of the circuit diagram (which lies somewhere between monument and document) and the electrotechnical infrastructure; in other words, it requires not only analytical knowledge, but also media-operative hermeneutics.

> There were already medium-wave transmitters at the beginning of radio history. This is why historical radios still work today and contemporary radios would have also been able to receive transmissions in the past. The technology of contemporary radios is similar to that of the first radio receivers. Both cases involve so-called tuned radio frequency receivers, which *do not use intermediate frequencies*—unlike Superheterodyne receivers, which later became standard.[80]

The "retro" radio is thus equiprimordially related to the "historical" model. A counter-question then follows: Why doesn't this "retro" radio also receive broadcasts from the past?

In the context of the historical discourse, the antiquarian value of a medium constantly increases; however, this value is based purely on subjective impressions. The historical dimension of the "people's receiver," for example, depends media-phenomenologically on the design, the form of the housing, which is immediately classified in a so-called historical context (i.e., the broadcasting policies of the Third Reich) learned from illustrations or radio museums. As such, it remains deeply embedded in the collective memory of Germans; it is only because of this context that in the early 1970s Quelle Versand was able to launch a remake of this radio in the old design but in colors and supplemented with FM reception. The de-shelling of the historical radio—in other words, the excavation of its chassis, components, and loudspeakers—affords an unsentimental insight (real media *theoría*) into this relation. What is paramount, therefore, is the technical function, the media-archeological time of the medium, which is only revealed in implementation. However, an excavation of the transistorized and thus minimized inner life of the remake of the "people's receiver," which is equipped with printed circuit boards, reveals nothing but a yawning emptiness.

Does the media-historical discourse depend on housings? It can be deduced from the design history of electronic apparatuses, and above all from the mass media of the second half of the twentieth century, that technical devices were gradually removed from furniture consoles. The first plain technical designs were open structures like the Loewe "local receiver" from 1926 (Ortsempfänger OE333) or subsequent radio cabinets (as literal "black boxes"). Special so-called "radio saws" made it possible for tinkerers to make their own custom-fit wooden housings around the radio chassis. Media analysis took a different course. The historical impression produced by an antiquarian radio largely depends on the form of the housing; when the rear panel is removed, the tubes become visible. "Open software" is a political implementation of current media criticism in the digital humanities, but "open hardware" applies to the world and to the time of electrophysically implemented *techné*, as it refers to the media-archeological attempt to see past the decorative casing and to critically analyze the electrophysics of technical media themselves. If the antiquarian radio is ultimately reduced to its chassis *plus* its peripheral equipment (loudspeakers, tuner, etc.), then it assumes a media-archeological condition and mediates or reveals another temporality that would be opposed to the external and anthropomorphic model of history. When an antiquarian radio has power and receives signals in the present, it is also undoubtedly a function of a special temporal configuration—not necessarily that of history, but rather sublated time. The technical assembly (tubes, etc.) reveals a time window that is only accessible to the media-archeological gaze, which (and insofar as it) is free from the trappings of history—both in terms of narrative and materiality (housing). While the media-historical approach focuses on the frontal view of an antique television, the media-archeological approach focuses on the reverse side after the removal of the protective rear panel; it thus provides a truly media-*theoretical* insight into the essence of the apparatus—namely, its circuitry. The technical *dispositif* is only revealed in its media-temporal construction.

If an old German radio like the "people's receiver" from the National Socialist era is brought back into operation, what one hears in the present is not the voice of Joseph Goebbels, minister of propaganda, but rather at best a current long- or medium-wave transmitter. An ancient Greek sculpture in an archeological museum retains the character of its appeal, as it is still able to immediately address people's aesthetic sense of proportions. With a "historical" radio, however, the creation of presence takes place under well-defined technological rules, which must remain stable over long periods of time in order to be effective. In contrast to the statue, the radio requires *active*, genuinely *media*-archeological-dynamic participation in order to reveal its temporal referential character. A purely phenomenological description of the outward appearance of a technical medium is not sufficient to grasp its operative being. A photograph of an activated radio shows everything except the medium in implementation; a descriptive reenactment of a medium requires a representation *in* time. Suppose that a radio receiver from the time after the end of World War II (shortly before the introduction of FM reception) carries a manufacturing stamp on its wooden housing. The apparatus can already be "historically" dated to that era based on the design of its visual appearance. When this housing is removed and the apparatus is thus media-archeologically reduced to the technological *dispositif* with its hand-wired components, it nevertheless suddenly conveys another, almost modern impression. The brutal three-dimensionality of the construction (the outstanding electron tubes, resistors, and capacitors) embodies one of three great eras in the history of electronic radio: the initial tube-based form, then the transistor-based form, and finally the integrated circuit-based form. Under power, however, this formerly "historical" appearance transforms into a functional one, which thus compresses the emphatic, deeply temporal, historical distance to a short-circuited and largely flat temporality. What is paramount from this point on is the actual reception, the astonishingly precise rectification and demodulation of current radio waves, which is not distracted by the nostalgic associations that would accompany the same sounds if they were heard coming from the radio in an intact housing. The wooden housing enriches the resonance of the loudspeakers with a specific tone color—the alleged "warm tone," which has become the acoustic signature of radio historicity. From an epistemological perspective, the media-archeological analysis of sound hears an entirely different message in the medium: the miracle of successful reception across a half century of temporal difference, and thus the kairotic moment with respect to the chronos of history.

There are two reasons for the fascination with antique radio transmitters and receivers. Insofar as the apparatus functions, even though it is historically dated, it behaves not historically but rather in the present; in its media-archeological archaism, however, such an apparatus is simultaneously reduced to its essential components and is thus analytically and fundamentally transparent. The decades of "historical," and thus political, cultural, and mental difference between the time when an antique radio was manufactured and the time of its reactivation in the present do not correspond to any actual technological difference; generations of media pulsate in other intervals and eras. In *Reading Capital*, Louis Althusser pleaded for another rhythm of historiography that would follow not the writing-induced aesthetic of linear progress, but rather the steps, upheavals, and short circuits of historical time and asynchronicities.

"Althusser's entire, and necessary, critique of the 'Hegelian' concept of history and of the notion of an expressive totality, etc., aims at showing that there is not one single history, a general history, but rather histories *different* in their type, rhythm, mode of inscription—intervallic, differentiated histories."[81] But is this still historiographically representable? A genuine media archeography, which is appropriate to high-tech media, results from the critique of the previous philosophy of history; media philosophy thus looks different. After two centuries of the emphatic philosophy of history, however, the Western perception of media time is still entangled in historical semantics. Successful reception with an antiquarian radio automatically conveys the sonic impression that past, "historical" time is streaming out of the radio, even though the reception itself is strictly in the present in terms of its form and content. Astonishingly, a radio receiver built in 1940 is still able to receive amplitude-modulated broadcasts in 2009. The technological being of the apparatus under power behaves the same with respect to the flow of historical time as long as it corresponds to the technological infrastructure: a power supply under 220 volts in 50 Hz cycles of alternating current, the laws of radio broadcasting, the form of modulation, etc.[82] The infrastructure of a given period, as soon as it gains acceptance as a standard, constitutes an *epoché* that behaves like a cliff with respect to the waves of entropic-linear time. When this infrastructure is no longer valid, however, the medium can no longer be implemented. The end of analog FM radio broadcasting is imminent; from this point on, analog FM receivers will become historical, as their reception will be suspended. On the other hand, they will remain in technological latency, ready at any time to once again process analog ultra-short waves in accordance with their configuration. The upstream connection of a converter for digital and analog FM signals provides them with a potential afterlife. The question remains as to whether the intervention of digital conversion (an intermediate medium) decouples the apparatus from its media historicity by transforming it into a function of mathematical worlds. Digital UK, the British agency responsible for promoting the shift from analog to digital television, advertised that antique televisions would still be able to receive digital channels. This is possible in England through the interposition of a special converter that translates digital television signals back into the earlier standard 441-line television; such a medium is then no longer functionally historical and invariant with respect to historical change. With the entropy of its hardware, however, it looks quite different. Obsolete capacitors, for example, must be replaced with functionally equivalent but physically new components, which leads to the critical question facing all curators of technology and media museums: Is a device restored in this way still an original? If an apparatus is only a medium in implementation, then a medium in implementation is always a dynamic original. However, a disruption in the historical consciousness occurs when a current program appears on an antique television. One answer is to establish a comfortable signal-technical and semiotic relationship between the technology and the program, such as showing only historical films from the 1930s and 1940s on a British Marconi television from 1936.

Jacob Roschy constructed a tube modulator for long- and medium-wave reception "in order to make sound sources audible on historic radio receivers."[83] This involves constructing a local home transmitter to convert low-frequency signal sources into high-frequency carrier waves so that they can be received by AM devices, while actual AM transmitters (long-wave, short-wave) are currently disappearing in the face of digital and FM broadcasts—thus a media-temporal outcrossing. The presence of "historical" radios in museum collections is thus replaced with an operative historicity. The modulator is built using classical tube technology in order to have the required affinity to antiquarian radio; its implementation is thus related to the medium (and not only the aesthetic impression of humans). Such local home transmitters are on the edge of broadcasting law. "The content of the transmitted signals also plays an important role. It is safest to transmit a radio program from another waveband, such as satellite or FM. This most likely ensures the character of the modulator as a pure converter,"[84] as the transmitter translates the present into another radio broadcast that is not experienced as artificially generated. Does such a modulator also allow the past to be translated (back) into the present? This opens up an intermediate realm of medial necromancy. "It could occur to someone that it would be appropriate to transmit a tirade of the so-called 'Führer' or the 'Minister of Propaganda' in order to recreate the authentic reception situation of a 'people's receiver.' However, an accidental listener could understand the meaning of this transmission entirely differently."[85] The former involves the technical modulation of a current radio broadcast for the purpose of its reception with an antique radio, while the latter involves—albeit in reverse—the reception of a historical recording through current technology. An experiment using technical media to realize the archive might look like this: a historical shellac record is employed as the signal source, such as the recording of a political speech from 1933. Because it is not simply played on a record player and understood as a historical source but is instead coupled to a contemporaneous "people's receiver," this historical recording is transformed into a broadcast; in other words, the device endows it with the qualities of an actual radio broadcast, which (in terms of its perception) always already includes the quality of nowness (*Jetzigkeit*). If signals are sent from a storage medium through a pure transmission medium, then the apparently archival recording is also endowed with the borrowed actuality of the "live" medium. The electronic transmission medium here functions as a modulator that converts the past into the present; at least this is how it appears to the senses, while cognitive consciousness seeks to preserve historical distance. This combination of storage and transmission media (instead of the traditional combination of storage and playback media for the purpose of historical research) establishes a unique, genuine media temporality *à la recherche du temps perdu*.

Unlike the museum presence of cultural-historical works, the temporality of high-tech artifacts only reveals itself in current implementation. Under power, a radio demodulates electromagnetic waves into audible signals. Although an antique radio can clearly be assigned to a specific moment in the past due to its technological condition as a manufactured product, it nevertheless behaves equiprimordially in the present when it receives a current medium-wave transmitter in amplitude modulation.

An advertisement for the reproduction of a historical crystal radio receiver released by the Berlin company Technische Requisiten Vorrath (TRV) featured the slogan "Listen to radio like in the year 1925 on the nostalgia wave." Taken literally, there is actually a reception mode that tunes into the present on resonant carrier frequencies of the past—a transitive concept of the "nostalgia wave." Walter Benjamin wrote in 1936: "The authenticity of a thing is the quintessence of all that is transmissible in it from its origin on, ranging from its physical duration to the historical testimony relating to it."[86] Yet, while the aura of the work of art "withers" in the age of its technological reproducibility,[87] antique radios not only have a testimonial character for a particular technical era due to their material survival, but they actually continue this era by processing signals in the present. Operational media from the past are transitively related to the so-called historical time; however, their passive exhibition as museum or collector's item remains intransitively related to it. Media history addresses such radios, but when they are once again brought into operation as techno-real artifacts they represent "time synthesizers"[88] through the concrete historicity of their individual components and their readable configurations as circuit plans. They not only "bear witness" to the technological past, but they also continuously generate it anew in the transmission and reception modes. The condition for this operative memory is the stability of the technical ensemble (and sometimes also the interposition of a corresponding converter). In the moment when an antique radio receives a current radio broadcast, the material apparatus resonates as the active presence of a past that is precisely *not* that of a museum—even if the signals to be processed are possibly played from a tape as a second order presence according to the law of media-technical reproducibility.

This analysis can be stated more precisely by means of a concrete scenario. On January 1, 2007, Deutschlandradio transmitted a performance of Handel's *Messiah* from the Philharmonie in Berlin. Media-induced disruptions of temporal relations often occur with such transmissions. Without additional information the listener is initially not able to determine whether it is actually a "live" transmission or a recent recording. A New Year's address by former German Chancellor Helmut Kohl on public television was legendary, as it later turned out that it was accidentally the video of the speech from the previous year, which most viewers did not notice. Whether it is "live" or "live on tape," both broadcasts occur in the same electromagnetic present. In storage media, the "historical" difference is only articulated through the indexical signs of their physical decay, such as "drop outs" in the case of magnetic tapes or scratches and cracks in the case of mechanical sound carriers.

Shifting to the perspective of the transmitter, the musicians and singers in the Philharmonie incorporate an archival document—namely, Handel's score. Through the culturally and technically familiar act of decoding a written text, this embodiment transforms the historical notation into the presentness of music. Is the work in this moment still from the past? The tradition of musical notation as aesthetic information is largely invariant with respect to the temporal intervention of a historical distance; encoding is a process designed to protect signals from the noise in the temporal transmission channel.

The philosophy of so-called historical performance practice for traditional music is also connected to the question of the historicity of the reception of current radio broadcasts with an antique radio. The use of historical instruments is a media-archeological argument concerning hardware, as they embody the physics of past sound worlds and thus offer the option of equiprimordially reviving these worlds. Does it make a difference if the concert transmitted from the Philharmonie is received by a historical "local receiver," whose electronic core consists of the legendary Loewe 3NF triple triode tube? This radio receiver went into production in 1926, as Heidegger's question concerning the entanglement of *Being and Time* was already in the epistemic air. The integrated circuit in the vacuum of the electron tube, designed by Manfred von Ardenne, was retrospectively declared the first integrated circuit in the world, and it thus implicitly already pointed beyond its own historicity; in archeological terms (read as a circuit), it transcended the era of discrete electrical engineering.

The concept of media time itself requires differentiation, which means distinguishing between chronological, logical, historical, and immanent time; the representation of time-technical procedures requires a real techno-mathematical historiography in the form of circuit plans, signal path diagrams, and network topologies (graphs). The ahistorical alternative to the reactivation of an antique electronic medium for the purpose of current signal reception—which is usually vetoed by curators in technology museums—is the reproduction of its circuitry. A document in a historical archive only interests historians, but according to archival studies an active archive represents a memory of those documents, which continues to remain valid in the present in a latent form (such as classical legal documents). However, the rereading of a memorable text from an archive differs fundamentally from the replay of a so-called "historical musical recording" on magnetic tape. A circuit plan handed down from the past is a text insofar as it remains largely invariant with respect to historical time and it can be read again as instructions for the technical reconstruction of the apparatus—a genuine media historiogram. On the basis of a historical circuit plan, active media archeology is able to bring the past of the electronic device into direct contact with the present—not only at the logical level but also at the operative, signal-processing level. This requires the active implementation of the underlying hardware. The realization of such a reproduction after decades is based not on the availability of historically original components, but rather on the possibility of reconstructing the essence of the medium using functionally identical elements. Functional equivalence is the crucial feature of an ahistorical relation to the media past. In contrast to classical archeology, the media-archeological principle depends not on material originals, but rather on equiprimordial implementation. Such a concrete realization of coded transmitted media knowledge from the past only reveals—and at the same time suspends—its historical index in operative implementation.

A unique combination of internal-technical time relations and historical time is present here at the media-temporal level. The so-called historical context is derived not from the proper logic of the technical medium, but rather from the circumstances of its production and thus an external parameter. Social conditions undoubtedly shape the technology used at a certain time (and vice versa). This relates to a key question in media studies. Stefan Andriopoulos has demonstrated the massive contribution of occult discourses to the development of the mass medium of television, from which he has derived an explicitly historical media archeology (which represents an oxymoron for radical ahistorical media theory). "Such an approach to the archaeology of media . . . addresses the complex relation between technology and culture by avoiding a simple determinism, which in its focus on a technological *a priori* threatens to reduce culture to a mere epiphenomenon."[89] For the actual functioning of technical media, however, immanent, literally techno-logical factors are most crucial. The question is how to write media history,[90] which always also provokes its negation: "How *not* to write media history?" This aims not simply at another mode of media historiography, but rather at a radical alternative to the imagination of "historical" development in medial genealogy. The repeated disappearance of sound film after it had already been made possible, for example, was neither technically nor economically motivated, but rather a result of the media-aesthetic identification of cinematography as silent film in its media-archeological uniqueness.[91]

The implicit message of this analysis is that technical media do not emerge historically, but rather they are reconfigured in rhythms, scansions, frequencies, and phase shifts that are closer to oscillatory tides than the transcendental signified "media history." How did a phantasm called television turn into a functioning technical apparatus? Joseph Hoppe assumes that the unfolding of media follows the logic of their proper time rather than that of cultural history in that he asks whether technical innovations "are not derived and developed inductively from technical standards that have already been achieved."[92] The concept of "induction" is appropriate here, as it has been part of the technical vocabulary of media worlds since Faraday. In the case of electronic image transmission, "it appears to have been inspired by factors that were technically immanent for a very long time."[93] It was not occult clairvoyance but rather Manfred von Ardenne's engineering skill that made television a high-tech and thus fully electronic medium. If it were left to mediumistic spiritualism, an image would never have appeared on the screen of the Braun tube as a function of the highly complex time integration of electrically converted light signals. Even Immanuel Kant, who dedicated a treatise to the *Dreams of a Spirit-Seer*, was persuaded by mathematician Leonhard Euler of the physical frequency theory of optical appearances.[94] The media-archeological process is able to reveal the epistemological core of a medium precisely by reducing it to its basic functional elements. "It is astonishing and moving how a few small and simply constructed components like tubes, which in principle 'only' represent a clever arrangement of metal parts in an evacuated glass bulb, are able to generate a 'live' video image"[95]—reducing the medium to its basic functional elements thus reveals its technical essence (the *arché*).

In the century of electronic mass media, the question of their "medial historiography" is fundamental. Isn't the still open-ended era of radio rather asymmetrically proportional to the narrative suggestion associated with concepts like "100 years of radio"? The medium appears to comply with the conceptual logic of historical eras, but it actually undermines this logic and introduces another temporal economy. When an original sound from the past is received by a radio, for example, its history is hardly audible unless it is contextualized discursively (in words). The only historical aspect that is audible is the quality of the recording (such as the noise of wax cylinders or shellac records), but apart from that radio practices a *compressed* time with regard to sensory perception (provided that this is not superimposed over the "historical sense," which cognitively corresponds not to the medium but rather to the logic of textual historiography). There is an indexical relation to the past of radio, but its original sound lies in the materiality of the technology itself rather than the macrotemporal frame of media history as a function of historiographic narrative. Antique technologies always exist in two temporal situations: as archeological monuments and as archival documents.

However, their presentness as media only occurs in the moment of technical signal implementation, such as when an Edison cylinder is played by a phonograph. This opens up a time window of operative presence. In the reading of an alphabetically coded text, the operation of realization depends solely on people; however, the grooves of a shellac record do not articulate sounds like a textual document in an archive until the record is played on a gramophone.

COMPUTER-ARCHEOLOGICAL REENACTMENTS: EMULATION AND SIMULATION

"Reenactment" refers to the reproduction of an event in a dramatic form. The concept is also plausible for high-tech media theater, such as Heinrich Hertz's experiments with electromagnetic waves. In the era of digital media culture, the ahistorical reproducibility of technical media yields entirely unique variants of reenactments. The Commodore 64 home computer, which came on the market in 1982 and was based on a relatively low-clocked 8-bit processor with 64 kilobytes of RAM, was graphics-capable and equipped with a sound chip. Its programs could be executed almost instantaneously, as the BASIC compiler made it possible to write byte values directly into storage cells and execute them with the speed of a machine language. How can computer games originally written for the C64 be reproduced again decades later? The restoration of a C64 along with its peripheral equipment, including datasette, joystick, and television monitor, is actually marked by signs of decay, as both the hardware and the bit strings stored on the magnetic tape of the datasette have become prone to error. Despite such historicism, it is nevertheless astonishing that a computer is still able to assiduously yield the same screen events and that it can still be used to interactively communicate decades later. The logic of hardware and software (microcodes in their fixed wiring and RAM and ROM data) proves astonishingly invariant with respect to techno-historical time and to the age of the electrophysical components. The medium is still able to position humans in the temporal relations that it prescribes.

The time-consuming process of loading a game program from the datasette into the working memory of a C64 articulates the dynamic temporal mode of electronic media reality, which is still recognizable. With ultra-high frequency processor clocks, however, the digital computer retains this temporal knowledge for itself.

When antique computer hardware is exhibited in a museum, its essential time- and *bit*-critical processes remain excluded. "In archaeological terms the operational continuity of contemporary culture cannot be assured."[96] However, software belongs—genuinely medially—to the class of generic objects.

The solution lies in transforming the material aspects of computer culture itself into software—that is, digitally *emulating* the computer hardware of the past. This suddenly results in a media culture that practices its own *un*reification—"logical replication as distinct from physical replication."[97] Operational media are actually materially embodied, but they themselves no longer represent "things." It is the principle of the universal Turing machine (UTM) that allows them to emulate all other machines with time-critical symbol operations. A new computer in the present *is* in the condition of a C64 when it emulates it—a new type of ahistoricity. However, this only applies *per definitionem* to the ahistorical—namely, logical-symbolic—level (the "archive" of statements); in order to become a real simulation, the current computer must take into account the earlier timing of the C64's electrical engineering for the purpose of "high temporal fidelity." This requires consideration of the algorithmic sequence of changes in condition (logical, mathematical-dramatic time) as well as the concrete microtemporal intervals that add up to the overall behavior of the system. The result is the resurrection of a "historical" computer as the physical world of its former time behavior—a unique kind of temporalization. Is the world of the C64 restored through the simulation of its temporal behavior? PegEm, the current emulation of the historic Pegasus computer developed by Ferranti Ltd. in the late 1950s, only claims that its simulation is limited. "Although no particular attempt was made to simulate correct timing, the similarity of the logic module to the actual hardware means that the various instruction times bear a reasonably correct ratio to each other."[98] The emulation thus models the logic of the Pegasus, but only the time behavior makes a replica exact. "The aim was to roughly model the gross timing behaviour of the original machine."[99] The computing cycles and time base of early computers must be handed down in order to be able to revive them in their temporal existence; current processors as "hosts" must take this into consideration in "reverse engineering" and media-active archeology.[100] Such technological anamneses are driven above all in the form of "retro computing," which practices anything but historical computer nostalgia.

The practical reproducibility of historical media technology represents an escalation of the question of the temporal modes of high-tech media. Emulation means functional equivalence in a time-invariant form.[101] This distinguishes symbolic machines from analog electronic media; the current reincarnation of a historic radio is only able to realize its model when the actual components are equivalently reused. Simulation is defined precisely by its temporal indexicality, which becomes its actual object. Since the era of the analog computer, the time-scale modeling of a physical process is familiar through electronic computing elements (time compression and time expansion); worldly time behavior was thus modeled by worldly means. However, the term "simulator operation" was reserved for a specific temporal mode: the representation of dynamic and electromechanical processes in the time window known as "real time." "One or several components of the physical device may take the place of the corresponding block in the computer itself. They may thus be tested, with the computer running on a 1:1 time scale"[102]— flight simulators are a prominent example. The process simulator is thus part of the being—that is, the being-in-time (*In-der-Zeit-Sein*)—of the real world. Unlike digital computers, which were initially computing-time intensive, the advantage of electronic analog computers lay especially in such time-critical applications; digital signal processing and "physical modeling" now perform precisely this. In digital signal processing, the strictly time-discrete binary computer models the continuous world at the level of its time behavior. Digital computers also calculate time series, which correspond to the protentions and retentions identified by Husserl as the phenomena of internal time consciousness in humans themselves. As an *a priori*, time no longer separates perception (Kant) from the coded, technological, mechanical "vulgar time" of the clock (Heidegger); rather, both of these temporal modes are sublated in computers as simulations of open-ended, non-linear processes, like weather and climate.

Directive 3633 from the Association of German Engineers (*Verein Deutscher Ingenieure*) defines simulation as "the reproduction of a system with its dynamic processes in an experimental model."[103] With the emulation of highly integrated chips, assembler programming makes it possible to inscribe this factor onto microcontrollers, as this language—in contrast to so-called "standard" languages—is machine-oriented. The assembler mnemocode preserves the proximity to the opcode (short for "operation code") on the most elementary level and thus to the actual temporal implementation of the computer. Depending on the operating system, it is actually possible to recreate the delicate network of timers, interrupts, sync signals to the VGA monitor, and thus the time behavior of the concrete hardware. In order to simulate the prehistoric electronic computer ENIAC (ca. 1945), for example, it is essential to reproduce its "clocking." Symbolic logic comes in the delicate "scheduling" (the allocation of time windows in temporal implementation), which culminates in preemptive systems and hard real-time systems. The smallest latency periods are here opened, and they each have their own "deadline": being-toward-death (*Sein-zum-Tode*) in the times of digital signal processing. "It was considered impossible to emulate an Amiga with other computers due to the complexity and . . . speed of its hardware. With the Fellow emulator for MS-DOS . . . it was already possible to reach the speed of an Amiga 500 with a 486 PC, which was only 100 MHz faster."[104] Simulation is a product of the emulator and its time-critical implementation. The computerization of matter upsets the philosophical dialectic of spirit and substance;

the computer brings forth a new model of memory *sui generis*. Running a classical programming language like BASIC on historical computer architecture, such as an early portable IBM computer under the DOS operating system, is by no means an act of pure nostalgia. In 1936, Turing defined the mathematical writing machine named after him above all as a logical operating sequence, which was largely independent of the concrete hardware version. However, simulation and emulation represent two sharply differing media-epistemological forms of chronotechnical modeling. The emulation of an old C64 computer on a contemporary computer is an illustration whose functionality (although realized with other algorithms) must strongly correspond to the logic of the original. If the computer is conceived as a technological sequence of steps, then the media state of the emulation is coessential with that of the original. A current emulation is thus not a historical citation or the invocation of a chapter from computer history; rather, an emulation on a new computer *is* the old computer in the moment of the corresponding configuration—in short, it is the imitation of a UTM on another UTM. In the face of this concept of state, the classical concept of historicity no longer applies. Emulation represents an operative form of media archeology, which discovers and reproduces the essential nature of earlier computers. The Computer History Simulation Project practices such media archeology. In the video documentation of a conference in 1999 on the occasion of the 50th anniversary of the first successful computation by the EDSAC computer in Cambridge, it was said that it is "so much easier preserving by simulation software rather than re-creating it in hardware." A computer from the past is emulated by a contemporary computer at the level of logic; however, historical time remains at work at the level of hardware, as the concrete physical environment underlies the transience of the material. Nevertheless, the electrophysical and mathematical laws on which the operation of the EDSAC was based remain the same in 1949 and 1999. They are natural laws of a second order, as they have been appropriated for technology—an entanglement of natural and cultural knowledge. Techno-mathematical relations can only be discursively negotiated up to a point. Historically relative knowledge competes with the proper time of the media frame (*Mediengestell*).

"HISTORICAL" COMPUTERS IN OPERATIVE PRESENCE

Only a profound familiarity with the theory and techno-mathematics of media processes makes it possible to elicit the knowledge implied in them. Human culture is thus constantly negotiating with the natural laws of the media it has brought forth. The analysis of the specific temporal processes in such technologies not only pertains to concrete media operations but also raises the epistemological question of the extent to which the time frame of media history is appropriate to the proper temporality of technical media.

What does it mean to deal with a computer "historically"? The Technikum computer museum in Kelkheim near Frankfurt am Main presents historical technologies with a firm emphasis on "active" exhibitions; indeed, technical media are always only in a truly medial state during moments of operative implementation. A historical computer is thus in a non-historical state as soon as it can be made to compute again. "Reenactment" is not only an experimental form of the human-performative recollection of historical events, as genuine, nonhuman media temporality comes into play in the "reenactment" of apparatuses. Past media technologies can only be understood in the moments when they are brought back into operation as a form of the operative presence of the past. At the Center for Art and Media in Karlsruhe, for example, veterans of the legendary Zuse Z22 computer (necessarily an electrician and a mathematician) wanted to bring it back into operation, and it has been computing again since 2005. The exhibition aesthetic of technical media is necessarily dramatic and active; this premise questions the traditional form of museum presentations in the context of the history of technology. What are the actual objects of media history: texts, circuit plans, mathematical formulas, or concrete material artifacts? Unlike paintings, such technologies are temporal objects in the present, and their implementation thus sublates museumality. When it is actually computing, the antique Z22 *in a state of being*;[105] the entanglement of being and time revealed by Heidegger thus culminates in the operative medium.

> The Z22 ... is considered the oldest fully functional and faithfully preserved tube computer in the world. As the centerpiece of the Center for Art and Media's "Algorithmic Revolution" exhibition ... the 1.5 ton heavy computer system is supposed to run continuously ... without being switched off. The "insides" of the machine will also be illuminated and viewed through a window set into the computer housing.[106]

By means of different generations of Zuse computers, it is possible to see how the computer principle remains largely invariant in the different worlds of its concrete embodiment (mechanical, electromechanical, and electronic). The concrete technical configuration is the historical content, but not the message of a digital computer exhibited in the museum. Its message is only revealed when it is presented not as a pure frame (*Gestell*), but rather as a computing operation; in this moment, however, it is precisely in a nonhistorical state. Its logical architecture (the symbolic regime) resists historical-linear time. Once developed, the variable electrotechnical elements of the technological infrastructure can also be functionally replaced (such as the replacement of electron tubes with transistors).

The electromechanical relays in the earliest digital computers are a reminder that the dialing mechanisms of telephonic step-switches already "computed" in an implicitly time-discrete way. The "Tunny" machine in Bletchley Park (England's cryptological establishment in World War II)[107] employed a central arithmetic unit using telephone dialing technology to emulate the cipher-wheels of the German Lorenz SZ42 cipher machine after its concrete setup was first ascertained in the Colossus computer. It thus involved the "reverse engineering" of one device through another—not in a historical, but rather in a functional, quasi real-time sense. The current operative reconstruction of this computer, which was dismantled after the war, thus represents not a historical, but rather a functional-equiprimordial equivalent. Coupled arrangements of operative media create a proper time field, which eludes thoroughly variable historical time for the duration of the entire era. Anthony Sale, the driving force behind the media-archeological revival of Bletchley Park, once unsuccessfully attempted in his youth to recreate the innovative Williams tube—an image storage tube for the electrostatic delay of data, which allowed the computer to monitor itself as an "interactive television." Decades later, as an expert in the restoration of historical computers, Sale came across a reproducible early technical description of the Williams tube in the Science Museum in London, which convinced him that in principle (and thus media-archeologically) early electronic systems could also be reconstructed on the basis of corresponding circuit plans. Since then, Bletchley Park has been computing again. A functional record sublates historical distance; the temporality of genuine media knowledge can no longer be called history.

NOTES

1. See Granino A. Korn and Theresa M. Korn, *Electronic Analog Computers (D-C Analog Computers)* (New York: McGraw-Hill, 1952).

2. "Drama means the controlled and observable flow of time." Hans-Thies Lehmann, *Postdramatisches Theater* (Frankfurt am Main: Verlag der Autoren, 1999), 61.

3. Plato, *Laws*, trans. R. G. Bury (Cambridge, MA: Harvard University Press, 1961), 1: 129.

4. Hans-Jörg Rheinberger, "Historialität, Spur, Dekonstruktion," in *Experiment, Differenz, Schrift. Zur Geschichte epistemischer Dinge* (Marburg: Basilisken, 1992), 51. Rheinberger here refers to Ilya Prigogine, *From Being to Becoming: Time and Complexity in the Physical Sciences* (San Francisco: W. H. Freeman, 1980).

5. Helmut Kowar, *Mechanische Musik. Eine Bibliographie und eine Einführung in systematische und kulturhistorische Aspekte mechanischer Musikinstrumente* (Vienna: Vom Pasqualatihaus, 1996), 46f.

6. Bernhard Siegert, "Spectres. Faradays Experimente 1830–1831," in *Apparaturen bewegter Bilder*, ed. Daniel Gethmann and Christoph B. Schulz (Münster: LIT Verlag, 2006), 47.

7. Sybille Krämer, "Friedrich Kittler. Kulturtechniken der Zeitachsenmanipulation," in *Medientheorien. Eine philosophische Einführung*, ed. Alice Lagaay and David Lauer (Frankfurt am Main: Campus Verlag, 2004), 223.

8. See, for example, Reinhart Koselleck, *Zeitschichten* (Frankfurt am Main: Suhrkamp, 2003).

9. Fernand Braudel, *The Mediterranean and the Mediterranean World in the Age of Philip II*, trans. Siân Reynolds (New York: Harper & Row, 1972), 1: 21.

10. In the late nineteenth century it was customary to employ frog legs as electrical instruments. For Heinrich Hertz they also served as indicators for electromagnetic waves ("the suddenness of the spark"). See Albrecht Fölsing, *Heinrich Hertz. Eine Biographie* (Hamburg: Hoffmann & Campe, 1997), 269 and 395.

11. See also Martin Heidegger, *History of the Concept of Time: Prolegomena*, trans. Theodore Kisiel (Bloomington: Indiana University Press, 1985).

12. This is the title of Abraham's contribution on musical timing and rhythm in the *Encyclopédie Français*.

13. Kay Kirchmann, *Verdichtung, Weltverlust und Zeitdruck. Grundzüge einer Theorie der Interdependenzen von Medien, Zeit und Geschwindigkeit im neuzeitlichen Zivilisationsprozeß* (Opladen: Leske & Budrich, 1998), 138.

14. See Barbara Burke Hubbard, *The World According to Wavelets: The Story of a Mathematical Technique in the Making* (Wellesley, MA: A. K. Peters, 1996).

15. This formulation explicitly follows Günther Stern's unpublished dissertation from 1930, in which he describes the musical ear as an "enclave" outside of historical and thus lost time. See Günther Stern, "Philosophische Untersuchungen zu musikalischen Situationen," Günther Anders fonds, Literature Archive of the Austrian National Library, Vienna. Qtd. in Veit Erlmann, *Reason and Resonance: A History of Aurality* (New York: Zone Books, 2010), 312.

16. Qtd. in Kassung, *Das Pendel*, 314.

17. Ibid., 322.

18. Heidegger, *History of the Concept of Time*, 129.

19. Marey, *Chronophotographie*, 32.

20. See Ludolf Herbst, *Komplexität und Chaos. Grundzüge einer Theorie der Geschichte* (Munich: C. H. Beck, 2004).

21. Objects must first pass through a "rubbish" phase before they are rediscovered as culturally valuable and thus become museum-worthy. See Michael Thompson, *Rubbish Theory: The Creation and Destruction of Value* (Oxford: Oxford University Press, 1979), 10.

22. "The information in the computer is clearly stored in a structural way." Heinz von Foerster, *Sicht und Einsicht. Versuche zu einer operativen Erkenntnistheorie* (Braunschweig: Vieweg, 1985), 134.

23. Florian Cramer and Elfriede Jelinek, eds., *Zauberhafte Klangmaschinen. Von der Sprechmaschine bis zur Soundkarte* (Mainz: Schott, 2008).

24. Jules David Prown, "The Truth of Material Culture: History or Fiction," in *History from Things: Essays in Material Culture*, ed. Steven Lubar and W. David Kingery (Washington: Smithsonian, 1993), 3.

25. Knut Ebeling, "Das technische Apriori," *Archiv für Mediengeschichte* 6 (2006): 21.

26. Ibid.

27. Heidegger, *Being and Time*, 347.

28. Martin Heidegger, "The Origin of the Work of Art," in *Off the Beaten Track*, trans. Julian Young and Kenneth Haynes (Cambridge: Cambridge University Press, 2002), 20–21.

29. Hans-Jörg Rheinberger, "Zeit und Biologie," in *Zeit-Zeichen. Aufschübe und Interferenzen zwischen Endzeit und Echtzeit*, ed. Georg Christoph Tholen and Michael O. Scholl (Weinheim: VCH/Acta Humaniora, 1990), 133.

30. Ibid.

31. Klaus Frerichs, *Begriffsbildung und Begriffsanwendung in der Vor- und Frühgeschichte. Zur logischen Analyse archäologischer Aussagen* (Frankfurt am Main: Lang, 1981), 73.

32. Wolfgang Hagen, *Gegenwartsvergessenheit. Lazarsfeld, Adorno, Innis, Luhmann* (Berlin: Merve, 2003), 8.

33. Ibid.

34. See Alex Preda, "The Stock Ticker," in Making Things Public: Atmospheres of Democracy, ed. Bruno Latour and Peter Weibel (Cambridge, MA: MIT Press, 2005), 622–27.

35. "The technical organization of the public image of the world . . . is the authentic form of the dominion of historicism." Martin Heidegger, "Anaximander's Saying," in *Off the Beaten Track*, trans. Julian Young and Kenneth Haynes (Cambridge: Cambridge University Press, 2002), 246.

36. Großklaus, *Medien-Zeit*, 48.

37. See Edgar Douglas Adrian, *The Mechanism of Nervous Action: Electrical Studies of the Neurone* (Philadelphia: University of Philadelphia Press, 1932).

38. Peter Berz, "Bitmapped Graphics," in *Zeitkritische Medien*, ed. Axel Volmar (Berlin: Kulturverlag Kadmos, 2009), 133.

39. See Mirjam Schaub, *Gilles Deleuze im Kino. Das Sichtbare und das Sagbare* (Munich: Fink, 2003), 202f.

40. W. Daniel Hillis, *The Pattern on the Stone: The Simple Ideas that Make Computers Work* (New York: Basic Books, 1998), 69.

41. Gilles Deleuze, *Proust and Signs*, trans. Richard Howard (Minneapolis: University of Minnesota Press, 2000), 58.

42. Heidegger, *Being and Time*, 348–49.

43. The verb form of the Low High German word for "time" (*tid* or *teid*) is "to move" (*tiden*).

44. Heinz von Foerster, *Sicht und Einsicht. Versuche zu einer operativen Erkenntnistheorie* (Braunschweig: Vieweg, 1985), 46.

45. Gilbert Simondon, "Discussion," *Les cahiers du centre culturel canadien* 4 (1976): 87.

46. See Ahasver von Brandt, *Werkzeug des Historikers* (Stuttgart: Kohlhammer, 1958), 48.

47. Frerichs, *Begriffsbildung und Begriffsanwendung in der Vor- und Frühgeschichte*, 73.

48. "The semiological change in the concept of the sign is the manifestation of a medial rupture." Michael Wetzel, "Verweisungen. Der semiologische Bruch im 19. Jahrhundert," in *Arsenale der Seele*, ed. Friedrich Kittler and Georg Christoph Tholen (Munich: Fink, 1989), 78.

49. See Michel de Certeau, "Writing vs. Time: History and Anthropology in the works of Lafitau," *Yale French Studies* 59 (1980): 37–64.

50. Hans-Georg Gadamer qtd. in Herbst, *Komplexität und Chaos*, 71.

51. Erich Hörl, "Abendländische Rekursionen," in *Rekursionen. Von Faltungen des Wissens*, ed. Ana Ofak and Philipp von Hilgers (Munich: Fink, 2010), 58.

52. See Sigmund Freud, "*A Note Upon the 'Mystic Writing-Pad*,'" in *The Standard Edition of the Complete Psychological Works of Sigmund Freud*, trans. James Strachey (London: Hogarth Press, 1961), 19: 227–32.

53. See Herbst, *Komplexität und Chaos*, 41.

54. Frerichs, *Begriffsbildung und Begriffsanwendung in der Vor- und Frühgeschichte*, 97.

55. See Inke Arns and Gabriele Horn, eds., *History Will Repeat Itself: Strategies of Reenactment in Contemporary (Media) Art and Performance* (Frankfurt am Main: Revolver, 2007).

56. "The Man with the Flower in his Mouth: The First British Television Play—1930," *The Dawn of TV: The Mechanical Era of British Television*, http://www.tvdawn.com/earliest-tv/the-man-with-the-flower-in-his-mouth.

57. See Christian Sichau, "Die Replikationsmethode. Zur Rekonstruktion historischer Experimente," in *Im Labor der Physikgeschichte. Zur Untersuchung historischer Experimentalpraxis*, ed. P. Heering, F. Rieß, and C. Sichau (Oldenburg: Bibliotheks- und Informationssystem der Universität Oldenburg, 2000), 60.

58. See Annette Bitsch, *Diskrete Gespenster. Die Genealogie des Unbewussten aus der Medientheorie und Philosophie der Zeit* (Bielefeld: transcript, 2009).

59. Hans-Ulrich Fuß, "Musik als Zeitverlauf. Prozeßorientierte Analyseverfahren in der amerikanischen Musiktheorie," *Zeitschrift der Gesellschaft für Musiktheorie* 2.2-3 (2005): 21–34.

60. See Arkady Pikorsky, Michael Rosenblum, and Jürgen Kurths, *Synchronization: A Universal Concept in Nonlinear Sciences* (Cambridge: Cambridge University Press, 2003). The historiography of the apparatus is also questioned.

61. Friedrich Kittler, *Kulturgeschichte der Kulturwissenschaft* (Munich: Fink, 2000), 235f.

62. See Wetzel, "Verweisungen," 71–95.

63. Wolfgang Schäffner, "Topologie der Medien. Descartes, Peirce, Shannon," in *Die Adresse des Mediums*, ed. Stefan Andriopoulos, Gabriele Schabacher, and Eckhard Schumacher (Cologne: DuMont, 2001), 87.

64. Ibid., 88.

65. See also Gotthard Günther, *Das Bewußtsein der Maschinen. Eine Metaphysik der Kybernetik* (Krefeld: Agis, 1963).

66. Gfeller, "Anmerkungen zum restauratorischen Hintergrund der Ausstellung," 124.

67. Ibid., 125.

68. Douglas Kahn, "Radio Was Discovered Before It Was Invented," in *Relating Radio. Communities, Aesthetics, Access. Beiträge zur Zukunft des Radios*, ed. Golo Földmer and Sven Theirmann (Leipzig: Spector Books, 2007).

69. David Gooding, "Picturing Experimental Practice," in *Experimental Essays. Versuche zum Experiment*, ed. Michael Heidelberger and Friedrich Steinle (Baden-Baden: Nomos, 1998), 312.

70. Wolfgang Schreier, *Die Entstehung der Funktechnik* (Munich: Deutsches Museum, 1995), 33.

71. Hagen, "Theorien des Radios."

72. Ibid.

73. Ibid.

74. "Arbeiten mit dem Oszilloskop," *ELEXS. Das Online-Magazin für Elektronik in Hobby und Ausbildung*, http://www.elexs.de/oszi3.htm.

75. See Jeffrey Sconce, "The Voice from the Void: Wireless, Modernity and the Distant Dead," *International Journal of Cultural Studies* 1.2 (1998): 211–32.

76. See Kittler, *Gramophone, Film, Typewriter*, 30–33.

77. Benjamin, *The Arcades Project*, 462.

78. See Wolfgang Ernst, "Merely the medium? Die operative Verschränkung von Logik und Materie," in *Was ist ein Medium?*, ed. Stefan Münker and Alexander Rösler (Frankfurt am Main: Suhrkamp, 2008), 158–84.

79. Burkhard Kainka, *Das "60 Jahre Franzis" Retroradio selber bauen. Das Komplettpaket mit Gehäuse und allen benötigten Bauteilen* (Poing: Franzis Verlag, 2008), 12.

80. Ibid., 2.

81. Jacques Derrida, *Positions*, trans. Alan Bass (Chicago: University of Chicago Press, 1981), 57–58.

82. This distinguishes high-tech devices from traditional kinds of cultural-technical tools like hammers, which as archeological finds can in principle also be brought back into operation and thus de-museumized.

83. See "Radio-Forum: Forum for Radio Collectors," *Radio Museum*, http://www.radiomuseum.org/forum/modulator_fuer_lang_und_mittelwelle.html.

84. Ibid.

85. Ibid.

86. Benjamin, "The Work of Art," 254.

87. Ibid.

88. A concept developed by Jussi Parikka on the occasion of his visit to the Media Archaeological Fundus in the Department of Musicology at Humboldt University in 2009.

89. Stefan Andriopoulos, "Psychic Television," *Critical Inquiry* 31 (2005): 622.

90. See Joseph Hoppe, "Wie das Fernsehen in die Apparate kam," in *TV-Kultur. Das Fernsehen in der Kunst seit 1879*, ed. Wulf Herzogenrath et al. (Dresden: Verlag der Kunst, 1997), 26. On the role played by industrial cartels in the delay of sound

film, which had already been technically possible for a long time, see Wolfgang Mühl-Bennighaus, *Das Ringen um den Tonfilm. Strategien der Elektro- und der Filmindustrie in den 20er und 30er Jahren* (Düsseldorf: Droste, 1999).

91. See Corinna Müller, *Vom Stummfilm zum Tonfilm* (Munich: Fink, 2003), 83.
92. Hoppe, "Wie das Fernsehen in die Apparate kam," 26.
93. Ibid.
94. See Jens Schröter, "Die Form der Farbe. Zu einem Parergon in Kants 'Kritik der Urteilskraft,'" in *Kants Schlüssel zur Kritik des Geschmacks*, ed. Ursula Franke (Hamburg: Meiner, 2000), 135–54.
95. Henry Westphal, "Klassische Fernsehtechnik: Anschauliche Darstellung des Prinzips und Einblicke in die Geschichte" (paper presented in the colloquium "Medien, die wir meinen," Humboldt University, Berlin, July 8, 2009).
96. Doron Swade, "Collecting Software: Preserving Information in an Object-Centred Culture," *History and Computing* 4.3 (1992): 209.
97. Ibid.
98. Christopher P. Burton, "Pegasus Personified: Simulation of an Historic Computer," *Computer Conservation Society*, http://ftp.cs.man.ac.uk/pub/CCS-Archive/Simulators.
99. Ibid.
100. See Jens-Martin Loebel, *Lost in Translation. Leistungsfähigkeit, Einsatz und Grenzen bei der Langzeitbewahrung digitaler multimedialer Objekte am Beispiel von Computerspielen* (Glückstadt: Verlag Werner Hülsbusch, 2014).
101. The Windows-compatible run-time environment for the WINE operating system says it already in the acronym "Wine Is Not an Emulator." It does not emulate every processor instruction, but it is sufficient to allow the executed programs to run as fast as they did under the original operating system.
102. Granino A. Korn and Theresa M. Korn, *Electronic Analog Computers (D-C Analog Computers)* (New York: McGraw-Hill, 1952), 9.
103. Qtd. in Claus Pias, "Klimasimulation," in *2°. Das Wetter, der Mensch und sein Klima*, ed. Petra Lutz and Thomas Macho (Göttingen: Wallstein, 2009), 112.
104. Volker Mohr, *Der Amiga. Die Geschichte einer Computerlegende* (Morschen: Scriptorium-Verlag, 2007), 93f. The antique Amiga had a clock pulse rate of 7 MHz, but it had a more effective graphics processor.
105. The Admiralty of the British Empire coined the term "*fleet in being*" to refer to a war fleet at sea. See Paul Virilio, *Speed and Politics*, trans. Marc Polizzotti (Los Angeles: Semiotext(e), 2006), 62.
106. Peter-Michael Ziegler, "'Die Zuse' läuft wieder," *heise online*, March 9, 2005, http://www.heise.de/newsticker/meldung/Die-Zuse-laeuft-wieder-142122.html.
107. See Ronald Lewin, *Entschied ULTRA den Krieg? Alliierte Funkaufklärung im 2. Weltkrieg* (Koblenz: Wehr & Wissen, 1981).

Chapter 9

Equitemporalities in Media Knowledge

ARCHÉ, ARCHAISM, AND THE EQUIPRIMORDIALITY OF TECHNICAL MEDIA

As initially technical and later technomathematical configurations, media embody physical (hardware) and logical (circuit plan and software) temporal modes, which are asymmetrically proportional to the constant flux that is called history. "When we replace the parameter *t*, which refers to time, with *t+T*, whereby T can be an arbitrary number, then all the equations of physics remain unchanged. As a result. . ., the laws governing a physical experiment are independent of the chosen point in time."[1] As matter that has been informed by cultural knowledge, media operate simultaneously in both physical and logical time—that is, they are based on matter and energy, yet at the same time, they are also removed from them. Media history is written in the space between these diachronic and synchronic axes. So where does the temporal index of media lie? From the perspective of linguistics, the syntagm is determined by the temporal extension and therefore it only produces order over time in the sense of narrative. In contrast, the paradigm is an actualization out of archival latency. The paradigmatic order is *in absentia*, while the actual articulation is *in praesentia* and thus unfolds in time.[2] In his interpretation of the ancient Greek concept of presence as "unconcealment" (*aletheia*), Heidegger surmised that being stood in the clearing of time in a concealed way.[3] Media archeology is an attempt to conceptualize this concealed temporal mode for technomathematics. According to Anaximander, the concept of *arché* refers to the emergence of the thing in a double sense: "It assumes a *substance* that is actually temporal and physical."[4] In media-archeological terms, this is "not a matter of grounding by deduction but rather of laying bare and exhibiting the ground."[5] As the underlying principle of technical media, the *arché* oscillates between a temporal-incipient and a functional-archaic meaning, and it is thus deemed historiographic as well as infrastructural. Both of these perspectives are thus mutually differentiable as well as integrable. This is apparent in the architecture of electronic microchips: the formerly developed version is also sublated in the current version as technical knowledge, and it is thus equiprimordial in the computative present.[6] This invokes the concept of asynchronicity. In opposition to the "false idea of one-dimensional historical stretches of time,"[7] Wilhelm Pinder once defined multidimensional space-time as "a coordinate system that arises from juxtaposition in the course of time (the time area), which is perpendicular to the course of life and parallel to every juxtaposition."[8] What Pinder defined as "time cubes" also applies to media-archeological artifacts in implementation: they are time structures, "although 'time' here is only coarsely meant as historical."[9]

If the model of technical-historical progress is replaced with configurations that are repeated but nevertheless more complex in each reactualization, then it is possible to discern a simple pattern. Media *archaism*—the return of medial artifacts and processes—thus refers to the continuous recourse to fundamental technograms rather than a search for origins. This gives rise to the fascination for basic electrotechnical experiments. Media archeology employs a virtually pre-Socratic method of reflecting on media-epistemic tools, as it reconciles nature and numbers in a way that was already familiar to natural philosophy.

In the moment of operative implementation, media belong neither to history nor to the future: "In the Moment, authentic existence is indifferent to both of these alternatives."[10] Technical events are the repetition of a continual knowledge—whether it involves materiality (*physis*) or logical machines. This is the culmination of equiprimordiality. An implicit knowledge is first revealed through high-tech media. Ever since frequencies were epistemologically conceptualized through the analysis of oscillations and the electromagnetic field was conceptualized through experimentation as well as differential calculus, media-technical phenomena have emerged from the shadows of ancient Greek thought; a world controlled by gods thus transformed into a world of temporalized matter,[11] which resulted in a new form of knowledge induced by technical processes themselves.

Technomathematical media theory attempts to explain unexpected experiences based on experimental results; these explanations are in turn verified through veritable theory experiments, which are specifically constructed for this purpose.[12] In high-tech terms, this epistemological syllogism proceeded from the experimental discovery of electromagnetic induction (Oersted) to its mathematical analysis (Maxwell's equations) to its "reentry" in the implementation of technical media (Hertz's "radio" installation *avant la lettre*). Hertz himself described this scientific process.

We form for ourselves images or symbols of external objects; and the form which we give them is such that the necessary consequents of the images in thought are always the images of the necessary consequents in nature of the things pictured. In order that this requirement may be satisfied, there must be a certain conformity between nature and our thought.[13]

A materialistic media epistemology researches technological things in their energetic contexts. The accuracy of these theoretical constructs (also understood in the mathematical sense as function, as "mapping," as the one-to-one allocation of values) is precisely *not* "at our discretion."[14] Hertz's insight became a media-technical reality in the analog computer. There is no direct analogy here between the apparatus and the world, but rather an equiprimordial relation founded on the mathematical analysis of a concrete physical situation that can be mapped in another corresponding physical medium—such as the mapping of a mechanical pendulum swing through an electronic oscillating circuit. This corresponds to "physical modeling" in digital signal processing—"an entirely poietic process"[15] that represents a drastic paradigm shift to the real time world of matter, as it is capable of reproducing the physical properties of sounds generated by instruments through symbolic computation. The concept of equiprimordiality here makes concrete media-archeological sense. This results in technical isomorphism with respect to the actual time event.

In his experiment with the monochord, Pythagoras discovered the equivalence of whole-numbered fractions and musical intervals, but not the periodicity of the actual oscillation event. Insight into the temporal structure of a sound, which could then in turn be written mathematically in frequencies, remained reserved for the modern era (Sauveur, Mersenne): "In the seventeenth century, the birth of modern science required a new mathematics, more fully equipped for the purpose of analysing the characteristics of vibratory existence. And now in the twentieth century we find physicists largely engaged in analysing the periodicities of atoms."[16] This leads to the crucial media-epistemological question: Is such knowledge inherent in operative things themselves or is it subject to a constant relativization in the sense of cultural discourses? According to Karl Popper, there is a third world that is apart from the world of nature and the world of culture. This world was created by humans, but at the same time, it "transcends" the knowledge of its creator.[17] This is an appropriate definition of the world of technological media: "One need only think of the effects of the transmission of electrical energy."[18] The concept of a "flash of inspiration" is here not only metaphorical. Forms of electrical discharge were already known through the rubbing of amber with dry textiles to generate sparks (and ancient radio broadcasts). However, it was not until the modern era that this moment was knowingly reproduced as precise "situation analysis."[19] Hero of Alexandria's sketch of a steam engine also remained inconsequential until its principle was recalled in the thermodynamically informed industrial revolution. Sometimes, technical media are able to invoke the knowledge of an earlier era in a nonlinear way. When the experience of such a technical moment "is too aberrant for the current state, it sinks to the bottom like a foreign body in order to be covered by the sediment of time; it is only later, or never, that it has an effect."[20] So there are latency times for knowledge and its embodiments in media, which are always invoked again in new technical implementations. The monochord was for a long time an epistemic medium of theory rather than a musical instrument. However, it is precisely because of its removal from culture that it retains its ahistorical validity: "As an ancestor of our piano, the memory of the monochord must still be appreciated today; the acoustic relations that it represented will be valid for all eternity."[21]

Knowledge is actually already implicit in (electro)technical relations—even if it is not yet explicitly articulated to the *logos*.[22] Joseph Henry describes this in connection with his discovery of the electrical resonant circuit: "A ready explanation is afforded by it of a number of phenomena which are to be found in the older works on electricity, but which have until this time remained unexplained."[23] For a long time, therefore, electrical resonance waited to be uncovered—not only implicitly as a phenomenon, but also explicitly as knowledge—in order for it to pass from a metahistorical event to a historically datable invention. This relation can be described as the continuous resonance of cultural knowledge and technical media physics, which is traced through "diagrammatic reasoning" (Peirce) rather than technical-historical narrative.

TECHNICAL KNOWLEDGE IN THE DOUBLE-BIND OF HISTORY AND INVARIANCE

On the one hand, technomathematical knowledge unfolds in the field of human-made culture, which is defined by Ernst Cassirer as the world of the symbolic and by Giambattista Vico as the world of history. In its epoch-making historicity, this knowledge is relative. On the other hand, however, the objects of this field—namely, electrophysics in its materiality and mathematics—always become operatively evident. It is true that the behavior implicit in these objects can only be discussed as explicit knowledge in discursive, historically identifiable contexts, yet the relation of latent media potentiality to human knowledge is not passive. There is a media-active urge that always appeals to the human desire for knowledge and that constantly reasserts itself across historical time, as the underlying laws are repeatedly brought to the fore. Media states are at the same time historically relative and scientifically invariant. "The paradoxical 'simultaneity' of invariance and variability corresponds to the general problem of structuring the autopoietic reproduction of systems."[24]

The media-archeological concept of *mathesis* (mediamatics) already indicates its proximity to mathematics. According to its essence (or self-understanding), however, mathematics is an ahistorical science, "in which the individuality of the mathematician and the historical specificity of his contribution can disappear in the final structure"[25]—validity versus genesis. The particular historicity of each technical medium is also sublated in the moment of operative or computational implementation. What is valid in the present takes priority over what is merely handed down.[26] Media often already practice such situations before their knowledge is possessed by humans. In 1879, Edward Hughes and George Gabriel Stokes observed that a distant spark was perceptible on a telephone as crackling: "Without knowing it, these physicists had experimented with rapid electrical oscillations and thus also with electromagnetic waves. . . . But they could not connect their observations to Maxwell's less well-known theory. They lacked the basis for further development."[27] In the incubation phase of new technologies, humans themselves sometimes become the "media" of a knowledge that anamnetically insists on manifesting. Technologically implicit knowledge is particularly compliant with cultural practices and human curiosity—much like the concept of the "meme," which makes use of humans and machines as hosts for the purpose of its transmission.[28] A knowledge that uses cultural tradition as a temporal channel in order to continuously transmit itself "is no mere *human* doing, even if the human belongs to the carrying out of such a [knowledge]."[29]

DIFFERENTIAL TIME, RECURSIVE TECHNO-LOGIC

The history of science consists not only of an inventory of artifacts and discourses; technical-mathematical laws are also part of the archive. Their adequate formulation requires not only a history of technology, but also a structure of algebraic formulas and technical diagrams.[30] Culture is not just a realm of signs in the sense of cultural semiotics (Jurij Lotman); media practices are also a realm of signals. Other time relations prevail here.[31] The event world of technological media is not exclusively the emphatic, semanticized time of historiography but also a web of microtemporal shifts in which the time-critical momentum below every historical sense is crucial. The being-in-time of technical media is not fully grasped by the historicization of its objects as scientific and cultural history.

Media archeology is certainly aware of the irreducible historicity of technical knowledge, yet it identifies this index not only in the particular discourse of each era but also in the technomathematical *dispositif* itself. The physics lecture hall of the Karlsruhe Institute of Technology facilitated Heinrich Hertz's experiments on the propagation of electric power. This room could be completely darkened and soundproofed in order to allow sparks to be seen and heard as indicators of electromagnetic waves; the dismantling of electroconductive substances like chandeliers and heating pipes also made it possible to create an interference-free electromagnetic field.[32] A cultural relativization of the discovery of that electromagnetic field as merely a function of discourses at the time runs into the danger of forgetting media reality itself. *There are* media in which something always occurs that does not depend on human-cultural knowledge. The modeling of electrotechnical knowledge may be variable with respect to concrete experimental settings, and it may be "epistemically, culturally, and even socially"[33] relative as an amalgam of hypotheses, mathematical methods, and measuring instruments, yet it is at the same time grounded in a virtually ahistorical and acultural *fundamentum in re.*

"Are we to consider that human acts are the sole factors of history?"[34] To some extent, actor network theory (ANT) also recognizes the nonhuman agents that play a role in this knowledge, yet this position fluctuates between sociological relativism and definitive media archeology. A system is spontaneously and recursively regenerated in the "organizational memory" of all technologies, and this system continuously yields surprises—the response of information theory seems more plausible than the teleological model of time.

IMPLICIT MACHINE AND MEDIA KNOWLEDGE: *AUTOMATHESIS* AND AGENCIES

It was already true of the pre-electronic era "that machines lived and moved according to their own needs rather than through the understanding and ability of humans."[35] This has intensified in the era of symbolic machines.

> When a . . . program . . . is first entered into a machine, the entire computing process runs automatically in the electron brain at a speed that seems impossible to grasp intellectually. The electron brain now works in "microprocesses" of time and compresses these processes into such short periods of time that they can be neither utilized nor even imagined through human actions and thoughts.[36]

Machine knowledge escalated in the era that appropriated the electromagnetic field for the purpose of telecommunications and computing. The copying telegraph, which was patented by Alexander Bain in 1843 but never actually realized, is a true media-archeological thing above all in its technological speakability. There are technical apparatuses "in which a certain pre-instrumental and pre-theoretical knowledge emerges long before it . . . cools down in a systematized form."[37] Cultural and nonhuman knowledge is stored in technical media, and it occasionally erupts in electrical suddenness instead of developing gradually. Television is not simply the temporal function of a development between phototelegraphy and the electronic image. The "from . . . to" indicates not merely a vector of technological history but also a nonlinear constellation—a short-circuited temporal mode as a shortcut of history.[38] Nonlinear media history means excluding historical contextuality in favor of another historicity. What is at issue, therefore, is not *another* way of writing media history, but rather *the other* of writing media history itself. Instead of expanding linear narrative history by adding the concept of contingency, nonlinear physics developed a concept of weak causality, according to which small changes of an initial state can induce large changes in future behavior[39]—a time-dilatory induction. Benoît Mandelbrot proved that the iterations of self-similar objects were based on non-whole number fractions. However, only the computer was able to demonstrate the numerical temporal mode of such fractals on the screen as an imaging process—an algorithmic, nonnarrative concept of event sequences that is similar in nature to the temporal modes of computing media themselves. The concept of "self-organized criticality"[40]—a time-critical processuality—applies to recursive dynamics. If these dynamics are conceived as unfolding in time, then they open up an alternative form of scientific historicity. Through nonsynchronized but coupled oscillators, such complex space-time dynamics can simulate how they emerge in vibratory media.

There is a techno-logic that constantly eludes the whims of historical change. In current mobile telephones, for example, electromagnetic radio waves continue to exist in another modulation, even though radio itself already belongs to media history. "Media are constant, forms alternate," writes Niklas Luhmann,[41] as the process of recursive reconstruction applies to the rhythm of technical media time: "Style respects tradition by deviating from it."[42] Peirce's operative diagram of an electrified circuit logic in 1886 remained as inconsequential as Paul Ehrenfest's coupling of algebraic logic and conducting paths charged with electricity in 1910, "so that as of 1935 the optimization of circuits required the analogy of Boolean algebra and electrical circuitry to be newly invented."[43] At the same time, however, the diagram enabled the logical reproducibility of technomathematical relations as real practice in the mode of symbolic reconfiguration.[44]

As the closed field of media-epistemological knowledge was opened up by Faraday and Boole, it developed nonlinearly in the play of latency and actualization. Electronic circuits—insofar as they are not interactive in the human-machine interfaces of the computer—yield processes that are largely independent of conscious decisions and subjective intentions[45]—"with the same necessity with which the spider spins."[46] The environment of technical media appears to be virtually automatic. To say this more pointedly: technical media bring forth structures that know about themselves—an internal mechanism that differs from everyday communication precisely because it is independent of its historical context.[47] According to Heidegger, therefore, "positionality [*Ge-Stell*], the essence of technology, cannot be anything merely human."[48] The miller thus stands in the structure of the mill and is framed by its operatively reified knowledge. On the other hand, this knowledge must be "assembled loosely enough"[49] that it can be coupled differently; the central computing and logic unit of the mechanical computer was thus derived from the "mill" in Babbage's analytical engine. Technical thought as the *a priori* of modern thought co-emerges with all concrete technologies.[50] Historian Leopold von Ranke's programmatic claim that he wanted to "extinguish" himself in his writing of history in order to allow the events to speak for themselves expresses an instinct for the nondiscursive side of this epistemology. The natural and technical sciences of the nineteenth century dreamed of minimizing the intervention of humans in physical measuring processes with the help of self-writing machines.[51] Media themselves thus become active agents in the creation of knowledge. What is writing itself here? Indicators and kymographs are epistemologically "hot" apparatuses, as they communicate a knowledge of nature that cannot be denied through discourse-analytical relativization and that only manifests in technical media.[52]

THESES ON THE HISTORIALITY OF TECHNICAL STRUCTURES

There is an implicit knowledge inherent in the implementation of every medium. "Otherwise, how could it be that certain things are in the air and almost want to be thought?"[53] The media-archeological time figure of equiprimordiality is an alternative to the cultural-historical concept of knowledge in so far as it assumes "that the knowledge and being of things [emerge] from one and the same source."[54] Unlike narrative history, *historiality* (according to Rheinberger) envisages "a kind of recurrence that precedes every consideration and thus interpretation." It claims "that recurrence is at work in the differential movement of the *system itself* and its time structure. What is called its history is 'subsequent' in a virtually constitutive sense: the recent is . . . the result of something that was not."[55] The emergence of things themselves serves as a model. "The historian goes from the beginning to the present so that contemporary science is to a certain degree always already signaled in the past. The epistemologist goes from the present back to the beginning so that only a part of what was considered science in the past appears to a certain degree to still be valid in the present."[56] In metrological and thus random structures, a temporality is at work that "fundamentally infiltrates"[57] the seemingly stable appearances of a historical era; this gap opens the possibility of entirely different chronologies. "Every system of material units and every system that deals with such units . . . would thus have its own *internal time*. This internal time is not a dimension of its existence in space and time. It characterizes the sequence of system states in so far as it can be comprehended in the form of cycles of non-identical replication."[58] However, this "ruptures the lavish imagination of history."[59] The alternative lies in concepts like the dissipative field; in the case of technological media, moments of stability are "a necessary but not sufficient condition of historial structures. Historiality assumes the possibility of the retention of knowledge in the form of its continuous material reactivation."[60] This corresponds to the phenomenon of resonances in physical acoustics, which serves as an appropriate time figure to describe media processes; they are of an ahistorical nature in so far as their interrelationship does not necessary involve "originally intertwined appearances."[61] The temporality revealed in sonic phenomena corresponds to the temporal modes of technical media in a virtually equiprimordial way. "If we sense that the description of sympathetic vibration bears some resemblance to radio broadcast, it is no coincidence, the same principle is at work. The processes of contemporary media systems are latent in the laws of nature."[62]

A media-archeological theory of time differs markedly from the traditional philosophy of history. The micromedial event space in the phenomenology of the electromagnetic field developed by Faraday and Maxwell also serves as an explanatory model of macrotemporal evolution in general. Alexander Gurvitsch's discovery of ultra-weak photoemissions during cell division was only made possible through measuring media, which were based on as well as reacting to the temporal modes of the electromagnetic field in a technophysical sense.[63] Morphogenetic fields are thus not only responsible for the development of structures in the present, but their natural frequencies also resonate with structures in the past. This isomorphism in the temporal domain is opposed to symbolic transcoding, which is reflected in the literality of alphanumeric recordings: Boltzmann's notion of entropy (as the static dispersion of thermodynamic particles in space) and Shannon's notion of entropy (as coded signal transmission) thus represent alternative temporal modes. In his book *The Presence of the Past* (1988), Rupert Sheldrake combined the mathematical temporal mode of Markov chains, where only the present state determines the probable occurrence of future states, with evolutionary macrohistory. What is central here is a communication process whose formulation is common to radio and telecommunications. "The morphic resonance of countless early organisms is a composite or average of the previous forms, which represents a probability structure."[64] If the relationship between dynamic models is transferred to the time axis, then the oscillation is at the same time both the subject and the object of time and its temporal operation is resonance in the form of harmonic oscillators. The operative reproduction of a musical or poetic melody also resonates with the corresponding past, which enables "lively" communication with the dead.[65] A transmission-oriented model (telecommunications in terms of space and tradition in terms of time) is thus replaced by the tunneling of transfer time; as stimuli, signals evoke almost immediate effects in resonance-capable receiving systems. Resonance corresponds to the dynamic temporal relations of an electromagnetic field rather than one-dimensional transmission in a linear channel. "We seek to strike a responsive chord. . ., not get a message across. This involves . . . examining how stored experiences are patterned in our brain, and how previous experiences condition us to perceive new stimuli."[66] This describes the nonlinear epistemology of the sonic time field,[67] which thus serves as a model for the nonhistorical temporal mode of technical media.

THE DISCOURSE OF TECHNOLOGICAL HISTORY VERSUS MEDIA-ARCHEOLOGICAL EQUIPRIMORDIALITY AND ITS SUBLATION IN THE CONCEPT OF INFORMATION

A field of compressed knowledge yields similar insights at different moments. Insight into anachronistic, ahistorical knowledge relations emerges from the shadows of knowledge-historical influences. Media-specific time analysis examines the emergence in different technological fields of modes of implementation that yield similar phenomena despite the fact that there is no direct exchange of communication between them. Influences thus do not simply form traditions in the sense of cultural evolution, but rather they are processed in preexisting structures.

As rules of mapping, logical structures initially dispel temporal momentum. As a generalization of repetition, however, recursive data structures constitute a subcategory of dynamic structures that change their structure during the execution of a program.[68] However, the concept of recursions, which was borrowed from algorithms, is usually still interpreted historiographically instead of being expressed algorithmically as the inherent logic of media-archeological time.[69] Gilbert Simondon chose the concept of "recurrent causality" as a variant of dynamic equiprimordiality in techno-cultural relations.[70] It is worth being independent of inventor biographies for a moment and to write an archeology of technological configurations in terms of their concrete media-technical modes of existence. Heidegger described being-in-the-world (*In-der-Welt-Sein*) as the "co-original" or "equiprimordial" (*gleichursprünglich*) character of being.[71] In the mechanical act of copying a text, the copyist participates in the operative moment of the technologizing of the word, which corresponds to repeatable signal processing in technical media. As machinic implementations of astronomical knowledge, the ancient Antikythera mechanism and the medieval astrolabe implement this code independently of their textual tradition; the techno-logic of mathematics also reproduces knowledge (such as Euclid's *Elements*) that is equiprimordially new. In his treatise on the *Origin of Geometry*, Husserl defines a geometrical existence that lies beyond physical, discursive, or cultural history, as it is actualized supertemporally at different points in time[72]—even though it is susceptible to misinterpretation. In its concrete technical instantiation, every ideal structure becomes a contingent individuality. Technical media are at the same time empirically concretized and metahistorically transcendent. At the level of actual implementation, there is no such thing as a pure, eternal tone; its concrete activation through a string instrument or an electronic synthesizer always causes transients to resonate as temporal contamination. A temporal (but not necessarily historical) index is thus already registered in the media event as a trace of the real.

The phonetic alphabet continuously carries within itself the trace of its equiprimordial orientation toward oral poetry;[73] nevertheless, the alpha-numeric code in which digital signal processing is implemented today is a character set designed for the purpose of telecommunications. In the Turing era, "writing disappears in machines as numbers. ... So we live as guests of a world in which culture accelerates the sound of nature."[74] The media-archeological approach does not treat such technical recursions within culture as historical figures, but rather it uses them to describe the autonomy of electrophysics, logic, and mathematics, which is negotiated in technical media. It is the genuine operativity of technomathematical media that suggests an alternative form of time-writing: algorithmic time, the temporal logic of programming in loops and conditions.

There is one method, one name, for this new way of writing: recursion. It looks for the return of the same—and actually in the same way that its history of being transformed. "We 'run backwards' in time, from today to the Greeks, but at the same time also forwards in time, from the first beginning to its repeated distortion. ... Sometimes the threads branch out like a fork; at other times separate threads become intertwined like a mesh."[75] Whether this still results in a form of history writing or rather its alternative remains to be seen. In any case, the current interest in such operative time figures indicates a growing discomfort with the earlier model of media history—the denial of which is represented by technical *chronopoiesis*. In the model of recursion, media time is negotiated as a continuously new configuration of isomorphic constellations. The cultural unconscious of technologies reveals the existence of laws whose logic is always repeatedly and equiprimordially compelling in different cultural-historical eras.

The novel *The Difference Engine,* in which Babbage's protocomputer in Victorian England functions as the protagonist, plays out an iterative chronotechnics as both subject and object of a narrative sequence.[76] Niklas Luhmann has in turn adopted the cybernetic concept of recursion to describe memory performance in autopoietic system reproduction.

> I use the term "memory" here neither in the sense of a possible return to the past nor in the sense of a store of data or information we can draw on as need be. I am talking about a function in constant but only present use that tests all incipient operations for consistency with what the system constructs as reality.[77]

Leibniz chose a chronotechnical form to describe the metaphysical construction of a prestable harmony. Leibniz's *Monadology* presents a world of the smallest automata, which he imagines (as a true disciple of Christian Huyghens) according to the model of communication between clockworks. "Though the monads reflect one another, the reflection does not consist in a transfer of the causal chain from one to another. They are actually as self-contained as, or rather more self-contained than, the passively dancing figures on top of a music box."[78] The cybernetic concept of informational equivalence is an extension of this idea. "When a speech is recorded on a gramophone record, on tape, printed, or transmitted over electrical radio waves, etc. the amount of semantically identical information remains the same, although its physical carrier is entirely different"[79]—as long as information is understood as discrete symbol sequences. The situation is different for the physical event of vocal articulation—in other words, the classical analog signal world. Georg Klaus's theorem is only absolutely valid again with the rise of digital "sampling." What remains invariant in technical transduction over time (from gramophone record to tape to printed newspaper article) is neither substance nor consciousness but rather form (*Gestalt*).

> The gramophone record generates sound waves, which are automatically transformed into electrical waves at a radio broadcasting station and then recorded again at the point of reception and transferred through a (technically constructed) device to the keys of a typewriter that finally puts the entire speech down on paper. Naturally a thinking consciousness must have constructed all of these devices, and a consciousness must have once created the content of this speech.[80]

In this respect, the typewriter and the text of the speech are equiprimordial. "Once the information is created, it acquires to a certain extent its own life."[81] This results in a technological theory of media-cultural transmission based on matter as well as symbolic coding.

At the end of the analysis of media-technical implementation, the traditional concept of time proves to be too rigid in its totalizing referentiality to grasp with sufficient elasticity the tempor(e)alities of technical media. Technical worlds cling to their concrete embodiments and are at the same time deconstructed by them; the horizon of microtemporal, mesotemporal, and macrotemporal media worlds, which was fostered by the primacy of historical time, is burst open. Media-induced time "is no longer a single fixed reference point that exists external to events. Time is now . . . choreographed directly,"[82] whether in the electrophysics of hardware, algorithmic software, or interfaces linked to the human sense of time. The media message of those temporal modes is something other than what we once called "time."

NOTES

1. Michel Serres and Nayla Farouki, eds., *Thesaurus der exakten Wissenschaften* (Frankfurt am Main: Zweitausendeins, 2001), 242.

2. See Andreas R. Becker, *Netzereignis—Ereignisnetz. Prozesse und Strukturen medialer Ereignisse im Internet* (Marburg: Schüren, 2009), 13.

3. von Falkenhayn, *Augenblick und Kairos*, 16.

4. Oswald Spengler, "Heraklit," in *Reden und Aufsätze* (Munich: Beck, 1937), 22.

5. Heidegger, *Being and Time*, 6. Heidegger also notes that "the 'iteration' (*Wiederholung*) of the similar is fundamentally different from the recovery of the relation to the *same* in retrieval (*Wieder-Holung*)"—a sentence that can be read as concretely media-archeological. Martin Heidegger, *Die Geschichte des Seyns* (Frankfurt am Main: Vittorio Klostermann, 1998), 22f.

6. "We need an eye which can see the past in its place with its definite differences from the present, and yet so lively that it shall be as present to us as the present." T. S. Eliot, *Selected Essays* (London: Faber & Faber, 1934), 64. See also Reinhold Hammerstein, "Musik als Komposition und Interpretation," *Deutsche Vierteljahresschrift für Literaturwissenschaft und Geistesgeschichte* 40.1 (1966): 23.

7. Wilhelm Pinder, *Das Problem der Generation in der Kunstgeschichte Europas* (Berlin: Frankfurter Verlags-Anstalt, 1928), 3

8. Ibid., 11.

9. Ibid., 12.

10. Heidegger, *Being and Time*, 353.

11. See Sebastian Döring, "Der offene Bereich des Geistes. Der Äther und seine Medien," in *sonArc::ion Projekt. Äther oder der domestizierte Blitz*, ed. Jan-Peter Sonntag (Berlin: Kulturverlag Kadmos, 2008), 167–255.

12. See B. L. van der Waerden, "Die Harmonielehre der Pythagoreer," *Hermes* 78.2 (1943): 177.

13. Heinrich Hertz, *The Principles of Mechanics*, trans. D. E. Jones and J. T. Walley (London: Macmillan, 1899), 1.

14. Friedrich Herneck, *Bahnbrecher des Atomzeitalters. Große Naturforscher von Maxwell bis Heisenberg* (Berlin: Buchverlag der Morgen, 1968), 64. The theory of numbers is perhaps already sufficient proof.

15. Michael Harenberg, "Virtuelle Instrumente zwischen Simulation und (De) Konstruktion," in *Soundcultures. Über elektronische und digitale Musik*, ed. Markus S. Kleiner and Achim Szepanski (Frankfurt am Main: Suhrkamp, 2003), 90.

16. Alfred North Whitehead, "Inapplicability of the Concept of Instant on the Quantum Level," in *The Concepts of Space and Time*, ed. Milic Capek (Dordrecht: Reidel, 1976), 537.

17. Karl R. Popper, *Objektive Erkenntnis. Ein evolutionärer Entwurf* (Hamburg:Hoffmann & Campe, 1984), 165.

18. Ibid.

19. Ibid., 194.

20. Wilhelm Ostwald, *Elektrochemie. Ihre Geschichte und ihre Lehre* (Leipzig: Veit & Company, 1896), 1f.

21. Sigfrid Wantzloeben, *Das Monochord als Instrument und als System* (Halle: Niemeyer, 1911), 128.

22. This media-archeological concept of knowledge differs decidedly from Michael Polanyi, *Implizites Wissen* (Frankfurt am Main: Suhrkamp, 1985).

23. Joseph Henry, "On Induction from Ordinary Electricity; and on the Oscillatory Discharge," in *Scientific Writings of Joseph Henry* (Washington: Smithsonian, 1886), 1: 200–3. Qtd. in Julian Blanchard, "The History of Electrical Resonance," *Bell System Technical Journal* 20.4 (1941): 416.

24. Luhmann, *Art as a Social System*, 129.

25. Herbert Breger, "Mathematik im Wandel der Geschichte," *Wechselwirkung* 5 (1982): 31. See also Carsten Busch, "Analyse und Bewertung des Informationsbegriffs nach C. E. Shannon" (MA diss, Technical University of Berlin, 1989), 70.

26. Gerhard Kropp, *Geschichte der Mathematik. Probleme und Gestalten* (Wiesbaden: Aula, 1994), 5. This is familiar from the Christian Eucharist: on the one hand, it is understood as a (supposed) historical moment (the Last Supper of Christ); on the other hand, it is also encapsulated in sublated time and revealed or realized continuously through liturgical repetition.

27. Wolfgang Schreier, *Die Entstehung der Funktechnik* (Munich: Deutsches Museum, 1997), 15. See also Ana Ofak and Philipp von Hilgers, introduction to *Rekursionen. Von Faltungen des Wissens*, ed. Ana Ofak and Philipp von Hilgers (Munich: Fink, 2010), 25–46.

28. See Susan Blackmore, *The Meme Machine* (Oxford: Oxford University Press, 1999).

29. Martin Heidegger, "Positionality," in *Bremen and Freiburg Lectures*, trans. Andrew J. Mitchell (Bloomington: Indiana University Press, 2012), 29.

30. "The set of possible statements (É) can be represented by a function ($f_É$) as follows: $f_É$: R x P x A x L → S. Here *R* is the class of 'referentials,' *P* is the class of 'subjective positions,' *A* is the class of 'associated domains,' *L* is the class of spatiotemporal locations, and *S* is the class of strings of signs." Martin Kusch, "Discursive Formations and Possible Worlds: A Reconstruction of Foucault's Archeology," *Science Studies* 1 (1989): 18.

31. See Wolfgang Ernst, "Von der Mediengeschichte zur Zeitkritik," *Archiv für Mediengeschichte* 6 (2006): 23–32.

32. Falk Rieß, "Erkenntnis durch Wiederholung. Eine Methode zur Geschichtsschreibung des Experiments," in *Experimental Essays. Versuche zum Experiment,* ed. Michael Heidelberger and Friedrich Steinle (Baden-Baden: Nomos, 1998), 165.

33. Hans-Jörg Rheinberger, *Experiment—Differenz—Schrift. Zur Geschichte epistemischer Dinge* (Marburg: Basilisken, 1992), 9. See also Klaus Hentschel, "Feinstruktur und Dynamik von Experimentalsystemen," in *Experimental Essays. Versuche zum Experiment,* ed. Michael Heidelberger and Friedrich Steinle (Baden-Baden: Nomos, 1998): 329.

34. Gabriel Tarde, *The Laws of Imitation,* trans. Elsie Clews Parsons (New York: Henry Holt, 1903), 1.

35. Andrej Platonow, *Tschewengur. Die Wanderung mit offenem Herzen* (Berlin: Volk & Welt, 1990), 44.

36. Rolf Strehl, *Die Roboter sind unter uns. Ein Tatsachenbericht* (Oldenburg: Gerhard Stalling, 1952), 26.

37. Christian Kassung and Albert Kümmel, "Synchronisationsprobleme," in *Signale der Störung,* ed. Albert Kümmel and Erhard Schüttpelz (Munich: Fink, 2003), 144.

38. See Franz Pichler, *Elektronische Bilder aus der Ferne. Technische Entwicklung von Bildtelegraphie und Fernsehen bis zum Jahre 1939. Von den mechanisch-optischen Anfängen zu den elektronischen Systemen* (Linz: Universitätsverlag Trauner, 2010).

39. Jürgen Kurths and Udo Schwarz, "Nichtlineare Wissenschaften—Neue Paradigmen und Konzepte," *Kunstforum International* 55 (2001): 66.

40. Ibid., 68.

41. Luhmann, *Art as a Social System,* 129.

42. Ibid., 131.

43. Schäffner, "Topologie der Medien," 89f.

44. See Charles Sanders Peirce, *Collected Papers, Volume II: Elements of Logic* (Cambridge, MA: Harvard University Press, 1932). See also Matthias Bauer and Christoph Ernst, *Diagrammatik. Einführung in ein kultur- und medienwissenschaftliches Forschungsfeld* (Bielefeld: transcript, 2010).

45. See Jens Gerrit Papenburg, "Der Synthesizer als Apriori: Körper und Maschinen in der Popmusik," *Paragrana. Internationale Zeitschrift für Historische Anthropologie* 14.2 (2005): 91–104.

46. Friedrich Nietzsche, "Truth and Lies in a Nonmoral Sense," in *Philosophy and Truth: Selections from Nietzsche's Notebooks of the Early 1870s,* trans. Daniel Breazeale (Atlantic Highlands, NJ: Humanities Press, 1990), 87.

47. Friedrich Kittler, *Optical Media: Berlin Lectures 1999*, trans. Anthony Enns (Cambridge: Polity Press, 2010), 44. Kittler here refers to Claude Shannon's theory of communication: "When Shannon explicitly says that we have no need for a communications system for eternal truths . . . because such truths must be continuously reproducible at different times and places without technical transmission, it becomes abundantly clear how the essence of media diverges from our everyday concept of faith."

48. Heidegger, "Positionality," 37. Qtd. in Wolfgang Hagen, *Das Radio. Zur Geschichte und Theorie des Hörfunks—Deutschland/USA* (Munich: Fink, 2005), 255.

49. Rheinberger, "Historialität, Spur, Dekonstruktion," 54.

50. See Peter-Paul Verbeek, *What Things Do: Philosophical Reflections on Technology, Agency, and Design* (University Park: Pennsylvania State University Press, 2005).

51. See Lorraine Daston and Peter Galison, "The Image of Objectivity," *Representations* 37 (1992): 67–106.

52. See Wolfgang Ernst, "Sekretärinnen ohne Chef (Mnemosyne, Klio, Schreibmaschinen)," in *Europa: Kultur der Sekretäre*, ed. Bernhard Siegert and Joseph Vogl (Zürich: diaphanes, 2003), 253–65.

53. Martin Donner, e-mail message to author, May 17, 2008. See also Martin Donner, "Rekursion und Wissen. Zur Emergenz technosozialer Netze," in *Rekursionen. Von Faltungen des Wissens*, ed. Ana Ofak and Philipp von Hilgers (Munich: Fink, 2010), 77–115.

54. This is how Rainer Bayreuther formulates Descartes and Leibniz's disentanglement from ancient Pythagorean musical arithmetic. See Rainer Bayreuther, "Mathematisches Denken in der Musik des 16. und 17. Jahrhunderts," in *Die mathematischen Wurzeln der Kultur. Mathematische Innovationen und ihre kulturellen Folgen*, ed. Jochen Brüning and Eberhard Knobloch (Munich: Fink, 2005), 137.

55. Rheinberger, "Historialität, Spur, Dekonstruktion," 48f.

56. Georges Canguilhem, "Die Geschichte der Wissenschaften im epistemologischen Werk Gaston Bachelards," in *Wissenschaftsgeschichte und Epistemologie*, trans. Michael Bischoff and Walter Seitter, ed. Wolf Lepenies (Frankfurt am Main: Suhrkamp, 1979), 12.

57. Rheinberger, "Historialität, Spur, Dekonstruktion," 50.

58. Ibid., 50f.

59. Ibid., 52.

60. Ibid., 54.

61. Hans-Jörg Rheinberger and Michael Hagner, "Experimentalsysteme," in *Die Experimentalisierung des Lebens*, ed. Hans-Jörg Rheinberger and Michael Hagner (Berlin: Akademie, 1993), 12.

62. Viola, "The Sound of One Line Scanning," 157.

63. Alexander Gurvitsch, "Über den Begriff des embryonalen Feldes," in *W. Roux' Archiv für Entwicklungsmechanik* (1922): 353–415. Another effect of subtle electrophysiology is the recent discovery of channel-free "field coupling" as a condition of time-critical communication between neurons.

64. Rupert Sheldrake, *Das Gedächtnis der Natur. Das Geheimnis der Entstehung der Formen in der Natur* (Munich: Piper, 1996), 158. See also Friedrich Cramer,

Symphonie des Lebendigen. Versuch einer allgemeinen Resonanztheorie (Frankfurt am Main: Insel, 1996).

65. William Stern, *Die differentielle Psychologie in ihren methodischen Grundlagen* (Leipzig: Barth, 1911), 146f. In part two of the same work, Stern determines this process strictly mathematically: variations and correlations.

66. Tony Schwartz, *The Responsive Chord* (Garden City, NY: Anchor Books, 1974), 27.

67. "In discussing electronically based communication processes, it is very helpful to use auditory terms . . . like *feedback* . . . *reverberation* . . . *tuning*." Ibid., 23.

68. Niklaus Wirth, *Algorithmen und Datenstrukturen* (Stuttgart: Teubner, 1975), 10.

69. See Shintaro Miyazaki, "Algorithmics: Understanding Micro-Temporality in Computational Cultures," *Computational Culture* 2 (2012), http://computationalculture.net.

70. Simondon, *On the Mode of Existence of Technical Objects*, 51–60.

71. Heidegger, *History of the Concept of Time*, 238.

72. Jacques Derrida, *Edmund Husserl's Origin of Geometry. An Introduction*, trans. John P. Leavey Jr. (Lincoln: University of Nebraska Press, 1989), 160.

73. Barry Powell, *Homer and the Origin of the Greek Alphabet* (Cambridge: Cambridge University Press, 1991).

74. Friedrich Kittler, *Musik und Mathematik I. Hellas 2: Eros* (Munich: Fink, 2009), 80.

75. Ibid., 245.

76. Gibson and Sterling, *The Difference Engine*.

77. Niklas Luhmann, *Theory of Society*, trans. Rhodes Barrett (Stanford: Stanford University Press, 2012), 1: 349.

78. Wiener, *Cybernetics*, 41.

79. Georg Klaus, *Kybernetik in philosophischer Sicht* (Berlin: Dietz, 1962), 77.

80. Ibid.

81. Ibid.

82. McLuhan and McLuhan, *Laws of Media*, 53.

Selected Bibliography

Aarseth, Espen. "Aporia of Epiphany in *Doom* and *The Speaking Clock*: The Temporality of Ergodic Art." In *Cyberspace Textuality: Computer Technology and Literary Theory*, edited by Marie-Laure Ryan, 31–42. Bloomington: Indiana University Press, 1999.

Adorno, Theodor W. *Current of Music: Elements of a Radio Theory*, edited by Robert Hullot-Kentor. Frankfurt am Main: Suhrkamp, 2006.

Adrian, Edgar Douglas. *The Mechanism of Nervous Action: Electrical Studies of the Neurone*. Philadelphia: University of Philadelphia Press, 1932.

Aiserman, Mark Aronowitsch et al. *Logik, Automaten, Algorithmen*. Munich: Oldenbourg, 1967.

Arns, Inke, and Gabriele Horn, eds. *History Will Repeat Itself: Strategies of Reenactment in Contemporary (Media) Art and Performance*. Frankfurt am Main: Revolver, 2007.

Aschoff, Volker et al. *Die Zeit. Dauer und Augenblick*. Munich: Piper, 1992.

Baird, John Logie. *Television and Me: The Memoirs of John Logie Baird*, edited by Malcolm Baird. Edinburgh: mercatpress, 2004.

Barthes, Roland. *Camera Lucida: Reflections on Photography*. Translated by Richard Howard. New York: Hill & Wang, 1981.

Becker, Andreas R. *Netzereignis—Ereignisnetz. Prozesse und Strukturen medialer Ereignisse im Internet*. Marburg: Schüren, 2009.

Bergson, Henri. *Creative Evolution*. Translated by Arthur Mitchell. London: Macmillan, 1922.

Berz, Peter. "Uhrwerk und Zeitgetriebe." In *Zeitreise. Bilder, Maschinen, Strategien, Rätsel*, edited by Georg Christoph Tholen, Michael Scholl, and Martin Heller, 171–88. Basel/Frankfurt am Main: Stroemfeld/Roter Stern, 1993.

Bitsch, Annette. *Diskrete Gespenster. Die Genealogie des Unbewussten aus der Medientheorie und Philosophie der Zeit*. Bielefeld: transcript, 2009.

Bolter, David. *Turing's Man: Western Culture in the Computer Age*. Chapel Hill: University of North Carolina Press, 1984.

Braun, Ferdinand. "Ueber ein Verfahren zur Demonstration und zum Studium des zeitlichen Verlaufes variabler Ströme." *Annalen der Physik und Chemie* 60.1 (1897): 552–59.

Collins, Nick, Alex McLean, Julian Rohrhuber, and Adrian Ward. "Live Coding in Laptop Performance." *Organised Sound* 8.3 (2003): 321–30.

Cramer, Florian, and Elfriede Jelinek, eds. *Zauberhafte Klangmaschinen. Von der Sprechmaschine bis zur Soundkarte.* Mainz: Schott, 2008.

Cramer, Friedrich. *Symphonie des Lebendigen. Versuch einer allgemeinen Resonanztheorie.* Frankfurt am Main: Insel, 1996.

Crivellari, Fabio, Kay Kirchmann, Marcus Sandl, and Rudolf Schlögl, eds. *Die Medien der Geschichte. Historizität und Medialität in interdisziplinärer Perspektive.* Konstanz: UVK, 2004.

Czech, Josef. *Der Elektronenstrahl-Oszillograf. Aufbau, Arbeitsweise, Meßtechnik.* Berlin-Borsigwalde: Verlag für Radio-Foto-Kinotechnik, 1955.

de Certeau, Michel. "Writing vs. Time: History and Anthropology in the works of Lafitau." *Yale French Studies* 59 (1980): 37–64.

de Chadarevian, Soraya. "Die 'Methode der Kurven' in der Physiologie zwischen 1850 und 1900." In *Die Experimentalisierung des Lebens. Experimentalsysteme in den biologischen Wissenschaften 1850/1950*, edited by Hans-Jörg Rheinberger and Michael Hagner, 28–49. Berlin: Akademischer Verlag, 1993.

de la Motte-Haber, Helga, and Günther Rötter, eds. *Musikpsychologie.* Laaber: Laaber-Verlag, 2005.

Deleuze, Gilles. *Cinema 2: The Time Image.* Translated by Hugh Tomlinson and Robert Galeta. Minneapolis: University of Minnesota Press, 1989.

Dennhardt, Robert. *Die Flipflop-Legende und das Digitale. Eine Vorgeschichte des Digitalcomputers vom Unterbrecherkontakt zur Röhrenelektronik 1837–1945.* Berlin: Kulturverlag Kadmos, 2009.

Dienst, Richard. *Still Life in Real Time: Theory after Television.* Durham, NC: Duke University Press, 1994.

Dolar, Mladen. *His Master's Voice. Eine Theorie der Stimme.* Frankfurt am Main: Suhrkamp, 2007.

Dotzler, Bernard. *Diskurs und Medium. Zur Archäologie der Computerkultur.* Munich: Fink, 2006.

Ebeling, Knut. "Das technische Apriori." *Archiv für Mediengeschichte* 6 (2006): 11–22.

Egly, Max. *Eintritt frei Fernsehen.* Lausanne: Rencontre, 1963.

Engel, Friedrich, Gerhard Kuper, and Frank Bell. *Zeitschichten. Magnetbandtechnik als Kulturträger.* Potsdam: Polzer Media Group, 2008.

Ernst, Wolfgang. "Im Reich von Δt. Medienprozesse als Spielfeld sonischer Zeit." In *Sound Studies. Traditionen—Methoden—Desiderate. Eine Einführung*, edited by Holger Schulze, 125–42. Bielefeld: transcript, 2008.

———. "Temporary Items. Die Beschleunigung des Archivs." In *ephemer_temporär_provisorisch*, edited by Immanuel Chi, Susanne Düchting, and Jens Schröter, 77–88. Essen: Klartext, 2002.

———. "Von der Mediengeschichte zur Zeitkritik." *Archiv für Mediengeschichte* 6 (2006): 23–32.

Selected Bibliography 271

Faulstich, Werner, and Christian Steininger. *Zeit in den Medien—Medien in der Zeit*. Munich: Fink, 2002.

Feddersen, Berend Wilhelm. *Entladung der Leidener Flasche, intermittierende, kontinuierliche, oszillatorische Entladung und dabei geltende Gesetze. Abhandlungen*. Edited by Theodor Des Courdres. Leipzig: Engelmann, 1908.

Frizot, Michel. "Analyse und Synthese der Bewegung. Étienne-Jules Mareys Methode." In *Apparaturen bewegter Bilder*, edited by Daniel Gethmann and Christoph B. Schulz, 141–54. Münster: Lit Verlag, 2006.

Gabor, Denis. "Acoustical Quanta and the Theory of Hearing." *Nature* 4044 (May 1947): 591–94.

Gelhard, Andreas, Ulf Schmidt, and Tanja Schultz, eds. *Stillstellen. Medien—Aufzeichnung—Zeit*. Schliegen: Argus, 2004.

Großklaus, Götz. *Medien-Zeit. Medien-Raum. Zum Wandel der raumzeitlichen Wahrnehmungen in der Moderne*. Frankfurt am Main: Suhrkamp, 1995.

Hagen, Wolfgang. "Die Entropie der Fotografie. Skizzen zur einer Genealogie der digital-elektronischen Bildaufzeichnung." In *Paradigma Fotografie. Fotokritik am Ende des fotografischen Zeitalters*, edited by Hertha Wolf, 1: 195–235. Frankfurt am Main: Suhrkamp, 2002.

Hayes, Harvey C. "Measuring Ocean Depths by Acoustical Methods." *Journal of the Franklin Institute* 197.3 (March 1924): 323–54.

Heidegger, Martin. *Being and Time*. Translated by Joan Stambaugh. Albany: State University of New York Press, 1996.

Heidelberger, Michael, and Friedrich Steinle, eds. *Experimental Essays. Versuche zum Experiment*. Baden-Baden: Nomos, 1998.

Heidenreich, Stefan. *FlipFlop. Digitale Datenströme und die Kultur des 21. Jahrhunderts*. Munich: Hanser, 2004.

Herbst, Ludolf. *Komplexität und Chaos. Grundzüge einer Theorie der Geschichte*. Munich: C. H. Beck, 2004.

Hertz, Heinrich. *Über sehr schnelle elektrische Schwingungen. Vier Arbeiten*. Leipzig: Akademische Verlagsgesellschaft, 1971.

Hoogstad, Jan Hein. "Time Tracks." PhD diss., Utrecht University, 2005.

Hubbard, Barbara Burke. *The World According to Wavelets: The Story of a Mathematical Technique in the Making*. Wellesley, MA: A. K. Peters, 1996.

Husserl, Edmund. *On the Phenomenology of the Consciousness of Internal Time*. Translated by John Barnett Brough. Dordrecht: Kluwer, 1991.

Jongmanns, Georg. "Gute Zeiten, schlechte Zeiten. Das Echtzeit-Real." In *Authentizität als Darstellung*, edited by Jan Berg, Hans-Otto Hügel, and Hajo Kurzenberger, 250–72. Hildesheim: University of Hildesheim, 1997.

Kassung, Christian, and Albert Kümmel. "Synchronisationsprobleme." In *Signale der Störung*, edited by Albert Kümmel and Erhard Schüttpelz, 143–65. Munich: Fink, 2003.

Kassung, Christian. *Das Pendel. Eine Wissensgeschichte*. Munich: Fink, 2007.

Kessler, Frank. "Bilder in Bewegung. Für eine nicht-teleologische Mediengeschichtsschreibung." In *Apparaturen bewegter Bilder*, edited by Daniel Gethmann and Christoph B. Schulz, 208–20. Münster: Lit Verlag, 2006.

Kittler, Friedrich. *Draculas Vermächtnis. Technische Schriften.* Leipzig: Reclam, 1993.

———. *Gramophone, Film, Typewriter.* Translated by Geoffrey Winthrop-Young and Michael Wutz. Stanford: Stanford University Press, 1999.

Kirchmann, Kay, *Verdichtung, Weltverlust und Zeitdruck. Grundzüge einer Theorie der Interdependenzen von Medien, Zeit und Geschwindigkeit im neuzeitlichen Zivilisationsprozeß.* Opladen: Leske & Budrich, 1998.

Kley, A., and G. Meyer-Brötz. "Analoge Rechenelemente als Abtaster, Speicher und Laufzeitglieder." *Elektronische Rechenanlagen* 3.3 (1961): 119–22.

Koselleck, Reinhart. *Zeitschichten.* Frankfurt am Main: Suhrkamp, 2003.

Lazzarato, Maurizio. *Videophilosophie. Zeitwahrnehmung im Postfordismus.* Berlin: b-books, 2002.

Lessing, Gotthold Ephraim. *Laocoon: An Essay on the Limits of Painting and Poetry.* Translated by Ellen Frothingham. Boston, MA: Roberts Brothers, 1887.

Levin, Thomas. "Rhetoric of the Temporal Index: Surveillant Narration and the Cinema of 'Real Time.'" In *CTRL[SPACE]: Rhetorics of Surveillance from Bentham to Big Brother,* edited by Ursula Frohne and Peter Weibel, 578–93. Cambridge, MA: MIT Press, 2002.

Lipfert, Kurt. *Das Fernsehen. Eine allgemeinverständliche Darstellung des neuesten Standes der Fernsehtechnik.* Munich: Lehmann, 1938.

Lovink, Geert. "Die Indifferenz der vernetzten Gegenwart. Über Internet-Zeit." In *Zero Comments,* edited Geert Lovink, 167–80. Bielefeld: transcript, 2008.

Lyotard, Jean-François. *The Inhuman: Reflections on Time.* Stanford: Stanford University Press, 1991.

Mach, Ernst. "Bemerkungen über wissenschaftliche Anwendung der Photographie." *Jahrbuch für Photographie und Reproductionstechnik* 2 (1888): 284–86.

Marey, Étienne-Jules. *Die Chronophotographie.* Translated by A. von Heydebreck. Berlin: Mayer & Müller, 1893.

McLean, Donald. *Restoring Baird's Image.* London: Institution of Electrical Engineers, 2000.

McLuhan, Marshall. *Understanding Media: The Extensions of Man.* New York: McGraw-Hill, 1964.

———, and Eric McLuhan. *Laws of Media: The New Science.* Toronto: University of Toronto Press, 1988.

Melitopoulos, Angela. "Timescapes." In *Lab. Jahrbuch 1996/97 für Künste und Apparate,* edited by the Kunsthochschule für Medien, 172–82. Cologne: Walther König, 1997.

Mindell, David A. *Between Human and Machine: Feedback, Control, and Computing before Cybernetics.* Baltimore: Johns Hopkins University Press, 2004.

Mulser, Peter. "Zeitlupe Computer. Analyse von Lichtimpulsen jenseits der Meßbarkeit." In *Simulation. Computer zwischen Experiment und Theorie,* edited by Valentin Braitenberg and Inga Hosp, 100–25. Reinbek bei Hamburg: Rowohlt, 1995.

Ofak, Ana, and Philipp von Hilgers, eds. *Rekursionen. Von Faltungen des Wissens.* Munich: Fink, 2010.

Paflik, Hannelore, ed. *Das Phänomen Zeit in Kunst und Wissenschaft.* Weinheim: VCH Acta Humaniora, 1987.

Selected Bibliography

Pattee, H. H. "Discrete and Continuous Processes in Computers and Brains." In *Physics and Mathematics of the Nervous System*, edited by M. Conrad et al., 128–48. Berlin: Springer, 1974.

Pflüger, Jörg. "Wo die Quantität in Qualität umschlägt. Notizen zum Verhältnis von Analogem und Digitalem." In *Hyperkult II. Zur Ortsbestimmung analoger und digitaler Medien*, edited by Martin Warnke, Wolfgang Coy, and Georg Christoph Tholen, 27–94. Bielefeld: transcript, 2005.

Pias, Claus, ed. *Cybernetics: The Macy Conferences 1946–1953*. Zürich: Diaphanes, 2003.

———. "Elektronenhirn und verbotene Zone. Zur kybernetischen Ökonomie des Digitalen." In *Analog/Digital—Opposition oder Kontinuum? Zur Theorie und Geschichte einer Unterscheidung*, edited by Jens Schröter and Alexander Böhnke, 295–309. Bielefeld: Transcript, 2004.

Rheinberger, Hans-Jörg. *Experiment, Differenz, Schrift. Zur Geschichte epistemischer Dinge*. Marburg: Basilisken, 1992.

Rieger, Stefan. *Kybernetische Anthropologie. Eine Geschichte der Virtualität*. Frankfurt am Main: Suhrkamp, 2003.

Roch, Axel. *Claude E. Shannon. Spielzeug, Leben und die geheime Geschichte seiner Theorie der Information*. Berlin: gegenstalt Verlag, 2009.

———, and Bernhard Siegert. "Maschinen, die Maschinen verfolgen. Über Claude E. Shannons und Norbert Wieners Flugabwehrsysteme." In *Konfigurationen. Zwischen Kunst und Medien*, edited by Sigrid Schade and Georg Christoph Tholen, 219–30. Munich: Fink, 2003.

Rosenberg, Daniel, and Anthony Grafton. *Cartographies of Time*. New York: Princeton Architectual Press, 2010.

Sandbothe, Mike, and Walther Ch. Zimmerli, eds. *Zeit—Medien—Wahrnehmung*. Darmstadt: Wissenschaftliche Buchgesellschaft, 1994.

Schlemmer, Mirjam. "Audiovisuelle Wahrnehmung. Die Konkurrenz und Ergänzungssituation von Auge und Ohr bei zeitlicher und räumlicher Wahrnehmung." In *Musikpsychologie*, edited by Helga de la Motte-Haber and Günther Rötter, 173–84. Laaber: Laaber-Verlag, 2005.

Schröter, Jens. "Die Form der Farbe. Zu einem Parergon in Kants 'Kritik der Urteilskraft.'" In *Kants Schlüssel zur Kritik des Geschmacks*, edited by Ursula Franke, 135–54. Hamburg: Meiner, 2000.

Schubiger, Irene, ed. *Schweizer Videokunst der 1970er und 1980er Jahre. Eine Rekonstruktion*. Zürich: JRP Ringier, 2009.

Sharpless, T. K. "Mercury Delay Lines as a Memory Unit." In *Proceedings of a Symposium on Large-Scale Digital Calculating Machinery*, edited by Howard Aiken, 103–9. Cambridge, MA: Harvard University Press, 1948.

Sheldrake, Rupert. *Das Gedächtnis der Natur. Das Geheimnis der Entstehung der Formen in der Natur*. Munich: Piper, 1996.

Shiff, Richard. "Something is Happening." *Art History* 28.5 (2005): 752–82.

Siegert, Bernhard. "Das Leben zählt nicht. Natur- und Geisteswissenschaften bei Dilthey aus mediengeschichtlicher Sicht." In *Medien. Dreizehn Vorträge zur Medienkultur*, edited by Claus Pias, 161–82. Weimar: Verlag und Datenbank für Geisteswissenschaft, 1999.

———. *Passage des Digitalen. Zeichenpraktiken der neuzeitlichen Wissenschaften 1500–1900*. Berlin: Brinkmann & Bose, 2003.

———. *Relays: Literature as an Epoch of the Postal System*. Translated by Kevin Repp. Stanford: Stanford University Press, 1999.

Simondon, Gilbert. *On the Mode of Existence of Technical Objects*. Translated by Ninian Mellamphy. London: University of Western Ontario, 1980.

Stangl, Burkhard. *Ethnologie im Ohr. Die Wirkungsgeschichte des Phonographen*. Vienna: WUV, 2000.

Stiegler, Bernard. *Technics and Time*. Translated by Stephen Barker. Stanford: Stanford University Press, 2009.

Stockhausen, Karlheinz. ". . .wie die Zeit vergeht. . ." *Die Reihe. Information über serielle Musik* 3 (1957): 13–42.

Swade, Doron. "Virtual Objects: Threat or Salvation?" In *Museums of Modern Science*, edited by S. Lindquist, M. Hedin, and U. Larsson, 139–47. Canton, MA: Science History Publications, 2000.

Tarde, Gabriel. *The Laws of Imitation*. Translated by Elsie Clews Parsons. New York: Henry Holt, 1903.

Tholen, Georg Christoph, and Michael O. Scholl, eds. *Zeit-Zeichen. Aufschübe und Interferenzen zwischen Endzeit und Echtzeit*. Weinheim: VCH/Acta Humaniora, 1990.

Vief, Bernhard. "Die Inflation der Igel. Versuch über die Medien." In *McLuhan neu lesen. Kritische Analysen zu Medien und Kultur im 21. Jahrhundert*, edited by Derrick de Kerckhove, Martina Leeker, and Kerstin Schmidt, 213–32. Bielefield: transcript, 2008.

Viola, Bill. *Reasons for Knocking at an Empty House: Writings 1973–1994*. Edited by Robert Violette. Cambridge, MA: MIT Press, 1998.

Volmar, Axel, ed. *Zeitkritische Medien*. Berlin: Kulturverlag Kadmos, 2009.

Völz, Horst. *Handbuch der Speicherung von Information*. Aachen: Shaker, 2005.

von Falkenhayn, Katharina. *Augenblick und Kairos. Zeitlichkeit im Frühwerk Martin Heideggers*. Berlin: Duncker & Humblot, 2003.

von Helmholtz, Hermann. "Über die Methoden kleinste Zeittheile zu messen und ihre Anwendung für physiologische Zwecke." *Königsberger naturwissenschaftliche Unterhaltungen* 2 (1851): 169–89.

von Hilgers, Philipp. *Kriegsspiele. Eine Geschichte der Ausnahmezustände und Unberechenbarkeiten*. Munich: Fink, 2008.

von Mihály, Dionys. *Das elektrische Fernsehen und das Telehor*. Berlin: Krayn, 1923.

Weber, Samuel. *Mass Mediauras: Form, Technics, Media*. Stanford: Stanford University Press, 1996.

Webers, Johannes. *Handbuch der Film- und Videotechnik. Die Aufnahme, Speicherung, Bearbeitung und Wiedergabe audio-visueller Programme*. Munich: Franzis, 1991.

Weyl, Hermann. "Zeitverhältnisse im Kosmos. Eigenzeit, gelebte Zeit und metaphysische Zeit." In *Proceedings of the Sixth International Congress of Philosophy*, edited by Edgar Sheffield Brightman, 54–58. New York: Longmans, Green and Company, 1927.

Wiemer, Carl. "Im Rauschen des Realen. 'La dernière bande'—Becketts medientechnologische Antwort auf Prousts *Recherche*." *Romanistische Zeitschrift für Literaturgeschichte* 25.1-2 (2001): 169–76.

Wiener, Norbert. *Cybernetics or Control and Communication in the Animal and the Machine*. Cambridge, MA: MIT Press, 1985.

———. "Time, Communication, and the Nervous System." *Annals of the New York Academy of Sciences* 50 (1948): 197–219.

Williams, Frederic C., and Tom Kilburn, "A Storage System for Use with Binary-Digital Computing Machines." *Proceedings of the Institution of Electrical Engineers* 96.40 (1949): 81–96.

Winckel, Fritz, ed. *Technik der Magnetspeiche*. Berlin: Springer, 1960.

Zenneck, J. "Eine Methode zur Demonstration und Photographie von Stromcurven." *Annalen der Physik* 305.12 (1899): 838–53.

Zielinski, Siegfried. *Deep Time of the Media: Toward an Archaeology of Hearing and Seeing by Technical Means*. Translated by Gloria Custance. Cambridge, MA: MIT Press, 2006.

———. *Zur Geschichte des Videorekorders*. Berlin: Wissenschaftsverlag Spiess, 1986.

Index

Adorno, Theodor, 103
Adrian, Edgar Douglas, 215
Althusser, Louis, 232–33
AM (amplitude modulation), 22, 141, 183, 233–34
Amiga 500, 241
Anaximander, 251
Andriopoulos, Stefan, 237
ANT (Actor Network Theory), 256
Antheil, George, 176
anti-aircraft predictor, 5, 30
Ardenne, Manfred von, 46, 129, 236, 238
Arezzos, Guido von, 40
Aristotle, 16, 19, 45, 48, 50, 65, 67, 70, 80, 84, 105, 141, 175, 206
Aristoxenus, 75, 139

Babbage, Charles, 66, 69, 71, 79, 144, 258, 262
Baer, Karl Ernst von, 9, 38
Bain, Alexander, 210, 257
Baird, John Logie, 124–27, 131, 134–35, 138–39, 149–50
Baran, Paul, 184
Barker, Timothy Scott, ix
Barthes, Roland, 124, 190
Barwise, John, 210
Baudot code, 177

Beckett, Samuel, 113–17
Behm, Alexander, 16
Bell, Alexander Graham, 101
Benjamin, Walter, 48, 114, 159, 178, 228, 235
Bennett, Charles Harper, 19
Bergson, Henri, ix, 45, 47–50, 86, 106, 112, 137, 151, 158–59, 192, 206, 218
Bertalanffy, Ludwig von, 190
Beuys, Joseph, 179
Bigelow, Julian, 75
Boccioni, Umberto, 50
Bode, Hendrik Wade, 185
Böhme, Harmut, 189
Bohr, Niels, 135
Boltzmann, Ludwig, 259
Boole, George, 257
Bragaglia, Anton Giulio, 50
Branly, Édouard, 225
Braudel, Fernand, 208
Braun, Ferdinand, 17–18, 55, 225
Braun tube, 17, 56, 124, 238
Bronk, Otto von, 101
Brouwer, Luitzen Egbertus Jan, 78
Bruch, Walter, 18
Burns, Ken, 40
Burton, Christopher, 75
Bush, Vannevar, 42, 66, 132, 193
Butler, Samuel, 41–42

Calahan, Edward Augustin, 195
camera obscura, 21, 41
Campbell, Jim, 153
Cantor, Georg, 86
Cassirer, Ernst, 254
Cerf, Vinton, 184
chronophotography, xx, 3, 11, 19, 43–48, 146
Church, Alonzo, 216
cinematography, xx, 19, 37, 39–40, 48–50, 74, 103, 107, 128, 131–32, 216–17
Colossus, 32, 80, 244
Commodore 64, 85, 239–42
computers, xx, 3, 9–12, 23–32, 42, 63–89, 102, 132, 152, 177, 190–94, 207, 210, 220, 239–44, 257
Cros, Charles, 100
Crosby, Bing, 22, 111
Cunningham, Merce, 163
cybernetics, 5, 30, 42, 77, 262

de Chateaubriand, François-René, 101
de Forest, Lee, 130
de France, Henri Georges, 18
Deleuze, Gilles, ix, 158, 216
Derrida, Jacques, 70 100, 151
Dieckmann, Max, 17
difference engine, 79, 262
DRM (Digital Radio Mondiale), 227
Droysen, Johann Gustav, 219–20
DSP (digital signal processing), 10, 64–65, 86, 157, 193
du Bois-Reymond, Emil, 42
Dumas, Alexandre, 195

echolot, 16–17
Eckert, Gerd, 104
Edison, Thomas Alva, 100, 116
EDSAC (Electronic Delay Storage Automatic Calculator), 242
Ehrenfest, Paul, 257
Einstein, Albert, 50
electronic music, 117–18
ENIAC (Electronical Numerical Integrator And Computer), 66, 81, 241

ergodicity, 177
ergograph, 54
Euclid, 261
Euler, Leonhard, 238

Faraday, Michael, 19, 49, 117, 160, 207–8, 225, 238, 257, 259
Feddersen, Berend Wilhelm, 53
Féré, Charles, 54
Fiedler, Ladislaus, 176, 210
Figgis, Mike, 156
film. See cinematography
Fizeau, Hippolyte, 51–52
Flusser, Vilém, 80, 117, 216
Fluxus, 162, 191
Fontana, Bill, 114
Foucault, Michel, xiv–xviii, 40, 84, 175, 182, 191, 216
Fournier d'Albe, Edward, 15
FM (frequency modulation), 22, 123, 177, 183, 210, 233–34
Freud, Sigmund, 50, 107, 150

Galilei, Galileo, 19–20, 66
Geisler, Michael, xv
Gelernter, David, 190
Geoghegan, Bernard Dionysius, xvii
Giedion, Sigfried, 208
Gilbreth, Frank Bunker Sr., 194
Gitelman, Lisa, xvi, xxiv
Goddard, Michael, xvii
Goethe, Johann Wolfgang von, 40, 57, 145
Google, 187
Graham, Dan, 162
gramophone, 104, 114, 126, 130, 134, 150, 152, 175, 214, 239, 262
Greenberg, Clement, 153
Grimmelshausen, Hans Jakob Christoffel von, 111
Großklaus, Götz, 215
Guattari, Félix, 216
Gurvitsch, Alexander, 259

Haase, Frank, 52
Halbwachs, Maurice, 191

Index

Hansen, Mark B. N., ix
Hegel, Georg Wilhelm Friedrich, 30, 100, 154, 192, 220, 226, 233
Heidegger, Martin, ix, 11, 144, 149, 161, 188, 209–10, 214–15, 218–24, 228, 236, 241, 243, 251, 258, 261
Heider, Fritz, 176
Heine, Heinrich, 182
Heisenberg, Werner, 46, 139
Helmholtz, Hermann von, 11, 29, 37, 39–42, 47, 52
Hennig, Ludger, 156
Henry, Joseph, 254
Heraclitus, 84, 154
Hering, Ewald, 41
Hero of Alexandria, 254
Herschel, John, 56
Hertz, Heinrich, 16, 225, 239, 252–53, 256
Hilbert, David, 78, 210, 223
Hiller, Lejaren, 119n28
Hillis, William Daniel, 216–17
Homer, 7, 110, 116, 164, 181, 193
Hoppe, Joseph, 238
Huffman, David Albert, 175
Hughes, Edward, 255
Huhtamo, Erkki, ix, xvii
Husserl, Edmund, 11, 56, 112, 158, 207, 222, 241, 261
Huyghens, Christian, 20, 37, 222, 262

Innis, Harold, 215
Internet, 4, 63, 73, 157, 176–82, 187–92, 223
IP (Internet Protocol), 184

Jacquard, Joseph Marie, 144
Jensen, Wilhelm, 50
Jünger, Ernst, 53

Kahn, Bob, 184
Kahn, Douglas, 225
Kant, Immanuel, 145, 238, 241
Kantorowicz, Ernst, 215
Keats, John, 101
Kipling, Rudyard, 101

Kirchmann, Kay, 209
Kittler, Friedrich, xiii–xxv, 6
Klaus, Georg, 262
Kluitenberg, Eric, xvii
Krajewski, Markus, xvi
Kracauer, Siegfried, 50, 152
Kriesche, Richard, 159
kymograph, 17, 38, 54, 57, 100, 125, 258

Lacan, Jacques, 87, 110, 114
Lafitau, Joseph-François, 220
Lamarr, Hedy, 176
laterna magica, 41
Leibniz, Gottfried Wilhelm, 6, 9–10, 16, 41, 47, 52, 65, 77, 85, 130, 212, 262
Lessing, Gotthold Ephraim, 11, 47, 50, 148, 159, 189
Lieben, Robert von, 73
Link, David, ix
Lipfert, Kurt, 130
Llull, Ramon, 80
Lorenz SZ42 cipher machine, 244
Lotman, Jurij, 255
Lucas, Keith, 215
Ludwig, Carl, 54
Luhmann, Niklas, 69, 164, 257, 262
Lullus, Raimundus, 71
Lyotard, Jean-François, 177

Mach, Ernst, 46
Maddox, Richard Leach, 19
magnetic tape, xx–xxi, 6, 21–24, 29, 66, 103–4, 107–17, 142, 150–51, 160, 236–39, 262
Mandelbrot, Benoît, 257
Mann, Thomas, 37, 44
Marconi, Guglielmo, 101, 225
Marey, Étienne-Jules, 17, 40–48, 53, 175
Marinetti, Filippo Tommaso, 50
Markov, Andrey Andreyevich, 174
Maxwell, James Clerk, 16, 19, 21, 108, 160, 225, 252, 255, 259
McCulloch, Warren Sturgis, 76, 142
McLean, Donald, 149–50

McLuhan, Marshall, 5, 11, 23, 56, 69, 74, 80, 103, 123, 132, 140, 142, 180, 190, 215–16
Meissner circuit, 72, 209
Melitopoulos, Angela, 163
memes, xxiv, 107, 255
Mersenne, Marin, 212, 253
Michelangelo, 50
Michelson, Albert, 20
Mills, Dave, 189
Minkowski, Hermann, 6, 159
Moles, Abraham, 38
Monet, Claude, 43, 153
monochord, 209, 212, 253–54
Morse code, 38, 174–76, 225
Müller, Johann Helfreich 77
Münsterberg, Hugo von, 107, 146
Mullin, Jack, 112
Mumford, Lewis, 69
Murray code, 144
Muybridge, Eadweard, 41, 48, 50, 53

Nelson, Theodor Holm, 190
NTP (Network Time Protocol), 189
Neumann, John von, 66, 68, 75, 81, 87, 152
Newton, Isaac, 9–10, 16, 47–48, 134
Niépce, Nicéphore, 19
Nipkow, Paul, 41, 46, 101, 128, 136
Nipkow disk, 44, 125, 137, 139, 142, 181
Nyquist, Harry, 185
Nyquist-Shannon sampling theorem, 84

Oersted, Hans Christian, 225, 252
Ong, Walter, 110
OSC (Open Sound Control), 189
Orwell, George, 125
oscilloscope, 17, 27–28, 49, 53–57, 143, 149, 227
Otth, Jean, 143

packet switching, 63, 184
Paik, Nam June, 162–63
PAM (pulse amplitude modulation), 183

Parikka, Jussi, ix, xiii, xvii, xxiii–xxiv
PCM (pulse code modulation), 183–86
Pegasus, 240
Peirce, Charles Sanders, 43, 173, 219, 223, 254, 257
phenakistoscope, 41
Phidias, 49
Phillips, Sam, 21, 115
phonautograph, 40, 100
phonograph, xx, 74, 100–2, 103, 108, 116, 212, 215, 239
photography, xx, 9, 19–21, 41, 56, 68, 73, 103, 123–25, 140
phototelegraphy, 123–24, 129, 132, 136, 140, 174, 176, 210, 257
Pias, Claus, 82
Pinder, Wilhelm, 252
Pirandello, Luigi, 221
Pitts, Walter, 75, 142
Plato, 206, 216
Plessner, Helmut, 53
Popov, Alexander Stepanovich, 225
Popper, Karl, 253
Post, Emil, 216
Postel, Jon, 187
Poulsen, Valdemar, 17, 99, 101, 118n1
PPM (pulse phase modulation), 184
Presley, Elvis, 21, 115
Prigogine, Ilya, 206
Prince, 111
print, 68, 132, 164, 174, 190–91, 207, 226, 262
Proust, Marcel, 29, 101–2, 113, 137
Ptolemy, 41
Pudovkin, Vsevolod, 131, 146
Pushkin, Alexander Sergeyevich, 174
Pythagoras, 209, 212, 253

radar, 20, 56, 82, 152, 162
radio, xx–xxiii, 22, 25, 38, 66, 74, 102–12, 130, 132, 183, 190, 210, 225–38, 262
radio photography, 55
RAM (random-access memory), 191, 218, 239
Ranke, Leopold von, 138, 258

Index

ROM (read-only memory), 191, 218, 239
Roschy, Jacob, 234
Reeves, Alec Harley, 183
Reis, Philipp, 101
Rhein, Eduard, 137
Rheinberger, Hans-Jörg, 206–7, 258
Robida, Albert, 103
Rodin, Auguste, 47

St. Augustine, 39, 67
Salcher, Peter, 46
Sale, Anthony, 244
Sattler, Felix, 156
Sauveur, Joseph, 253
Schabowski, Günter, 50
Schaeffer, Pierre, 115
Schiller, Friedrich, 7
Schimana, Elisabeth, 118
Schröter, Fritz, 140
Schüller, Eduard 112
Schum, Gerry, 159
Scott, Édouard-Léon, 40, 100
SECAM (SÉquentiel Couleur A Mémoire), 18
Serres, Michel, 185
Shannon, Claude, 17, 30, 44, 73, 77, 80, 128, 145, 174, 177, 185, 259
Sheldrake, Rupert, 259–60
Shelley, Mary, 110
Shiff, Richard, 153
Siegert, Bernhard, xv–xvi, xix
Sieveking, Lance, 127, 221
Simondon, Gilbert, ix, 4, 144, 219, 260
sonar, 20
Spencer-Brown, George, 78
Spitz, Samuel, 17
Sterling, Bruce, xix, 150
Stern, William, 15
Stibitz, George Robert, 79
Stiegler, Bernard, ix
stock ticker, 195–96
Stockhausen, Karlheinz, 115
Stokes, George Gabriel, 255
Stroud, John, 75

Talbot, Henry Fox, 140
tape recorder, 21–22, 109–17, 130, 213
TBC (Time Base Corrector), 138, 141, 154
TCP (Transmission Control Protocol), 184, 187–89
Teilhard de Chardin, Pierre, 132
telecinematography, 129
telegraphy, 38, 100–2, 138–39, 173–78, 182, 195–96, 210, 257
telegraphone, 17, 99, 101
telephone, 100–3, 183–86, 255, 257
teletext, 143
television, xx, 3, 22–23, 37, 41, 56, 66, 74, 103–9, 128–61, 190, 213, 216–18, 221, 224, 227–28, 233–38, 257
Tholen, Georg Christoph, 80
TTL (Time To Live), 180
Turing, Alan, 66, 70, 75, 77, 79–80, 83–86, 210, 216, 223, 242
typewriter, 144, 174, 263

Uexküll, Jakob von, 39
UDP (user datagram protocol), 187
UTM (Universal Turing Machine), 65, 70, 72, 77–78, 83, 86–87, 240, 242

Vico, Giambattista, 205, 254
video art, 162–64, 191
video compression, 157
video recorder, 22, 130, 142, 154, 161, 217
Viola, Bill, 42, 183
Virilio, Paul, 50, 136, 182
Vismann, Cornelia, xvi
Vogel, Hermann Wilhelm, 9

Warburg, Aby, 191
Watts, Donald, 184
Weibel, Peter, 159
Wellbery, David E., xiv
Weyl, Hermann, 65
Whitehead, Alfred North, 154
Wiener, Norbert, 5, 12, 25, 65, 68, 73, 137, 142, 162

Wikipedia, 190–91
Williams, Raymond, 151–52
Williams tube, 24, 56, 244
Williams-Kilburn tube, 25–26, 31
Winthrop-Young, Geoffrey, xiv–xvii
wireless telegraphy, 101, 176, 183
World Wide Web, 178, 191

Zelwies, Juliane, 146
Zenneck, Jonathan, 55
Zielinski, Siegfried, ix, 189
zoetrope, 47
Zuse, Konrad, 12, 65, 81
Zuse Z22, 76, 243
Zworykin, Vladimir Kosmich, 148

About the Contributors

Wolfgang Ernst, Full Professor for Media Theories at the Institute of Musicology and Media Studies, Humboldt-University, Berlin. Studied history, Latin philology, and (partially) classical archaeology. PhD thesis on historicism and museology; habilitation on the technical and symbolic infrastructures of national memory. Teaching experience and guest professorships at several universities (Kassel, Leipzig, Cologne, Weimar, Bochum, Paderborn). Other publications in English: *Digital Memory and the Archive* (2013); *Sonic Time Machines* (forthcoming).

Anthony Enns is Associate Professor of Contemporary Culture at Dalhousie University in Halifax, Nova Scotia. His edited collections include *Screening Disability* (2001), *Sonic Mediations* (2008), and *Vibratory Modernism* (2013). His work in media studies has appeared in such journals as *Screen, Senses and Society; Culture, Theory & Critique; Journal of Sonic Studies; Journal of Popular Film and Television; Quarterly Review of Film and Video; Popular Culture Review;* and *Studies in Popular Culture*, as well as the anthologies *Sounds of Modern History: Auditory Cultures in the 19th and 20th Century* (2014) and *Hörstürze: Akustik und Gewalt im 20. Jahrhundert* (2005). He has been a guest lecturer at several universities, including the University of Amsterdam and the University of Kiel. His other translations include Friedrich Kittler's *Optical Media* (2010) and Sybille Krämer's *Medium, Messenger, Transmission: An Approach to Media Philosophy* (2015).